HORMONES AND CELL REGULATION

HORMONES AND CELL REGULATION

European Symposium Volume 9

HORMONES AND CELL REGULATION

Proceedings of the Ninth INSERM European Symposium on Hormones and Cell Regulation, held at Sainte Odile (France), 24-27 September, 1984.
Sponsored by Institut National de la Santé et de la Recherche Médicale.

Edited by:

J. E. DUMONT, B. HAMPRECHT and J. NUNEZ

VOLUME 9

Scientific Committee:

E. Carafoli	B. Hamprecht	F. Morel
R. M. Denton	R. J. B. King	J. Nunez
J. E. Dumont	H. J. van der Molen	G. Schultz

1985

ELSEVIER SCIENCE PUBLISHERS
AMSTERDAM · NEW YORK · OXFORD

CONTENTS

VI

VIII

FOREWORD

As in previous years, the program of the 9th meeting on 'Hormones and Cell Regulation' may appear rather eclectic since it covers several general topics which, at first glance, are little or not at all interrelated. The reader will find papers on the molecular biology of cancer cells, the oncogenes and growth factors, different proteins of the cytoskeleton and their functions, the cyclic AMP model, the ion channels and finally different growth factors operating during neural development. Such diversity has been present since 1976, when the first meeting was organized. At that time, the organizers felt that there was a need for a high level, non-specialized, European meeting in the field of hormone action. The idea was that such a meeting should provide not only the most recent information on the mechanisms of hormone action but also on the new concepts and techniques of cell and molecular biology.

During this ten year period, rapid progress has been accomplished in the field precisely by the combination of a variety of experimental approaches and the advances achieved in different scientific domains. The concept of what is a hormone, a neurotransmitter or a growth factor has itself markedly changed.The cyclic AMP model, which is now much better understood, can be considered as one of the regulatory pathways, among others, operating at the membrane level. The derivatives of phosphatidyl inositol metabolism and calcium ions obviously represent very important regulatory systems operating in a variety of cells. The transfer of information triggered by the different signals from the membrane to cell is mediated by a number of proteins and enzymes which have been isolated and characterized (such as the different protein kinases, the Ca^{2+}-binding proteins, several components of the cytoskeleton, etc.). Moreover, increasing evidence suggests that these different regulatory pathways are closely interconnected among themselves. One of the most exciting examples is that of the relationship which appears to exist between some hormones or growth factors, their receptors, the oncogenes and cell division. Very much remains to be learned about the relationships existing between these regulatory systems and the differentiating cell, but the conceptual and technical bases of future progress in this field are probably already in hand.

We hope that this book will reflect such an evolution of the concepts in the field of cell regulation an on the mechanism of action of the hormonal, growth and nervous signals which modulate the activity and the expression of the cells of eukaryotic organisms.

J. NUNEZ

LIST OF PARTICIPANTS

BARDE, Y., Max-Planck Institut, Dept. of Neurochemistry, D-8033 Martinsried, F.R.G.

BIDEN, T., University of Geneva, Institut de Biochimie Clinique, Sentier de la Roseraie, CH-1211 Geneva, Switzerland.

BOONSTRA, J., University of Utrecht, Dept. of Molecular Cell Biology, Padualaan 8, NL-3508 TB Utrecht, The Netherlands.

BRIDOUX, A-M., INSERM U 282, Hôpital Henri Mondor, Créteil, France.

BRUNT, R., University of Bath, Dept. of Biochemistry, Claverton Down, Bath, U.K.

BURGER, A., Unité de Thyroide Lab. 4-767, Hôpital Cantonal, CH-1211 Geneva 4, Switzerland.

CARAFOLI, E., Laboratorium für Biochemie, ETH, CH-8092 Zürich, Switzerland.

CASTAGNA, M., Institut de Recherches Scientifiques sur le Cancer, B.P. 8, F-94802 Villejuif Cedex, France.

CLARET, M., Unité de Recherches INSERM U 274, CNRS, Bât. 443 Université Paris-Sud, F-91405 Orsay, France.

COOKE, B.A. Royal Free Hospital, School of Medicine, Dept. of Biochemistry, Rowland Hill Street, London NW3 2PF, U.K.

COUCHIE, D. INSERM U 282, Hôpital Henri Mondor, Créteil, France.

De BRABANDER, M., Janssen Pharmaceutica Research Laboratoria, Turnhoutseweg, 30, B-2340 Beerse, Belgium.

De LAAT, S., Hubrecht Laboratory, International Embryological Institute, Uppsalalaan 6, NL-3584 CT Utrecht, The Netherlands.

DENTON, R.M., University of Bristol, Medical School, Dept. of Biochemistry, University Walk, Bristol BS8 1TD, U.K.

DOERFLER, W., Institut für Genetik der Universität Köln, Weyertal 121, D-5000 Köln 41, F.R.G.

DROSDOWSKY, M., Département de Biochimie, C.H.U., Avenue Côte de Nacre, F-14040 Caen Cedex, France.

DUMONT, J.E., Faculté de Médecine, Campus Hôpital Erasme, Route de Lennik, 808, B-1070 Brussels, Belgium.

EBENDAL, T., University of Uppsala, Institute of Zoology, Box 561, S-75122 Uppsala, Sweden.

ERNEUX, C., Institut de Recherche Interdisciplinaire, C.P. 602, Faculté de Médecine, Route de Lennik, 808, B-1070 Brussels, Belgium.

FRANKE, W.W., Institut für Zell- und Tumorbiologie, Deutsches Krebsforschungszentrum, D-6900 Heidelberg, F.R.G.

FRELIN, C., Centre de Biochimie du CNRS, Université de Nice, Parc Valrose, F-06034 Nice Cedex, France.

FRIEDL, A., Physiologisch-Chemisches Institut der Universität Würzburg, Koellikerstr. 2, D-8700 Würzburg, F.R.G.

FUKADA, K., California Institute of Technology, Division of Biology 216-76, Pasadena, California 91125, U.S.A.

GERSTER, T., Institut für Molekularbiologie II, Universität Zürich, Hönggerberg, CH-8093 Zürich, Switzerland.

GHYSDAEL, J., INSERM U 186 Laboratoire d'Oncologie Moléculaire, Institut Pasteur de Lille, P.B. 3415, F-59019 Lille Cedex, France.

GIERLICH, D., Pharmazeutisches Institut der Universität Tübingen, Auf der Morgenstelle 8, D-7400 Tübingen, F.R.G.

GILL, G.N., University of California, School of Medicine, M-013, La Jolla, Ca. 92093, U.S.A.

GULLICK, W.G., Imperial Cancer Research Fund Laboratories, P.O.B. 123, Lincoln's Inn Fields, London, WC2A 3PX, U.K.

HAMPRECHT, B., Physiologisch-Chemisches Institut der Universität, Koellikerstr. 2, D-8700 Würzburg, F.R.G.

HEATHERS, G., University of Bath, Dept. of Biochemistry, Bath, Avon, U.K.

HEMMINGS, B., Friedrich-Miescher-Institut, Postfach 2543, CH-4002 Basel, Switzerland.

HERMANN, J., Université Pierre et Marie Curie, Physiologie de la Reproduction (Groupe Steroides), 9 quai Saint-Bernard, F-75005 Paris, France.

HESKETH, J., Rowett Research Institute, Dept. of Cell Biochemistry, Greenburn Rd., Bucksburn, Aberdeen, U.K.

HUCHON, D., Université Pierre et Marie Curie, Physiologie de la Reproduction, 9 quai Saint-Bernard, F-75005 Paris, France.

JESSUS, C., Université Pierre et Marie Curie, Physiologie de la Reproduction, 9 quai Saint-Bernard, F-75005 Paris, France.

JOB, D., INSERM U 244, Université scientifique et médicale de Grenoble, B.P. 54X, F-38041 Grenoble Cedex, France.

JOCKUSCH, B., Universität Bielefeld, Lehrstuhl für Entwicklungsbiologie, Postfach 8640, D-4800 Bielefeld, F.R.G.

KING, R. Imperial Cancer Research Fund, Lincoln's Inn Fields, P.O.B. 123, London WC2A 3PX, U.K.

LAMPRECHT, S., Soroka Medical Center, Dept. of Gastroenterology, Beer-Sheva 84105, Israel.

LA'IY, F., Faculté de Médecine, IRIBHN, Route de Lennik 808, B-1070 Brussels, Belgium.

LETERRIER, J.-F., INSERM U 44 Centre de Neurochimie, CNRS, 5 rue Blaise Pascal, F-67084 Strasbourg Cedex, France.

LEW, D., Division de Maladies Infectieuses, Hôpital Cantonal, CH-1211 Geneva, Switzerland.

LOEFFLER, J.-P., Max-Planck-Institut für Psychiatrie, Dept. of Neuropharmacology, Am Klopferspitz 18a, D-8033 Martinsried, F.R.G.

MAIRESSE, N., IRIBHN, Campus Erasme, Route de Lennik 808, B-1070 Brussels, Belgium.

MANCINI, A., Catholic University School of Medicine, Dept. of Internal Medicine, Division of Endocrinology, Via F. Filelfo 5, I-00135 Rome, Italy.

MARTIN, R., University of Cambridge, Dept. of Biochemistry, Tennis Court Rd., Cambridge CB2 1QW, U.K.

MARTINI, O.H.W., Physiologisch-Chemisches Institut der Universität Würzburg, Koellikerstr. 2, D-8700 Würzburg, F.R.G.

MAUGER, J.P., INSERM U 274, Bât. 443, Université Paris-Sud, F-91405 Orsay Cedex, France.

MELDOLESI, J., University of Milano, Dept. of Medical Pharmacology, Via Vanvitelli, 32, I-20129 Milano, Italy.

MERTENS-STRIJTHAGEN, J., Facultés Universitaires Notre Dame de la Paix, Laboratoire de Physiologie, rue de Bruxelles 61, B-5000 Namur, Belgium.

MIESKES, G., Universität Göttingen, Abteilung Klinische Biochemie, Innere Medizin, Robert-Koch-Str. 40, D-3400 Göttingen, F.R.G.

MONARD, D., Friedrich-Miescher-Institut, Postfach 273, CH-4002 Basel, Switzerland.

MOREL, F., Collège de France, 11, place Marcelin Berthelot, F-75231 Paris Cedex 05, France.

MORGAN, M.J., The Wellcome Trust, 1, Park Square West, London NW1 4LJ U.K.

MORIN, L., Université National du Rwanda, Dept. of Biology, B.P. 508, Butará, Rwanda.

MÜLLER, R., Universität Göttingen, Abt. Klinische Biochemie, Innere Medizin, Robert-Koch-Str. 40, D-3400 Göttingen, F.R.G.

NOVAK-HOFER, I., Friedrich-Miescher-Institut, P.O.B. 2543, CH-4002 Basel, Switzerland.

NUNEZ, J., INSERM U 282, Hôpital Henri Mondor, F-94101 Créteil, France.

OSTERRIEDER, H., Hoffmann-LaRoche, Abt. Pharmazeut. Forschung, Bau 70/426, Grenzacher Str. 12, CH-4002 Basel, Switzerland.

PADEL, U. Universität Göttingen, Abt. Klinische Biochemie, Innere Medizin, Robert-Koch-Str. 40, D-3400 Göttingen, F.R.G.

PASSAREIRO, H., ULB School of Medicine, IRIBHN, Route de Lennik 808, B-1070 Brussels, Belgium.

PAVOINE, C., Institut de Recherches Scientifiques sur le Cancer, B.P. 8, F-94802 Villejuif Cedex, France.

PETERSEN, O.H., University of Liverpool, Physiological Laboratory, Brownlow Hill, P.O. Box 147, Liverpool L69 3BX, U.K.

POZZAN, T., University of Padova, Institute of General Pathology, Via Loredan 16, I-35131 Padova, Italy.

REINHART, P.H., Physiologisch-Chemisches Institut der Universität Würzburg, Koellikerstr. 2, D-8700 Würzburg, F.R.G.

REISER, G. Physiologisch-Chemisches Institut der Universtät Tubingen, Hoppe-Seyler-Str. 1, D-7400 Tübingen, F.R.G.

RINK, T.J., Physiological Laboratory, University of Cambridge, Downing Street, Cambridge, CB2 3EG, U.K.

ROGER, P., ULB School of Medicine, IRIBHN, Route de Lennik 808, B-1070 Brussels, Belgium.

ROSENBERG, A., Loyola University Medical Center, Dept. of Biochemistry and Biophysics, 2160 South First Aven., Maywood, Illinois 60153, U.S.A.

ROSSIER, B.C., Institut de Pharmacologie de l'Université de Lausanne, rue du Bugnon 21, CH-1011 Lausanne-CHUV, Switzerland.

SAEZ, J.M., INSERM U 162, Hôpital Debrousse, Chemin Soeur Bouvier, F-69322 Lyon Cedex, France.

SAINTENY, F., Unité de Recherches sur la Cinétique Cellulaire, INSERM U 250, Institut Gustave-Roussy, Pavillon de Recherches, Rue Camille Desmoulins- F-94805 Villejuif Cedex, France.

SENSENBRENNER, M., Centre de Neurochimie du CNRS, 5, rue Blaise Pascal. F-67084 Strasbourg Cedex, France.

SEVERNE, Y., Free University of Brussels, Dept. of Protein Chemistry, 65 Paardenstreet, B-1640 St. Genesius Rode, Belgium.

SORGATO, M.C., University of Padova, Institute of Biochemistry, Via Marzolo, 3, I-35131 Padova, Italy.

SCHLEGEL, W., Fondation pour Recherches Médicales, Dept. de Médecine, 64, av. de la Roseraie, CH-1211 Geneva 4, Switzerland.

SCHNACKERZ, K., Physiologisch Chemisches Institut der Universität Würzburg, Koellikerstr. 2, D-8700 Würzburg, F.R.G.

SCHÜTZE, S., Universität Göttingen, Abteilung Klinische Biochemie, Robert-Koch-Str. 40, D-3400 Göttingen, F.R.G.

SCHULTZ, G., Freie Universität Berlin, Institut für Pharmakologie, Thielallee 69/73, D-1000 Berlin 33, Germany

SCHULTZ, I., Max-Planck-Institut für Biophysik, Kennedy-Allee 70, D-6000 Frankfurt a. M. 70, F.R.G.

TAFT, D., Alfreo Hospital, 21, Stoke Ave. Kew, Victoria 3101, Australia.

TARDY, M., INSERM U 282, Hôpital Henri Mondor, F-94010 Créteil, France.

TANAKA, J., University of Pennsylvania, Dept. of Neurology and of Biochemistry, School of Medicine, Philadelphia, Penn., U.S.A.

THOMOPOULOS, P., INSERM U 282, Hôpital Henri Mondor, F-94010 Créteil, France.

VAN DER MOLEN, H.J., Erasmus Universiteit Rotterdam, Dept. of Biochemistry, Postbus 1738, NL-3000 DR Rotterdam, The Netherlands.

VAN HEUVERSWYN, B., Université Libre de Bruxelles, IRIBHN, Campus Erasme, Route de Lennik 808, B-1070 Brussels, Belgium.

VICENTINI, L., University of Milano, Dept. of Pharmacology, Via Vanvitelli 32, I-20129 Milano, Italy.

WEBER, W., Institut für Physiologische Chemie, Universitätskrankenhaus Eppendorf, Martinistr. 52, D-2000 Hamburg 20, F.R.G.

WEEDS, A.G., MRC Laboratory of Molecular Biology, University Medical School, Hills Road, Cambridge CB2 2QH, U.K.

WILLIAMS, H.K., Royal College of Surgeons, Dept. of Dental Sciences, Lincoln's Inn Fields, London WC2 3PX, U.K.

WILMES, G., IRIBHN, Faculty of Medicine, University of Brussels. Campus Hôpital Erasme, Route de Lennik, 808, B-1070 Brussels, Belgium.

WOLLHEIM, C., Institut de Biochimie Clinique, University of Geneva, Sentier de la Roseraie, CH-1211 Geneva 4, Switzerland.

WYKE, J. Imperial Cancer Research Fund Laboratories, St. Bartholemew's Hospital, Dominion House, Bartholemew Close, London EC1A 7BE, U.K.

MOLECULAR BIOLOGY OF CANCER AND CELL DIFFERENTIATION

BIOLOGIE MOLECULAIRE DU CANCER ET DE LA DIFFERENCIATION CELLULAIRE

RETROVIRUSES WITH TWO CELL DERIVED ONCOGENES : MODEL SYSTEMS TO CELLULAR ONCOGENES COOPERATION

J. GHYSDAEL, F. DENHEZ, F. FERRE and D. STEHELIN*

*INSERM U 186 Oncologie Moléculaire - Institut Pasteur -

15, rue Camille Guérin - B.P. 245 - 59019 LILLE CEDEX

The factors involved in the initiation, maintenance and progression of tumors are multiple. A major achievement in the understanding of tumor development has been the discovery that, regardless of the diversity of these factors, their genetic substrates may be common. This conclusion stems from two main series of observations. First, the discovery that the oncogenes of acutely transforming retroviruses are of cellular origin. Second, the increasingly growing body of evidence that implicates the cellular progenitors of retroviral oncogenes in the genesis of non-viral tumors, including human tumors.

In this short review, we will briefly summarize our present knowledge of the structure and function of viral oncogenes and their cellular progenitors and discuss the prospects that follow from these discoveries in the understanding of tumor development.

1. The retroviral oncogenes

Retroviruses are the causative agents of many different tumors (carcinomas, sarcomas, leukemias) in different vertebrate species, including man (reviewed in ref. 1). Based on their oncogenic properties and their genetic structure, these viruses have been classified into two large classes.

The first class includes viruses inducing various neoplasms (mainly leukemias) long after their infection of susceptible hosts. The genome of these viruses contain three genes essential to viral replication. The gag gene codes for a precursor to the internal non glycosylated core proteins of the virion, the pol gene codes for the virion reverse transcriptase and the env gene codes for the viral envelope glycoproteins. The order of these genes is 5'- gag-pol-env - 3'. Details of the replication cycle of retroviruses have been the subject of comprehensive reviews (1-3) and will not be considered here. The second class of retroviruses includes viruses that induce a broad spectrum of malignant tumors after a short latency period.

Experimental study of these acutely oncogenic retroviruses has been considerably helped by the ability of these viruses to transform various cells in tissue culture (fibroblasts, epithelial cells, bone marrow cells). These transformation assays have thus allowed first the biological cloning of the transforming viruses and, more recently, the molecular cloning of the corresponding proviruses. The genome of these viruses contains sequences unrelated to those of the gag, pol and env genes, and responsible for the induction of neoplastic transformation in vivo and in vitro. These sequences have therefore been named viral oncogenes (v-onc genes). Viral oncogenes appear to have been acquired by recombinaison with cellular sequences since related sequences are found in all uninfected vertebrate cellular DNA (reviewed in ref.1).

The cellular progenitors of v-onc genes have all the properties of normal cellular genes : they have a constant chromosomal position in a given animal species, are present in all the members of that species and segregate as true mendelian loci. The cellular progenitors of v-onc genes are referred to as c-onc genes.

About 20 different v-onc / c-onc genes have now been indentified in all vertebrate species and, in some instances, in invertebrates (Table 1).

2. Additional cellular oncogenes

Most of the cellular oncogenes characterized so far have been discovered as the progenitors of the v-onc genes of retroviruses.

Recently however additional cellular oncogenes have been detected in various tumors by virtue of the ability of the DNA of these tumors to induce neoplastic transformation after its introduction in the form of a calcium phosphate coprecipitate (transfection) into a mouse fibroblastic cell-line in vitro (4). Although in most cases it appeared that these oncogenes were activated versions of the already known cellular homologs of the mouse sarcoma viruses ras genes, novel c-onc genes have been identified using this transfection assay (Table 2).

3. Transformation by viral and cellular oncogenes

Cellular transformation by retroviruses results in the modification of numerous morphological and biochemical parameters (the so-called transformation parameters, ref. 2 and 3). To decide which of these modifications are etiological in nature or the mere consequence of cellular transformation is unresolved in most cases. It is clear however that expression of v-onc genes interferes with both cell proliferation and differentiation. For example, erythroblasts transformed by the ES 4 strain of AEV (see Section 7) are at a differentiation stage close to that of CFU-E cells (5,6). However, unlike normal CFU-E cells, AEV transformed cells have acquired an increased potential for self-renewal and have lost their ability to differentiate terminally. As described in Section 7, this complex phenotype requires the coordinate expression of the two oncogenes of AEV, v-erb A and v-erb B.

Cellular oncogenes are by themselves not capable of inducing cellular transformation. However they can acquire a transforming potential ('activation') by several mechanisms including modifications of their structure (mutations), changes of either the levels of their expression (gene dosage) or the cell type in which they are expressed (ectopic expression).

Understanding the mode of action of v-onc genes or activated c-onc genes imply thus to discover the ways the products of these genes interfere with both cellular proliferation and differentiation.

4. Families of viral and cellular oncogenes. Biochemical functions of the products encoded by oncogenes.

As a first approach as to how, in molecular terms, viral and cellular oncogenes interfere with cellular metabolism, these genes have been grouped into different classes. The classification presented here is based on group structural homology and to homology to previously defined cellular genes. It also takes into account the subcellular localization of the proteins encoded by oncogenes.

The first recognized class of oncogenes encompasses those related to the src gene of Rous sarcoma virus and its cellular progenitor, c-src (7). The translation product of both v-src and c-src are 60,000 m.w. proteins that are protein kinases specific for tyrosine residues (8,9). These proteins are located on the cytoplasmic side of the plasma membrane, especially in adhesion plaques (10).

The protein products of five other oncogenes have been shown to have this enzymatic activity and a similar subcellular location. These proteins are the products of the yes, fps/fes, abl, fgr and ros oncogenes. These products share with $p60^{src}$ a domain of about 250 amino-acids responsible for catalyzing the transfer of phosphate to substrate proteins. Four additional oncogenes (fms, mil/raf, mos, and v-erbB) have been shown to encode amino-acid sequences related to the protein kinase domain of $p60^{src}$. Whether the translation products of these three genes encode bona fide protein kinases or have some functions similar to that of protein kinases still needs to be tested. The v-erb B gene

Table 1 - Oncogenes transduced by retroviruses

Viral oncogene	Species of origin[1]	Tumor(s)	Oncogene product Size[2]	Intracellular location[3]	Biochemical properties
v-src	CK	Sarcoma	60 Ksrc	PM,C	Tyr Kinase
v-fps/fes	CK/C	Sarcoma	85-170 Kgas-fps	PM,C	Tyr Kinase
v-yes	CK	Sarcoma	80-90 Kgag-yes	PM,C	Tyr Kinase
v-ros	CK	Sarcoma	68 Kgag-ros	PM,C	Tyr Kinase
v-mil)	CK/M	Myelocytomatosis Carcinoma Sarcoma	100 Kgag-mil	C	
v-myc)	CK		57-200 Kx-myc	N (matrix)	DNA binding, in vitro
v-ets)	CK	Erythoblastosis Myeloblastosis	135 Kgag-myb-ets	N	
v-myb)	CK		48 Kenv-myb		
v-erbA)	CK	Erythoblastosis Sarcoma	75 Kgag-erbA	C	EGF receptor homology
v-erbB)			74 KerbB	PM	
v-ski	CK	Sarcoma	125 Kgag-ski	N	
v-rel	TK	Reticuloendothe-liosis	58 Krel	N	
v-abl	M	B-cell leukemia	90-160 Kgag-abl	PM	Tyr Kinase
v-H-ras / v-K-ras	RA	Erythroleukemia Sarcoma	21 Kras	PM	GTP - binding
v-mos	M	Sarcoma	37 Kenv-mos	C	
v-fos	M	Osteosarcoma	P55fos	N	

Table 1 (continued)

v-fms	C	Sarcoma	120-180gag-fms	PM	Glycoprotein
v-fgr	C	Sarcoma	70 Kgag-fgr	n.d.	Tyrosine Kinase
v-sis	MO	Sarcoma	28 Kenv-sis	secreted	Homology to PDGF

1 - CK, chicken ; C, cat ; M, mouse ; TK, turkey ; RA, rat ; MO, monkey.
2 - Different sizes of product are from different isolates with different genetic structures
3 - PM, plasma membrane ; C, cytoplasm ; N, nucleus.

Table 2 - Oncogenes defined by transfection experiments

Cellular oncogene	Species/Tumor	Oncogene product		Biochemical properties
		Size	Intracellular location	
c-Ha-ras-1	Human	21 Kd	PM	GTP binding
c-Ki-ras-2	Human	21 Kd	PM	GTP binding
N-ras	Human/sarcomas, neuroblastomas Leukemias	18 Kd	PM	GTP binding
B-lym-1	Avian/Human B. lymphoma "inter-mediate stage"	ND		Homology to transferin gene family
T-lym-1	Human T-lymphoma "intermediate stage"	ND		

ND, not determined

7

product has been shown to be highly homologous to at least part of the human receptor for Epidermal growth factor (EGF), thus suggesting that c-erb B might encode the EGF receptor (11). The purified EGF receptor, as other growth factor receptors (PDGF receptor, insulin-like growth factor receptor) is associated with a tyrosine-specific protein kinase the activity of which is increased upon EGF binding (12).

It is believed that expression of the oncogenes of this class in some way interferes with cellular growth and differentiation by altering the balance of phosphorylation - dephosphorylation of specific target proteins involved in cell growth regulation. The exact nature of these targets still needs to be defined although several candidates have been proposed (reviewed in ref.13).

The second class of oncogenes is composed of the murine sarcoma viruses v-Ha-ras and v-Ki-ras and their cellular progenitors as well as the N-ras gene found to be activated in several human tumors by transfection experiments. These genes encode plasma membrane-associated proteins of m.w. 21 000 ($p21^{ras}$). These proteins bind guanine nucleotides with high affinity and viral $p21^{ras}$ can undergo autophosphorylation on threonine residues (14,15). The significance of these biochemical properties of $p21^{ras}$ are still obscure.

The third class of oncogenes includes those encoding nuclear proteins. The products of the unrelated myc, myb, fos and ski oncogenes (Table 1 and 3) have been found to be localized in the cell nucleus, most probably in the nuclear matrix, a cellular compartment specialized in DNA replication. Evidence has been presented that the 110 000 m.w. translation product of the MC 29 v-myc gene binds to DNA in vitro (16) and that this activity correlates with macrophage transformation by MC 29 (17).

Table 3 - Families of oncogenes products

1. - Tyrosine kinase family - src, fps/fes, yes, ros, mil/raf, erbB, abl, mos, fgr, fms

2. - GTP-binding proteins - Ha-ras, Ki-ras, N-ras

3. - Nuclear proteins - myc, myb, fos, ski

4. - Proteins related to growth factors - sis, B-lym-1

5. - Unclassified - erbA, ets, rel, T-lym-1

The fourth class of oncogenes includes those encoding polypeptides related or identical to growth factors. The prototype of these is the v-sis gene of the SSV simian sarcoma virus. The aminoacid sequence of v-sis is virtualy identical to that of one of the polypeptide component of platelet derived growth factor (PDGF ; ref. 18,19).

It is hypothesized in this case that increased cell growth proliferation would be the result of autocrine stimulation of cells synthesising and secreting a PDGF-like growth factor.

Several oncogenes (erb A, ets, rel) de not group into a defined class yet mainly because of the lack of information about the biochemical function and localization of their product.

5. Function of cellular oncogenes in normal cells

The high phylogenetic conservation of the progenitors of viral oncogenes and of the cellular oncogenes defined by the transfection assay has usually been considered as evidence for a central role of their products in the control of vital cellular functions. Consistent with this hypothesis, the products of several oncogenes have been found highly homologous or identical to previously known proteins involved in the control of cell proliferation in animal cells (c-erb B vs. EGF receptor, c-sis vs PDGF, see section 4) or yeast. Indeed, the products of several yeast genes known to be involved in the control of cell division (CDC genes) have been shown to share statistically significant structural homology with those of several oncogenes e.g. the product of CDC 28 and those of the src gene family (20) or the products of CDC 4 and CDC 36 and that of the ets gene of avian leukemia virus E26 (21).

The involvement of c-onc gene expression during development is based on limited studies indicating time-specific or lineage-specific expression of these genes during pre- and post-natal development. The numerous but still disperse observations made along these lines have been comprehensively reviewed recently and will not be detailed here (22).

6. Involvement of cellular oncogenes in tumor cells

A considerable body of evidence indicates that altered versions of cellular oncogenes ('activated oncogenes') are probably involved in the genesis of tumors with no retroviral etiology.

The best documented case is without any doubt the B cell tumors of mouse and human origins where one of the alleles of the c-myc gene is translocated from its normal position on chromosome 15 (mouse) or 8 (man) to the immunoglobulin gene locus on chromosome 12 (mouse) or 14 (man). The result of this transloca-tion is either increased transcriptional activity of the translocated c-myc gene, generation of aberrant c-myc transcripts or of mutated myc gene products (23, 25-26).

In addition to translocations, cellular oncogene amplification and rearrangements have been observed in many different tumors : c-myc amplification in a human promyelocytic cell line, HL 60 (27) and in highly malignant variants of small cell lung carcinomas, c-Ki-ras amplification in a mouse adrenocortical tumor (28) , c-abl amplification in a human chronic myelogenous leukemia cell line, K562 (29), N-myc amplification in neuroblastomas and retinoblastomas (30-31). In all these cases, amplification of the oncogene results in an increased expression of the genes.

As already mentionned, transfection experiments on mouse NIH 3T3 cells with DNA extracted from human tumor tissues or tumor cell lines has allowed the definition of activated cellular oncogenes in these tumors. Using this technique, all members of the c-ras gene family have been shown to be activated in about 10 % - 20 % of the carcinomas tested as well as in some sarcomas, leukemias and neuroblastomas.

In several of these cases, the lesion responsible for the activation of the transforming potential of the c-ras genes has been demonstrated to be a point mutation resulting in a single amino acid substitution in the $p21^{ras}$ protein (32,33). Genes activated so far only in B and T cell lymphomas have also been defined by transfection experiments (34,35).

Epidemiological and pathological studies indicate that tumor development is a multi-step proscess. The discovery of the involvement of cellular oncogenes in tumor development has led to the suggestion that several of these ill-defined steps might in fact represent the activation of distinct oncogenes.
Suggestive evidence along these lines has been provided by the analysis of several different tumors in which two separate oncogenes were indeed found activated. The HL 60 cell line and Burkitt lymphoma cells, in addition to the increased transcription of c-myc as the result of amplification or chromosomal translocations, express altered versions of N-ras and B-lym respectively (34, 36). Similarly B lymphomas induced by Avian leukosis viruses (ALVs, retroviruses with no onc gene in their genome, see Section 1), in addition to increased c-myc transcription as the result of integration of an ALV provirus next to c-myc (37,38), express an activated version of B-lym as detected by DNA transfection experiments (39,40).
Direct experimental evidence for the cooperating effect of two oncogenes in vitro to generate a fully transformed phenotype has been provided recently (41, 42). In these experiments, full transformation of primary embryo fibroblasts could only been achieved by the cotransfection of two activated oncogenes (e.g. ras and myc). Transfection of either of these genes alone resulted in either no effect (myc) or partial transformation (ras).

In all naturally occuring tumors containing one or more activated oncogenes it remains to be established whether these modifications are etiological or associated with the progression of the tumors to a more malignant phenotype.

7. Cooperating oncogenes : retroviruses as model systems

The genome of three avian acutely transforming retroviruses has recently been shown to contain two distinct and unrelated oncogenes. Avian erythroblastosis virus contains the oncogenes v-erb A and v-erb B (43,44), Avian acute leukemia virus E26 contains the oncogenes v-myb and v-ets (43,45) and Avian carcinoma virus MH2 contains the oncogenes v-myc and v-mil (43,46).
Delineating the respective functions of each of these oncogenes in these viruses and of their cellular counterparts in normal cells is a major goal of our laboratory and will cleary help to understand the significance of the cooperation of two oncogenes to the final phenotype of the tumors induced by these viruses.

1. AEV (strain ES4)

The v-erb A and v-erb B genes of AEV are expressed as two independent transcription units into a cytoplasmic 75 000 m.w. gag-erb A fusion protein and a 74 000 m.w. plasma membrane glycoprotein, respectively (47,48).

AEV induces erythroblastosis in vivo and transform bone marrow cells into CFU-E-like erythroblasts in vitro. However, unlike normal CFU-E cells, AEV transformed cells are capable of extensive self-renewal and do not differentiate into erythrocytes. Deletion mutants of either v-erb A or v-erb B have been generated in vitro from a molecularly cloned AEV (49). Studies of the in vitro and in vivo oncogenic potential of these mutants has indicated that although the v-erb A$^+$ v-erb B$^-$ mutant lacked detectable transforming activity, the v-erb A$^-$ v-erb B$^+$ mutant induced erythroblasts transformation. However, unlike wild type AEV transformed erythroblasts, cells transformed by the v-erb A$^-$ v-erb B$^+$ mutant grew only in complex culture medium and have the potential to differentiate spontaneously. Thus, although v-erb A alone is not transforming, its expression is required together with that of v-erb B to block the differentiation of transformed cells and increase their self-renewal capacity. The v-erb A gene has been sequenced recently and shown to have significant homology to the family of the carbonic anhydrases (50).

2. E 26

The v-myb and v-ets genes of E 26 are expressed as a triple fusion polyprotein of m.w. 135 000, located in the nucleus of transformed cells (51, 52).

The virus induces both myeloblastosis and erythroblastosis _in vivo_ and tranforms bone marrow cells into erythroblasts and myeloblasts _in vitro_. Unlike AEV erythroblasts, E 26 erythroblasts are strictly dependent of erythropoietin-like factors for growth and have the potential to differentiate spontaneously (53). Similarly growth of E26 myeloblasts is strictly dependent upon the presence of a growth factor present in the medium of concanavalin A - stimulated spleen cells (53). The respective contributions of v-myb and v-ets in these properties still needs to be assessed.

3. M H 2

The v-mil and v-myc genes of MH2 are expressed as independent transcription units into a cytoplasmic 100 000 m.w. gag - mil fusion protein and a 61 000 m.w. v-myc nuclear protein (54,55).
MH2 induces carcinoma _in vivo_ and transforms fibroblasts, macrophages and epithelial cells _in vitro_. The respective role of v-myc and v-mil in the transforming potential of the virus is poorly understood. It appears clear that expression of v-myc alone is sufficent to induce _in vitro_ fibroblast and macrophage transformation (unpublished observations and T. Graf, pers. comm.). Expression of v-mil however seems to be required to dramatically increase the proliferative capacity of neuroretina cells transformed by MH2 (56) and MH2 transformed macrophages (57). Whether this increased growth potential requires cooperation between the products of v-myc and v-mil and how these _in vitro_ observations relate to the oncogenic properties of MH2 _in vivo_ are still open questions.

Acknowledgements
    The authors thank Mrs. MEURILLON and Mrs. LECLERQ for patient typing of the manuscript.

References
1. RNA Tumor Viruses - R. Weiss, N. Teich, H. Varmus and J. Coffin, Ed., Cold Spring Harbor Laboratory, 1982.

2. Hanafusa, H. Cell Transformation by RNA Tumor Viruses - In : Comprehensive Virology, H. Fraenckel Conrat and R. Wagner Ed., 1977, 8, 401-488.

3. Vogt, P.K. - Genetics of RNA Tumor Viruses - In : Comprehensive Virology, H. Fraenckel Conrat and R. Wagner, Ed., 1977, 9, 341-430.

4. Cooper, G.M. - 1982 - Science 218, 801-806.

5. Beug,H., Kirchbach,A., Doderlein,G., Conscience,J.F. and Graf, T - 1979 - Cell 18, 375-390.

6. Gazzolo, L., Samarut,J., Bouabdelli, M. and Blanchet,J.P. - 1980 - Cell 22, 683-691.

7. Stehelin, D., Varmus,H.E., Bishop,J.M. and Vogt, P.K. - 1976 - Nature 260, 170-173.

8. Collett, M.S and Erikson,R.L. - 1979 - Proc Nath. Acad. Sci. USA 75, 2021-2024.

9. Oppermanh,H., Levinson,A.D. , Varmus, H.E. , Levintow, L. and Bishop, J.M. - 1979 - Proc Natl. Acad. Sci. 76, 1804-1808.

10. Rohrschneider,L.R 1980 Proc Natl. Acad. Sci. USA 77, 3514-3518.

11. Downnard J., Yarden Y., Mayes E., Scrace G., Totty N., Stockwell P., Ullrich A., Schlessinger J. and Waterfield M.G., 1984 - Nature 307, 521-527.

12. Ushiro H, and Cohen S. (1980) J. Biol. Chem. 255, 8368.

13. Ghysdael J. - In : Gene expression in normal and transformed cells. J.E. Celis and R. Bravo Ed. Nato-Asi Series, Plenum Press, NY - London (1983).

14. Langbeheim H., Shih T.Y. and Scolnick E.M. (1980) Virology 106, 292-300.

15. Shih T.Y., Papageorge A.G., Stokes P.E., Weeks M.O. and Scolnick E.M., 1980 - Nature 287, 686-691.

16. Donner P., Greiser M.I. and Moelling K. - 1982. Nature 296, 262-266.

17. Donner P., Bunte T., Greiser-Wilke I and Moelling K. - 1983 - Proc. Natl. Acad. Sci. USA 80, 2861-2865.

18. Doolittle R.F., Hunkapiller M.W., Hood L.E., Devare S.G., Robbins K.C., Aaronson S.A. and Antoniades H.N. - 1983 - Science 221, 275-277.

19. Waterfield M.D., Scarce G.T., Whittle N., Stroobant P., Johnsson A., Wasteson A., Westermark, B., - 1983 - Nature 304, 35-39.

20. Lörincz A, and Reed S.I. - 1984 - Nature 307, 183.

21. Peterson T., Yochem J., Byers B., Nunn M., Duesberg P.H., Doolittle R.F. and Reed S.I. - 1984 - Nature 309, 556-558.

22. Müller R. and Verma I.M. - 1984 - Curr. Top. Microb. Immunol. 112, 74-100.

23. Shen-Ong G., Keath E., Piccoli S.P. and Cole M.D. - 1982 - cell 31, 443-452.

24. Taub R., Kirsch I., Morton C., Lenoir C., Swan D., Tronick S., Aaronson S. and Leder P. - 1982 - Proc. Nath. Acad. Sci. U.S.A., 79, 7837-7841.

25. Adams J., Gerondakis S., Webb E., Corcoran L.M. and Cory S. - 1983 - Proc. Natl. Acad. Sci U.S.A., 80, 1982-1986.

26. Stanton L.W., Watt R. and Marcu K.B. - 1983 - Nature 303, 401-406.

27. Dalla-Favera R., Wong-Staal F., Gallo R.C. - 1982 - Nature 299, 61-63.

28. Schwab M., Alitalo K., Varmus H.E., Bishop J.M. and George D. - 1983 - Nature 303, 497-501.

29. Collins S. and Groudine M. - 1982 - Nature 298, 679-681.

30. Brodeur G.M., Seeger R.C., Schwab M., Varmus H.E. and Bishop J.M. - 1984 - Science 224, 1121-1124.

31. Lee W.H., Murphee A.L. and Benedict W.F. - 1984 - Nature 309, 458.

32. Tabin C.J., Bradley S.M., Bargmann I., Weinberg R.A., Papageorge A.G., Scolnick E.M., Dhar R., Lowy D.R. and Chang E.M. - 1982 - Nature 300, 143-149.

33. Yasa Y., Srivastava S.K., Dunn C.Y., Rhimm J.C., Reddy E.P. and Aaronson S.A. - 1983 - Nature 303, 775-779.

34. Diamond A., Cooper G.A., Ritz J. and Lane M.A. - 1983 - Nature 305, 112.

35. Lane M.A., Sainten A., Doherty K.M. and Cooper G.M. - 1984 - Proc. Natl. Acad. Sci. U.S.A. 81, 2227-2231.

36. Murray M.J., Cunningham J.M., Parada L.F., Dautry F., Lebowitz P., Weinberg R.A. - 1983 - Cell 33, 749-757.

37. Neel B.G., Hayward W.S., Robinson H.L., Fang J. and Astrin S.H. - 1981 - Cell 23, 323-334.

38. Payne G.S., Bishop J.M. and Varmus H.E. - 1982 - Nature 295, 209-214.

39. Cooper G.M. and Neiman P.E. - 1981 - Nature 292, 857-758.

40. Goubin G., Goldman D.S., Luce J., Neiman P.E. and Cooper G.M. - 1983 - Nature 302, 114-119.

41. Land H., Parada L.F. and Weinberg R.A. - 1983 - Nature 304, 596-602.

42. Ruley H.E. - 1983 - Nature 304, 602-606.

43. Roussel M., Saule S., Lagrou C. Rommens C., Beug H., Graf T. and Stehelin D. - 1979 - Nature 281, 452-455.

44. Vennström B. and Bishop J.M. - 1982 - Cell 28, 135-143.

45. Leprince D., Gegonne A., Coll J., De Taisne C., Schneeberger A., Lagrou C. and Stehelin D. - 1983 - Nature 306, 395-397.

46. Coll J., Righi M., De Taisne C., Dissous C., Gegonne A. and Stehelin D. - 1983 - Embo J. 2, 2189-2194.

47. Hayman M.J., Royer-Pokora B. and Graf T. - 1979 - Virology 92, 31-45.

48. Hayman M.J., Ramsay G.M., Savin K., Kitchener G., Graf T. and Beug H. - 1983 - Cell 32, 579-588.

49. Frykberg L., Palmieri S., Beug H., Graf T., Hayman M.J. and Vennstrom B. - 1983 - Cell 32, 227-238.

50. Debuire B., Henry C., Benaissa M, Biserte G, Claverie J.M., Saule S., Martin P. and Stehelin D. - 1984 - Science 224, 1456-1459.

51. Bister K., Nunn M., Moscovici C., Perbal B., Baluda M. and Duesberg P. - 1982 - Proc. Natl. Acad. Sci U.S.A. 79, 2677-3681.

52. Klempnauer K.H., Symonds G., Evan G.I. and Bishop J.M. - 1984 - Cell 37, 537-547.

53. Radke K., Beug H. and Graf T. - 1982 - Cell 31, 643-653.

54. Hu S., Moscovici C. and Vogt P.K. - 1978 - Virology 89, 162-178.

55. Hann S.R., Abrams H.D., Rohrschneider L.R. and Eisenman R.N. - 1983 - Cell 34, 789-798.

56. Bechade C., Pessac B., Calothy G., Coll J., Martin P., Saule S., Ghysdael J. and Stehelin D. - 1984 - Submitted for publication.

57. Adkins B., Leutz A. and Graf T. - 1984 - Submitted for publication.

RESUME

Les propriétés oncogènes des rétrovirus transformants résultent de l'expression d'oncogènes viraux (v-onc). Les oncogènes viraux sont dérivés de gènes cellulaires (proto-oncogènes ou gènes c-onc) Une vingtaine de gènes v-onc/c-onc ont été identifiés et isolés. Les tumeurs humaines, dont la plupart ne possèdent pas d'étiologie rétrovirale, montrent fréquemment des réarrangements ou une expression inapropriée de gènes c-onc. Nous résumons ici le statut des connaissances actuelles de la structure et fonctions des gènes v-onc/c-onc ainsi que l'implication de ces gènes dans le contrôle multigénique du développement tumoral

Hormones and Cell Regulation, Volume 9
INSERM European Symposium
J.E. Dumont, B. Hamprecht and J. Nunez editors
© 1985 Elsevier Science Publishers B.V. (Biomedical Division)

THE MOLECULAR BASIS FOR PHENOTYPIC MODULATION IN CELLS CONTAINING
AN INTEGRATED VIRAL src ONCOGENE

JOHN A. WYKE   MINA J. BISSELL* DAVID A. F. GILLESPIE AND
PEDROS LEVANTIS

Imperial Cancer Research Fund Laboratories, St. Bartholomew's
Hospital, Dominion House, Bartholomew Close, London EC1A 7BE.

INTRODUCTION

Neoplastic conversion is generally an irreversible process and
the cellular changes underlying it are stable. Two mechanisms
embody this necessary constancy.

1) Changes in the cell DNA sequence, leading to alterations in
gene products or in the pattern of gene expression. Two examples
of this have been detected so far in tumour cells and might play a
part in neoplasia; point mutations in proto-oncogenes and gross
translocations of chromosomal material. It is quite feasible that
between extremes of fine and coarse disruptions of DNA sequence
there lies a range of mutational events of varying degrees that have
yet to be discerned.

2) Epigenetic alterations that do not affect the DNA sequence but
permanently change the pattern of gene expression. Such modulations
presumably occur during normal metazoan growth and differentiation
and, if they play a role in neoplasia, this implies a perversion of
their usual function in development. However, we are still largely
ignorant of the precise mechanisms by which various epigenetic
changes affect gene activity and their significance in neoplasia
remains uncertain. Moreover, where reversion of neoplasia occurs
in vivo it is not known whether an underlying genetic change has
been eliminated or whether gene expression has been modulated.

Our attention has been directed to these considerations by our
studies on Rous sarcoma virus (RSV). In common with other retroviruses, RSV integrates into the DNA of infected cells as a provirus
whose structure resembles that of a transposable element, the coding
regions of the virus being flanked by long terminal repeats (LTR).
These repeats contain promoter and enhancer elements that efficiently direct transcription of viral genes (1-3), including the oncogene
v-src whose activity is responsible for tumour induction in chickens
and transformation of a number of cell types in tissue culture.
Some years ago we set out to study the behaviour of the integrated

RSV proviruses in rat cells, using morphological transformation by
v-src as a marker for proviral expression. The choice of rat rather
than chick as host permitted an analysis of individual proviral-cell
interactions, for not only were the cells readily cloned but they
are non-permissive for virus replication, precluding virus spread
through the infected population. However, another reason for studying RSV in rat cells, and the stimulus behind the venture, was the
intriguing variety of behaviour exhibited by RSV proviruses in
mammalian hosts (4-6). B77 strain RSV efficiently penetrates cells
of the line Rat-1 and integrates its provirus into the cell genome,
yet only about one in a thousand infected cells becomes transformed.
Moreover, the phenotype of many (but not all) transformed clones is
unstable. They segregate morphologically normal (revertant)
daughter clones that are themselves unstable, reconverting to a
transformed morphology at a low but detectable frequency. These
behavioural peculiarities raise two major sets of questions.

1) Why is only a small minority of integrated RSV proviruses in
Rat-1 cells expressed and what determines that the majority are
phenotypically silent? Since the RSV LTR efficiently directs transcription in chick cells (and, indeed, in short term assays in
mammalian cells (2)), the factors that dictate its functions when
integrated in Rat-1 DNA are of considerable interest.

2) What mechanisms regulate the reversion and re-transformation of
RSV-transformed Rat-1 cells? Suppression of the transformed phenotype is an appealing observation for cancer researchers, but we also
hoped that these phenotypic modulations would provide malleable
materials with which to learn more about the mechanisms regulating
gene activity in general.

These two groups of questions may, of course, overlap considerably,
since failure to transform ab initio and reversion of the transformed phenotype could be different manifestations of the same
regulatory processes. However our data do not yet lead to this
conclusion, so we will deal with these questions separately, returning
to discuss the significance of our findings so far for the concepts
posed at the beginning of this article.

METHODS, RESULTS AND DISCUSSIONS
Why are only a small minority of RSV-infected Rat-1 cells transformed?

Early studies showed that transformation was not a property of a rare, highly transforming virus, nor of a rare highly transformable cell. The evidence can be summarized as follows. 1) Virus rescued from transformed rat cells (by fusion with permissive chick cells) was not markedly better at transforming rat cells than were virus clones that had never been passaged in rat (4,6). 2) Subclones of a transformed cell line that were morphologically normal because of mutations in v-src were no more susceptible to transformation than rat cells that had never been infected (6,7). 3) Morphological revertants of the kind described below can be re-transformed by fresh infection with RSV. Their susceptibility to transformation is the same as that of uninfected Rat-1, even when the transforming virus is a rescued descendant of the proviral genome that they already bear (8). Thus, transformation depends on factors peculiar to individual virus-cell interactions, not on heritable character-istics of the virus or the host. The last experiment in particular shows that separately integrated, but genetically identical, proviral genomes in the same host cell are affected by different factors.

The possible importance of integration. What are these factors? One feature of RSV infection that shows the prerequisite variability at infection and stability thereafter is the site in the host genome at which the provirus inserts. Several groups have shown that even in the small proportion of mammalian cells that become transformed, the number of detected integration sites is very large (6,9,10). Indeed, it is possible, but not proven, that the provirus integrates at random in the cell DNA. Unfortunately, this plethora of inte-gration sites makes it difficult to test whether insertion at a subset of them is crucial for transformation since we cannot yet direct proviral integration to specific locations. However, our examination of the structure of a number of integrated proviruses in transformed rat cells provides indirect evidence for the import-ance of integration site, as well as intriguing demonstrations of the plasticity of the mammalian genome.

Rat-1 cells were infected with very low multiplicities of B77 strain RSV (0.01 to 0.1 infectious units per cell) and transformed clones were subsequently isolated by micromanipulation. Restriction mapping of high molecular weight DNA from these clones showed that in most of them proviral sequences were integrated at single locations in the host genome, but their arrangement was often complex (Gillespie et al., submitted for publication; S. Searle,

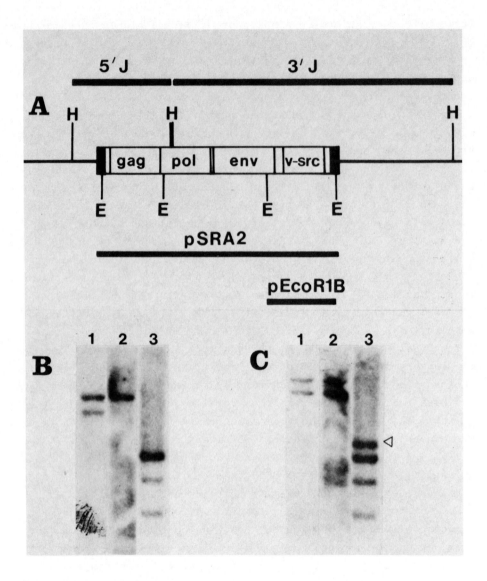

Fig. 1. Structure of integrated RSV proviruses with and without
adjacent genomic rearrangements. Panel A is a diagram of an integrated provirus, comprising v-src and the replicative genes, gag,
pol and env (open boxes) flanked by LTRs (solid boxes). The
restriction enzyme Hind III (H) cleaves the provirus at two closely
adjacent sites to generate, for each integrated provirus, two
discernible fragments whose size depends on the location of Hind III
sites in flanking rat DNA. 5' and 3' Hind III junction fragments
(5'J and 3'J) have been cloned from the infected lines A11 and B31.
Their use is shown in Table 1. Eco RI (E) cleaves the provirus in
the LTRs and twice more to generate 3 characteristic fragments from

(Fig. 1 legend; continued) a complete virus. Panels B and C show
DNA transfer experiments with two transformed clones digested with
<u>Hind</u> III (lanes 1 and 2) or <u>Eco</u> R1 (lane 3). Virus specific
fragments are seen by hybridization to probes prepared from cloned
unintegrated RSV provirus (11) and comprising either the whole
provirus (pSRA2; lanes 1 and 3) or subgenomic regions such as the
<u>Eco</u> RI fragment encompassing <u>v-src</u> (pEcoRIB; lane 2). The provirus
in Panel B shows no rearrangement whilst that in Panel C has an
additional <u>Eco</u> RI fragment (arrow head). Since the probe pEcoRIB
reveals both 5' and 3' Hind III junction fragments, the rearrangement is 5' to the provirus and contains sequences from its 3'
portion.

A. Green and M.J.B., unpublished data). All of 15 transformed
clones examined contained an apparently complete RSV provirus, but
9 clones showed, in addition, adjacent duplications of proviral
sequences (Fig.1). In six clones investigated further the duplications were invariably 5' to the integral provirus and in one clone,
A11, studied in detail, the 5' duplication includes flanking host
as well as proviral sequences (Gillespie et al., submitted for
publication; Fig. 2).

<u>What is the purpose of proviral rearrangements</u>? As far as we
know, such duplications in RSV infected rat cells have not been
reported before. Their frequent occurrence in these transformed
clones and their consistent positioning 5' to the integrated
provirus are thus significant. Moreover, rearrangements of the sort
seen in clone A11 must result from several recombination events and
are unlikely to occur commonly unless the rearrangement confers some
advantage on the infected cell. Since we have selected only for
morphological transformation, it is probable that the rearrangements
favour the transformed phenotype and there are two possible reasons
for this phenomenon. Either the rearrangement places the provirus
near a positive regulatory element or it alleviates some feature
of the integration site that is inimical to provirus expression.
Two sets of data are informative on this point.

1) Molecular cloning of the integrated provirus in the transformed line A11 (Gillespie et al., submitted for publication) has
provided a probe for flanking cell DNA that has been used to examine
and clone the corresponding single copy sequences from the
unoccupied integration site in uninfected Rat-1 cells (Fig.2).
In the uninfected cell these sequences are transcribed at a low
level but are resistant to digestion with pancreatic DNase 1,
suggesting that they are in a relatively inactive region of
chromatin. In A11 these sequences show an increase in both

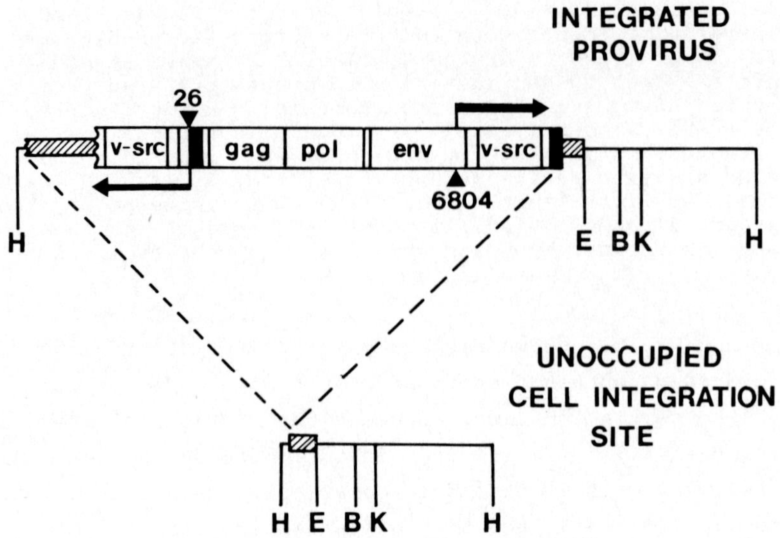

INTEGRATED PROVIRUS

26

v-src | gag | pol | env | v-src

6804

H   E BK   H

UNOCCUPIED CELL INTEGRATION SITE

H E BK   H

Fig. 2. The structure of the provirus in clone A11 and of the cell location at which it integrated. The insertion comprises a provirus and 5' flanking duplications of 1) sequences from the 3' portion of the provirus including v-src and part of env and 2) sequences from the cell immediately to the right of the integrated provirus (the hatched box between the 3' LTR and the adjacent cell Eco RI site shows the region used as probe to detect cell duplications). The 3' LTR has not been duplicated. The junction of the provirus and duplicated sequences is precise, occurring between the 5' LTR at a point 26 nucleotides from its 5' boundary and nucleotide 6804 in env (vertical arrow heads). The duplicated provirus sequences are in the opposite orientation to their native configuration (horizontal arrows). We have not yet determined the orientation and detailed structure of the duplicated cell sequences. For simplicity, only restriction enzyme sites that flank the insertion are shown; H, Hind III; B, Bam HI; K, Kpn I; E, Eco RI.

transcription and DNase 1 sensitivity, presumably as a consequence of proviral integration and rearrangement. Comparable experiments with a transformed line, B31, in which the provirus is not rearranged, show the cellular integration site to be highly sensitive to DNase 1 in uninfected Rat-1 cells, although not detectably transcribed (data not shown). These findings suggest that rearrangements 5' to the provirus can mitigate the negative influence of an unfavourable integration site, although we cannot rule out the possibility that integration occurred in one of a minority of cells in which this region happened to be in a favourable state.

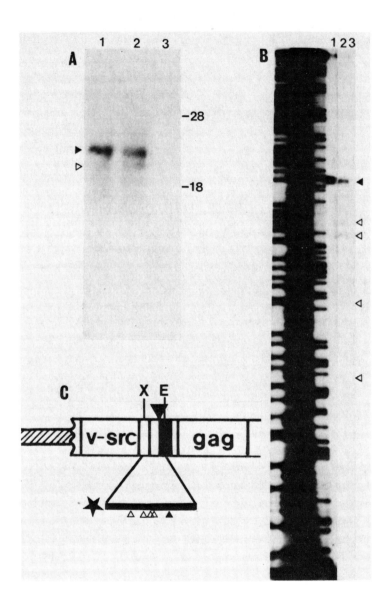

Fig. 3. Transcription of v-src sequences in RSV infected rat cells.
Panel A; transfer of poly A-selected RNA using a probe specific for
the coding region of v-src. Panel B; nuclease S1 protection
experiment in which total cell RNA protects an Xho I - Eco RI fragment from line A11 (sequence) whose location and position of radioactive label (*) is shown in Panel C. Lanes 1 show the unrearranged
transformed line B31, lanes 2 the transformed line A11 and lanes
3 a revertant of A11. Total RNA in B31 and A11 protects a major

(Fig. 3 legend; continued) fragment extending from the recombination between LTR and env in A11 (see Fig 2) to the Xho I site (Panels B and C, solid arrow heads). Their major mRNA is the same size as typical RSV v-src message (Panel A, solid arrow head; positions of 18s and 28s ribosomal RNAs are shown) and is initiated at the same site (data not shown). A11 contains another, far less abundant mRNA (Panel A, open arrow head) that is initiated at several sites in the env region of the proviral duplication (Panels B and C, open arrow heads). No transcription from the revertant line is detected in either Panels A or B.

2) The rearrangements studied so far do not consistently duplicate any particular region of the provirus, including the LTRs that contain proviral promoters and enhancers. Moreover, the 5' location of the duplications tends to distance most of the provirus from any positive regulatory elements in the cells. However, in A11 and some other (but not all) transformed cells examined the duplications involve v-src and it can be postulated that transformation in these lines results from expression of duplicated v-src genes. Studies on transcription in A11 by both RNA "blotting" and S1 nuclease experiments argue against this hypothesis. The major src-specific mRNA in the cell is of the same size, and is initiated at the same site, as the authentic viral src message (Fig.3). A transcript from the duplicated v-src gene, initiated from cryptic promoters in the duplicated region of the env gene (Fig.2) is present at far lower intensity (Fig.3).

It can still be argued that the unrearranged v-src gene in line A11 is defective in some way. The ready rescue of transforming RSV from this clone makes this unlikely, but to pursue the point we examined the ability of DNA from A11 to transform NIH 3T3 cells, comparing molecular clones containing either the unrearranged or duplicated v-src sequences with the src gene from B31, a transformed line with no proviral duplication in which the identity of the transforming gene is thus unequivocal (Table 1). Clones containing the 3' half of the A11 provirus (including v-src) and flanking cell DNA behaved identically to equivalent clones from B31. Neither alone detectably transformed NIH 3T3 (unlike a molecular clone of a comparable region of unintegrated RSV provirus) but both produced transformants when co-transfected with a clone containing 5' sequences from B31 which provided a proviral LTR in the appropriate position and orientation. Thus, v-src in the integral provirus is functional. More surprisingly, the clone containing only the duplicated A11 v-src sequence transforms with high frequency and

TABLE 1

TRANSFORMATION OF NIH3T3 CELLS BY CLONED PROVIRAL DNA

Donor DNA*	Foci per pmole v-src (x10^{-3})	Time to observed transformation (days)
A11, 3'J	<0.01	–
B31, 3'J	<0.01	–
pEco RIB	0.6	7-10
A11, 3'J + B31, 5'J	0.07	7-10
B31, 3'J + B31, 5'J	0.07	7-10
pSRA2	7.0	7-10
A11, 5'J	10.0	3-6

* The structure of donor DNA is shown in Fig.1.

unusual rapidity. This contrasts markedly with the low level of transcription of the duplicated v-src in A11 cells, adding further weight to the concept that the cell environment is unfavourable to proviral expression but the duplicated region 5' to the provirus bears the brunt of this inhibitory effect. Removal of cell sequences immediately flanking the duplicated v-src gene does not further enhance its transforming ability (Fig.4) so it is likely that the postulated inhibitory portions of the cell chromatin are outside the 20 kilobases of genome (including integrated provirus) that we have examined so far. Our attention is now turning to regions further 5' to the virus in the hope of detecting cellular elements that can inhibit RSV provirus expression over long distances.

Low frequency activation of "silent" proviruses does not involve rearrangement. The experiments considered so far provide circumstantial evidence for the importance of duplications 5' to the provirus in determing its expression. The concept would be strenghtened if we could show that inactive normal proviruses become active as a consequence of duplication and rearrangement. Other workers had previously shown that the majority of rat cells infected but not transformed by RSV harboured proviruses (4) and that these cells would segregate transformed derivatives at a low frequency, suggestive of a somatic mutation (12). We examined these claims by studying Rat-1 cells cloned by micromanipulation

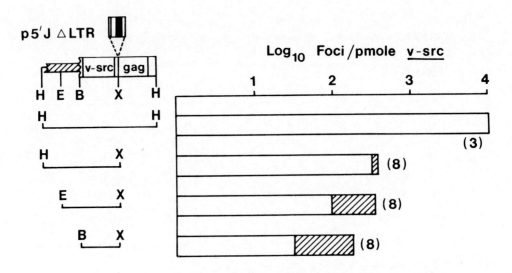

Fig. 4.   Transforming ability of cloned DNAs containing the
duplicated v-src gene of rat clone A11.   On the left of the figure
are shown the structure of the plasmid p5'ΔLTR and the fragments
derived from it that were used in transfection of NIH3T3 cells.   On
the right are shown the transformation efficiencies scored after 12
days (open bars) and 18 days (cross-hatched bars).   Numbers in
brackets are the day on which transformation was first detectable.
p5'ΔLTR is a derivative of the 5'J clone (Fig. 1) from which the
LTR has been removed by cleavage and religation at adjacent Xho1
sites (X).   Its transforming efficiency equals that of 5'J (compare
with Table 1) showing that elements in the LTR do not apparently
influence in-vitro transformation by the duplicated v-src.   Removal
of DNA to the right of X (in the gag region of the provirus) reduced
transformation 20-30 fold.   Further removal of DNA to the left of
v-src delays the detection of transformation but does not greatly
impair the final efficiency.   No fragment transforms better than
5'J.

16 hours after infection with RSV, extending to these clones a

molecular analysis of the integrated proviruses.   Logistics dictated

that we use a higher multiplicity of RSV infection than we used to

obtain transformed clones, and we chose 0.8 infectious units per

cell, a level that should maximise the number of clones containing

single integrated proviruses.   Table 2 shows that this was the case

and a number of clones containing 1 or 2 proviruses were identified

and maintained in culture for several months with the following

findings.

    1) Although several clones contained defective proviruses and no

rescuable RSV, only one of the clones that bore a complete provirus

showed evidence for a possible additional rearrangement of proviral

TABLE 2

CHARACTERISTICS OF CELL CLONES CONTAINING "SILENT" PROVIRUSES

Cells cloned one day after infection; all were initially of normal morphology

Number of Integrated Proviruses	Number of Clones	Clone Designation	Morphological Stability*	Provirus Structure**	Recovery of Virus***
>2	3	-	-	-	-
2	8	114,118,132, 135,149,151	N	-	-
		119,141	T	-	-
1	21	32,42,113	-	-	-
		1,6,19,123	N	D	O
		47,116	N	-	-
		2	N	U	5
		17	N	U	1
		26	N	U	60
		31	N	U	35
		33	N	U	37
		35	N	D	0
		16	T	R	-
		20	T	D	-
		22	T	U	40
		103	T	U	60
		105,130	T	U	-
0	62	-	NOT APPLICABLE		

*   N, no transformation after 17 weeks continuous culture; T, one or more transformed foci appear 3-12 weeks post infection; - not determined.

** U, complete provirus, no rearrangement; R, complete provirus + proviral rearrangement; D, provirus with internal defects.

*** Foci produced after fusing 10^5 rat cells (killed with mitomycin - C) to permissive chick cells.

sequences.

2) All clones were initially morphologically normal and in contrast to other reports (12) not all subsequently segregated transformed derivatives. Among those that did, transformation was usually infrequent and the status of the integrated provirus was identical in both the normal cells and their transformed derivatives. Thus no proviral rearrangements accompanied or preceded the rare provirus activations that occurred in these clones.

Three categories of integrated provirus. Our data so far suggest that competent RSV proviruses exist in rat cells in one of three states (defective proviruses with internal alterations comprise a fourth category).

1) Unrearranged proviruses that are transcriptionally inactive and stably maintain this phenotype. Provirus activation is rare, and since no detectable proviral rearrangements attend this event it could result from point mutations or epigenetic modulations of gene expression. Since transforming RSV can be rescued from cells harbouring single "silent" proviruses (Table 2) we presume that cell sequences or viral regulatory elements, and not v-src, are the targets for such alterations.

It would be of great interest to determine why the vast majority of RSV infected rat cells falls into this category, whereas such events have not been reported in RSV infection of chick cells. If we postulate that the site of integration is important and that proviruses can integrate at very many, possible random, sites then we must explain why only about 0.1% of integration sites permit provirus expression. If this level is less than the proportion of transcriptionally active DNA in the cell, an unfavourable chromatin structure may not be the sole arbiter of proviral "silence". A second possibility is that integration in a transcriptionally active region may impose a "transcriptional interference" on provirus expression (13). In the All clone the integration site is relatively inactive as judged by chromatin structure yet it is apparently transcribed at a low level, so either possibility could apply. We are probing these questions further by comparing the nature of cellular integration sites that have acquired both active and inactive proviruses.

2) Unrearranged proviruses that are transcriptionally active. These comprise a significant minority of transformed clones in our hands and all the transformed clones detected by others (reasons for

this discrepancy are discussed in the next paragraph). In common
with infected but untransformed cells the phenotypes of the clones
in this category that we have studied are stable, reversion occurring infrequently by mutation in v-src (7; J.A.W. unpublished data).
This stable activity presumably reflects integration at particularly
favoured sites, such as the B31 provirus integration site that is in
an active chromatin configuration but not detectably transcribed.

   3) Proviruses that are transcriptionally active and that show
rearrangements of proviral, and sometimes cell, DNA 5' to the integrated provirus. We find these comprise the majority of transformed
cells that contain single proviruses but it is not clear why they
are peculiar to our studies. In part this may be because they are
only revealed by a fairly detailed restriction enzyme analysis but,
since several other investigations have been very thorough, we
suspect their detection reflects our experimental protocol. We
used a very low multiplicity of infection to transform cells,
obtaining only about one transformed focus in each confluent tissue
culture dish. Thus, most transformed foci seen were cloned for
study and there was no unwitting selection for the more vigorous
transformed colonies. Indeed, the transformed areas we saw were
often small, or of mixed normal and transformed morphology, and all
clones subsequently obtained were phenotypically relatively unstable
and segregated morphologically normal revertants. This phenomenon
of reversion raised the next major question in our study.

   What regulates the reversion and re-transformation of RSV -
   transformed rat cells? Modulation in the morphology of RSV-
transformed Rat-1 cells is reversible, it does not involve any
detectable further alterations in the proviral restriction map
(although it is most frequent in clones that have already undergone
rearrangements flanking the provirus) and it is accompanied by
marked fluctuations in the levels of viral mRNA (8). Levels of RNA
transcribed from cell sequences flanking the provirus can show
concomitant alterations (D.A.F.G. and K. Hart, unpublished data),
all of which suggest that the observed morphological fluctuations
reflect cell mediated changes in transcription of the provirus and
its environs. These phenomena seemed to offer a particularly
versatile system to investigate gene regulation in mammalian cells
and we examined the mechanism of reversion, concentrating on
derivatives of the clone A11.

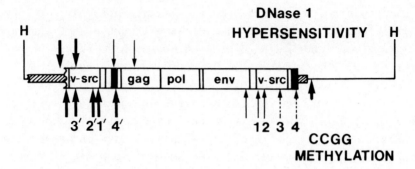

Fig. 5.   Chromatin structure in the proviral region of transformed
and revertant derivatives of A11.   The entire region between the
Hind III sites (H) is sensitive to DNase 1 in transformed cells and
relatively resistant in revertants.   Downward pointing arrows show
DNase 1 hypersensitive sites, heavy arrows being sites found in
both transformed and revertant cells, light arrows being those
found only in transformed clones.   Upward pointing arrows show Msp
1 sites (CCGG) whose methylation is examined by digestion with the
isoschizomer Hpa II.   Heavy arrows show unmethylated sites, light
arrows shows sites methylated in revertant but not transformed
clones and arrows with dashed shafts show sites methylated in some
clones irrespective of phenotype.   Numbers identify sites in
sequences that are duplicated at either end of the insert.

    Several studies have shown that the proviruses and flanking DNA
in revertant cells adopt changes characteristic of transcriptional
inertness in eukaryote cells.   Thus, there is an overall reduction
in the DNase 1 sensitivity of the provirus and its neighbouring
chromatin with a loss of some, but not all, sites that are hypersensitive to the nuclease (14,15; see Fig.5).   One hypersensitive
site that disappears is notable because it is located in the 5' LTR
of the provirus; comparable sites are found in the 5' (but not 3')
LTRs in three separate RSV-transformed Rat-1 clones (Dyson et al.,
submitted for publication).   Further evidence for alterations in
chromatin structure come from studies on the proximity of the integrated provirus to the nuclear "matrix", proviral sequences being
close to the matrix in transformed derivatives of A11 and relatively
remote in revertants (16).   Finally, proviral and flanking sequences
in revertants show an increase in methylation of cytosine in CpG
doublets as detected by cleavage with methylation sensitive
restriction enzymes (14,17; see Fig.5).   The presence of LTR
sequences at either end of the unrearranged provirus and of proviral
duplications in A11 derivatives allow us to conclude that identical
nucleotide tracts in different parts of the proviral region can

TABLE 3

TRANSFORMING EFFICIENCY OF GENOMIC DNA OF A11 DERIVATIVES

Donor DNA	Foci per µg DNA on NIH3T3 cells	HAT-resistant Colonies per µg DNA of LMTK cells*
A11 - transformed clone	0.10	0.11
A11 - revertant clone	<0.001	0.45
A11 - retransformed clone induced by 5 - azacytidine	0.07	Not tested
EJ**	0.10	0.15

*   Mouse L Cells lacking functional thymidine kinase
** A human bladder carcinoma cell line containing a transforming H-ras oncogene.

differ in CpG methylation as they can in DNase I hypersensitivity (Fig.5). Nonetheless, specific hypermethylation may play a role in reversion, for it shows a consistent pattern in clonal analyses and treatment with the methylation antagonist 5-azacytidine can induce a low level of re-transformation (17).

Cis and Trans Controls. In studying further the mechanisms regulating these changes in the integrated provirus we postulated that flanking cell sequences act in cis to determine levels of proviral activity. Transfection with high molecular weight genomic DNA from A11 derivatives supports this hypothesis; DNA from transformed cells readily transforms NIH 3T3 cells, whilst DNA from a revertant has never done so, although in testing another gene (thymidine kinase) both DNAs transform with equal facility (Table 3). The putative cis acting regulatory elements probably function by recognising the presence or absence of a trans acting factor, presumably a protein. Evidence for this comes from experiments in which certain RSV-transformed Rat-1 cells form morphologically normal hybrids when fused to normal rat or mouse cells. The trans acting regulatory factor donated by the normal parent in the hybrid not only suppresses proviral RNA levels (18) but also imposes on the provirus the changes in DNase 1 sensitivity, cytosine methylation and remoteness from the nuclear matrix that are characteristic of spontaneous morphological reversion (Dyson et al., submitted for publication).

Stratagems foiled. We reasoned that the proposed cell elements mediating reversion might be located by shearing revertant DNA and checking whether it regained transforming activity. These experiments have so far produced no positive results, for reasons that are not clear. Problems may be technical, or an inactive chromatin configuration may be inherited in a stable form after transfection, particularly if separated from elements that we conjecture propagate the signals mediating change.

As an alternative approach we tried to obtain molecular clones of the proviral region from revertant cells and have again met problems. The provirus from a transformed A11 cell is readily cloned, whereas that from a sibling revertant has never been obtained (Gillespie et al., submitted for publication; D.A.F.G. unpublished data). In contrast the unoccupied allele of the A11 proviral integration site is cloned with equal ease from both revertant and transformed cells. This intriguing problem is a set-back to immediate plans but may, in the long term, prove very informative. Its generality is being tested in attempts to clone proviruses from revertants of other transformed lines.

CONCLUSIONS AND IMPLICATIONS

The Rous sarcoma provirus is a self-sufficient genetic entity, bearing in its LTR sequence elements that promote and enhance the transcription of viral genes, including v-src. Despite the potency of these regulatory elements their effect is usually overridden when the provirus integrates in rat cells, but this subjugation is avoided in cells in which genomic rearrangements have occurred 5' to the provirus, or in cells where mutagenic or epigenetic alterations have occurred in the absence of detectable genomic rearrangements. The level of provirus expression is probably a consequence of its site of integration in the cell genome for, where examined, changes affecting the chromatin structure and activity of the provirus extend into flanking cell DNA. Since RSV proviruses can integrate at many sites, the transforming activity of v-src makes them useful sensors to detect elements in the cell that regulate gene expression but are not otherwise amenable to analysis. Our findings so far suggest that such elements operate in cis to regulate provirus activity but are affected by trans acting factors. We reason that transformation with genomic DNA should locate such regulatory regions whose DNA sequences can then be isolated by molecular cloning and tested and analysed by _in vitro_ reconstruction experiments.

We are still far from realising these aims but our work so far has a number of conceptual and practical implications.

1) Genomic rearrangements have so far only been found in transformed clones isolated 1 to 2 weeks after infection. We have not detected them in the delayed transformants segregated from phenotypically normal infected cells much later after infection. It is thus likely that their genesis is connected with events attending or soon after provirus integration. However, the structure of the proviral region in clone All shows that several recombination events must have occurred, involving cell as well as proviral sequences, and their time span is unknown.

The ability of cell sequences to participate in genomic rearrangements that potentially attenuate cell regulation of gene expression raises the possibility that the "spontaneous" occurrence of such events could lead to neoplasia. Such changes need not be gross karyological abnormalities nor even (as our findings show) alterations readily discerned in a simple restriction fragment analysis. They also need not involve changes in the coding region of the genes affected and thus may escape detection by present methods of identifying genetic alterations in neoplasia.

2) The modulations in gene activity typical of reversion, retransformation and perhaps some cases of delayed transformation are also sufficiently stable to contribute to neoplasia, particularly under the selective conditions of tumour evolution. No genetic alteration is apparent in these modulations and, although they should be identified in DNA transfer tests for activated genes (see Table 4), the behavioral differences between normal and transforming genes may not survive molecular cloning and analysis of the relevant sequences. Nonetheless, despite problems in implicating such alterations, their possible role in neoplasia should not be ignored.

3) The behaviour of the duplicated DNA in clone All also provides a _caveat_ on the use of DNA transfer to detect activated genes in cancer. NIH 3T3 transformants induced by All genomic DNA contain both unrearranged and duplicated v-src genes, but these frequently segregate independently in secondary transfections (P.L. unpublished data). Moreover, an analysis of cloned DNA would identify the duplicated v-src as the transforming gene in the cell, whereas the transcription pattern in the whole cell shows this is not the case.

The behaviour of transforming DNA isolated from its cellular context is clearly potentially artefactual.

4) The efficiency of provirus integration and the regulatory competence of the LTR have made retroviruses popular candidates as vectors in gene transfer experiments (19). However, as our studies show, expression of integrated proviruses is strongly subject to position effects and these may not parallel the desired expression patterns of the introduced genes. Other work has emphasised this point, notably the elegant investigations of Jaenisch and collaborators on the regulation during mouse development of murine leukaemia virus genomes introduced into the germ line (20). Much is known about the sequences in the vicinity of coding DNA that define it as a transcriptional unit, but we are largely ignorant of the elements that act at longer range to determine the temporal and spatial levels of gene activity. Studies on the expression of integrated proviruses should help to elucidate these controls.

ACKNOWLEDGEMENTS

We thank Tony Green, Kevin Hart and Sian Searle for helpful discussions and the provision of unpublished data and Carole Gardner and Joyce Newton for secretarial assistance.

REFERENCES

1. Yamamoto T, de Crombrugghe B, Pastan I, (1980) Cell 22;787
2. Gorman C M, Moffat L F, Howard B H (1982) Mol Cell Biol 2;1044
3. Luciw P A, Bishop J M, Varmus H E, Capecchi M R (1983) Cell 33;705
4. Boettiger D (1974) Cell 3:41
5. Varmus H E, Guntaka R V, Deng C T, Bishop J M (1975) Cold Spring Harbor Symp Quant Biol 3:987
6. Wyke J A, Quade K (1980) Virology 106:217
7. Varmus H E, Quintrell N, Wyke J (1981) Virology 108:28
8. Chiswell D J, Enrietto P J, Evans S, Quade K, Wyke J A (1982) Virology 116:428
9. Hughes S H, Shank P R, Spector D H, Kung H J, Bishop J M, Varmus H E, Vogt P K, Breitman M L (1978) Cell 15:1397
10. Collins C J, Boettiger D, Green T L, Burgess M B, Devlin B H, Parsons J T (1980) J Virol 33:760
11. DeLorbe W J, Luciw P A, Goodman H M, Varmus H E, Bishop J M (1980) J Virol 36:50
12. Turek L P, Oppermann H (1980) J Virol 35:466
13. Cullen B R, Lomedico P T, Ju G (1984) Nature 307:241

14. Chiswell D J, Gillespie D A, Wyke J A (1982) Nucl Acids Res 10:3967

15. Nicolas R H, Wright C A, Cockerill P N, Wyke J A, Goodwin G H (1983) Nucl Acids Res 11:753

16. Cook P R, Lang J, Hayday A, Lania L, Fried M, Chiswell D J, Wyke J A (1982) Embo J 1:447

17. Searle S, Gillespie D A F, Chiswell D J, Wyke J A (1984) Nucl Acids Res 12:5193

18. Dyson P J, Quade K, Wyke J A (1982) Cell 30:491

19. Williams D A, Lemischka I R, Nathan D G, Mulligan R C (1984) Nature 310:476

20. Jaenisch R, Jahner D, Nobis P, Simon I, Lohler J, Harbers K, Grotkopp D (1981) Cell 24:519

* M.J.B., was on leave of absence from Lab. Cell Biology, Lawrence Berkeley Laboratory, UC, Berkeley USA. She was supported by U.S. DOE and a Fogarty Senior Fellowship from the NIH.

RESUME

Le virus du sarcome de Rous est une entité génétique autosuffisante, qui porte dans sa séquence répétée terminale (LTR) des éléments qui induisent et augmentent la transcription de gènes viraux, notamment de l'oncogène v-src. Malgré le pouvoir de ces éléments de régulation, leur effet est habituellement annulé lorsque le provirus s'intègre dans des cellules de rat, mais cette subjugation est évitée dans les cellules où des réarrangements génomiques ont eu lieu en 5' par rapport au provirus, ou dans des cellules où des altérations mutagènes ou épigénétiques ont eu lieu en l'absence de réarrangements génomiques détectables. Le niveau d'expression du provirus est probablement une conséquence de son site d'intégration dans le génome cellulaire et, comme les provirus RSV peuvent s'intégrer en de nombreux endroits, l'activité transformante de v-src les rend utiles comme signaux permettant de détecter les éléments qui, dans la cellule, règlent l'expression des gènes mais qui ne sont pas analysables par d'autres moyens. De tels éléments opèrent apparamment en cis pour régler l'activité du provirus mais sont affectés par des facteurs agissant en trans. Une transformation par du DNA génomique devrait localiser ces régions de régulation présumées dont les séquences de DNA peuvent alors être isolées par clonage moléculaire et testées et analysées par des expériences de reconstruction in vitro. Jusqu'à présent, notre travail a des implications générales pour les mécanismes moléculaires de la néoplasie, pour les techniques employées pour définir ces mécanismes et pour l'utilisation des rétrovirus comme vecteurs pour la transduction de gènes.

Hormones and Cell Regulation, Volume 9
INSERM European Symposium
J.E. Dumont, B. Hamprecht and J. Nunez editors
© 1985 Elsevier Science Publishers B.V. (Biomedical Division)

THE PRODUCTION AND USE OF ANTIBODIES TO THE EPIDERMAL GROWTH
FACTOR RECEPTOR USING NATURAL AND SYNTHETIC IMMUNOGENS

W. J. GULLICK, J. J. MARSDEN AND M. D. WATERFIELD

Protein Chemistry Laboratory, Imperial Cancer Research Fund,
Lincoln's Inn Fields, London WC2A 3PX.

INTRODUCTION

Defining the normal pathways and mechanisms of control of cell division
by mitogenic factors is an important and fundamental area of research in cell
biology. The discovery that a growth factor (1,2) and the receptor for a
growth factor (3) are encoded by the oncogenes sis and erb-B has shown that
inappropriate expression of molecules involved in transduction of normal growth
factor signals may be involved in the process of cell transformation and neoplasia.   Much effort is now being devoted to the elucidation of the mechanism
of action of these normal and aberrant systems. Antibodies against oncogene
and proto-oncogene products have proved invaluable in many diverse applications,
and the development of more defined and refined reagents will undoubtedly
prove to be useful.

In this article we describe approaches, both from our own and other laboratories, to the production of antibodies to the epidermal growth factor receptor
(EGF receptor), we describe their use in receptor identification, assay and
purification and for studies of the biosynthesis and function of the receptor.
Currently the EGF receptor is one of the most well studied growth factor
receptors, primarily due to the fact that its ligand, EGF, can be purified
easily from an extremely rich source - the male mouse salivary gland. In
addition, the EGF receptor is expressed in extraordinarily high levels on the
human carcinoma cell line A431 providing a convenient source for study. Other
growth factor systems to date suffer either from a lack of a rich source of
receptor, (e.g., the insulin receptor) or from a dearth of both ligand and
receptor (e.g., platelet derived growth factor and its receptor). Consequently
more antisera and monoclonal antibodies (mAbs) have been generated against the
EGF receptor than with any other growth factor system. The results obtained
to date and future studies of the EGF receptor should provide an archetype
for work on other growth factor receptor molecules.

POLYCLONAL ANTISERA

Polyclonal polyspecific antisera

The discovery that the human vulval carcinoma cell line A431 expresses
approximately 2×10^{6} EGF receptors per cell on its plasma membrane allowed

the early production of polyclonal antisera containing antibodies to the EGF receptor. Haigler and Carpenter[4] prepared cell membranes from A431 cells with which they immunized rabbits. The IgG fraction from this serum inhibited the binding of iodinated EGF to human and mouse cells, cell membranes and solubilized EGF receptors. The antibodies also blocked EGF induced DNA synthesis. However these antisera were not mono-specific since several proteins were recognized in immunodiffusion experiments using solubilized A431 cell membranes. Horsch et al.,[5] have used a similar serum to examine the appearance of EGF receptors during mouse and rat embryogenesis. In this case the authors exploited the autophosphorylating property of the EGF receptor to identify it from the mixture of proteins precipitated and separated by gel electrophoresis.

Polyclonal monospecific antisera

A monospecific antiserum to the EGF receptor is clearly a more useful reagent and several approaches have been taken in order to obtain it. Decker[6] phosphorylated membranes from A431 cells and then separated the mixture of proteins on a preparative SDS gel. The 170,000 dalton major phosphorylated region was cut out of the gel and eluted and then chromatographed on a hydroxylapatite column run in a buffer containing SDS. The fraction containing the EGF receptor was rerun on a second preparative gel and then used for immunizing rabbits. The antiserum prepared to the SDS-denatured EGF receptor also reacted specifically with native methionine labelled EGF receptor from cell lysates of human, rat and mouse cells. Clearly there are therefore immunogenic determinants on the denatured molecule which generate antibodies that recognize structures on the native molecule. Whether in these sera they are directed against the protein or carbohydrate moieties of the EGF receptor is not clear but as discussed below both are immunogenic. Most likely the sera contain both anticarbohydrate and antiprotein sequence-type antibodies.

A second approach to obtain monospecific antireceptor serum has been reported by Stoscheck and Carpenter[7]. Five antisera to both the full length 170,000 dalton EGF receptor and the 150,000 dalton form which is produced by digestion with the calcium activated endogenous protease were obtained by immunizing rabbits with material purified by EGF affinity chromatography, in some cases followed by gel electrophoresis. Either approach, gave antibodies that precipitated the native form of the receptor. Three of the antisera inhibited EGF binding but none affected the kinase activity of A431 cell membranes in vitro. It is therefore possible to generate antibodies that either sterically or allosterically inhibit EGF binding and this property has been used by others to screen for hybridomas producing anti EGF receptor mAbs.[19]

A third method of antibody production employed by Das et al.,[8] takes advantage of cell line Cl 21 which is a mouse-human somatic cell hybrid containing human chromosome 7 as the only human chromosome. The EGF receptor has been mapped to the p13-q22 region of chromosome 7 (9). C57BL/6 mice were immunized with Cl 21 cells and the resulting antiserum absorbed on MC57G cells. The absorbed IgG fraction was labelled with iodine and allowed to bind to human WI-38 fibroblast cells expressing EGF receptors and the specifically bound antibody collected. The antibodies obtained by this method are consequently only directed against epitopes present on the extracellular region of the receptor. They did not crossreact with mouse EGF receptor and displayed no mitogenic activity on human cells, even after crosslinking with rabbit antimouse IgG. The antibodies blocked EGF binding but EGF did not inhibit a significant fraction of antibody binding, suggesting that the majority of the antibodies were directed at regions away from the ligand binding site. The immune IgG was found to slightly enhance EGF receptor kinase activity in the absence of EGF.

Finally, in our laboratory we have recently purified substantial quantities of EGF receptor from A431 cells using a mAb affinity column in order to determine its partial protein sequence (3). The material obtained was greater than 90% pure (Fig. 1A) and used to immunize rabbits. Animals were alternately immunized with native and SDS denatured EGF receptor to provide both antinative and antidenatured receptor reactive antibodies. The serum titer development with time was determined by ELISA assay using plates coated with purified EGF receptor (Fig. 1B).

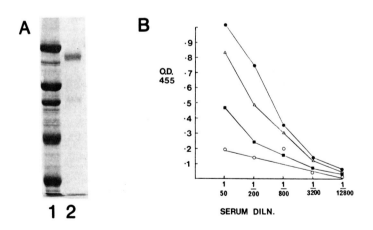

Fig. 1. A: Lane 1, molecular weight marker proteins, 200K, myosin; 116K, β-galactosidase; 92.5K, phosphorylase B; 66K, bovine serum albumin; 45K, ovalbumin. Lane 2, EGF receptor from A431 cells isolated by immunoaffinity chromatography on mAbEGFR1-Affigel 10. B: ELISA assay on serial bleeds from an individual male while NZ rabbit on plates coated with purified EGF receptor.

The antibodies obtained were specific for the EGF receptor (Fig. 2A) and crossreacted with native EGF receptor from several species including human, rat and chicken (data not shown) and were capable of recognizing denatured EGF receptor transferred to DPT paper by Western blotting (Fig. 2B).

MONOCLONAL ANTIBODIES

Antibodies raised to A431 cells or cell membranes

Until very recently the only reported method for generating mAbs to the EGF receptor has been to immunize BALB/c mice with either whole, usually trypsinized, A431 cells or with membranes prepared from A431 cells. The methods and schedules of immunization have varied but without obvious effect on the results. The methods for screening supernatants from hybridoma containing cells have included ELISA, solid phase radioimmunoassay and immunoprecipitation, but most often by using inhibition of EGF binding on whole cells. The latter approach selects not only for direct steric inhibition but

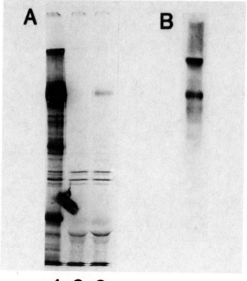

1 2 3

Fig. 2. A: Immunoprecipitate of EGF receptor labelled with ^{35}S-methionine by, lane 1, EGFR1 mAb; lane 2, preimmune rabbit serum; lane 3, immune serum from a rabbit immunized with immunoaffinity purified EGF receptor. B: EGF receptor from A431 cells transferred to nitrocellulose and probed with polyclonal rabbit anti EGF receptor serum followed by iodinated protein A.

also, potentially at least, for indirect allosteric inhibitors (and of course for antibodies to the extracellular domain of the molecule). There are now reports of approximately twenty mAbs which have been produced using this general approach (Table 1). The mAbs fall into two general categories, those against the protein backbone of the molecule (2G2, EGFR1, 528 and 255) and those directed against covalently attached carbohydrate (TL5, 1A-14A, 101 and EGF/G49).

On the face of it, it seems peculiar that there have been so many IgG3 subclass antibodies generated to the EGF receptor (10-15). This isotype in the mouse is however almost always directed against carbohydrate antigens (16). A431 cell membranes are particularly rich in blood group A antigen, mainly in the form of glycolipids, but also present covalently attached to the EGF receptor. The approximate mole ratio of blood group A antigen is 9:1 in favour of the glycolipid structure (V. Ginsberg, personal communication) and it is therefore quite possible that it is the glycolipid form that acts predominantly as the immunizing agent and that the antibodies produced crossreact, rather fortuitously, with the same structure attached to the EGF receptor. These antibodies have somewhat limited usefulness in that they do not recognize EGF receptors from normal human tissue or cells (13) and may crossreact with other proteins or glycolipids expressing the blood group A antigen. One of them however has proved useful as a means of purifying EGF receptor in an active form, from A431 cell lysate (13). This procedure involved incubating the lysate with an anti blood group A mAb immobilized on Affigel-10, washing off the unbound material, and then eluting with a buffer containing 250 mM N-acetyl galactosamine. The receptor obtained was 70-80% pure and retained its EGF stimulated autophosphorylating activity. This preparation is currently being used in attempts to reconstitute functional EGF receptors into artificial liposomes.

Gregoriou and Rees[12] have also raised a mAb directed to carbohydrate determinants on the EGF receptor. They suggest that this mAb inhibits EGF binding to receptors on A431 cells by selectively decreasing the affinity of the major species of low affinity EGF receptors while not influencing high affinity binding sites. The mAb also induces morphological changes in the cells and clustering and internalization of the receptor, thus mimicking EGF activity, but unlike the 2G2 mAb discussed below it did not stimulate the EGF receptor kinase.

There are presently only a few mAbs which have been raised by immunizing animals with whole EGF receptor that bind to the protein part of the molecule (Table 1). One of the most intensely studied is 2G2 which is an IgM (11,17). Unfortunately this antibody became unstable in long term culture, perhaps due to its isotype, which is disappointing since it possessed unique properties. The 2G2 mAb competed for EGF binding to A431 cells, human foreskin fibroblasts

42

TABLE I

SUMMARY OF MONOCLONAL ANTIBODIES THAT REACT WITH THE EGF RECEPTOR

mAb	Subclass	Specificity	Reference
2G2	IgM	Inhibits EGF binding to A431 cells, human foreskin, fibroblasts, Swiss 3T3 cells	Schreiber et al. (1981)
EGFR1	IgG2b	Antiprotein binds to the external domain of the EGF receptor, reacts with human, rat, mouse, chicken	Waterfield et al. (1982)
528	IgG2a	Competes for EGF binding to HeLa-S cells and A431 cells	Kawamoto et al. (1983)
255	IgG1	"	Kawamoto et al. (1983)
455	IgG1	Binds to A431 EGF receptor	Sato et al. (submitted)
F4	IgG1	Antisynthetic peptide from near the C terminus of EGF receptor	Gullick et al. (unpublished)
D10	IgG1	Antisynthetic peptide from near the C terminus of EGF receptor	Gullick et al. (unpublished)

---oo0oo---

mAb	Subclass	Specificity	Reference
TL5	IgG3	Blood group A antigen	Gooi et al. (1983)
1A,3A,4A,9A 10A,11A,13A,14A	IgG3	Blood group A antigen	Parker et al. (1983)
2A,8A	IgM	Blood group A antigen	Parker et al. (1983)
101	IgG3	Human blood group H type 1 antigen	Fredman et al. (1983)
EGF/G49	IgG3	Anti carbohydrate on A431 EGF receptor	Gregoriou and Rees (1984)

and mouse 3T3 fibroblasts. It appeared to mimic all the early and late biological effects mediated by EGF including causing membrane ruffling in A431
cells, enhancement of EGF receptor autophosphorylation, and DNA synthesis in
fibroblasts (11). In further studies (17) the mAb was shown to induce receptor
clustering and internalization. However an Fab preparation from 2G2 did not
induce EGF receptor clustering and was not mitogenic but did retain its ability
to enhance receptor autophosphorylation. Crosslinking the Fab molecules with
a second antiserum restored their stimulation of mitogenesis. Additionally
the 2G2 mAb increased the production of prolactin in rat pituitary cells,
another property in common with EGF (18).

Another mAb with very useful properties is EGFR1 (19). This antibody does not compete for EGF binding but does bind to the protein moiety of the receptor since it can precipitate early biosynthetic intermediates of the EGF receptor which lack carbohydrate side chains (20). In fact this property allowed a detailed description of EGF receptor biosynthesis showing that the major receptor species had a protein molecular weight of 138,000 to which approximately 11 N-linked high mannose-type oligosaccharide chains were added (20). This work also revealed the first evidence that truncated external domains of EGF receptors were produced by A431 cells, a finding subsequently confirmed by cDNA cloning and sequencing (21).

The EGFR1 mAb has also been used to develop a radioimmunoassay for detergent solubilized EGF receptor (22). The assay involved labelling the receptor with a saturating concentration of iodinated EGF of known specific activity and then separating this complex from free ligand by immobilization on EGFR1 mAb bound to Protein A Sepharose. The assay proved more reproducible than the polyethylene glycol precipitation method and also had lower non-specific background binding values. The assay was used to monitor the purification of the EGF receptor which was performed by immunoaffinity chromatography using the same mAb attached to Affigel-10. This provided a system in which manipulations that reduce the efficiency of binding to the column was rapidly revealed by the RIA in solution. The subsequent direct protein sequencing of the EGF receptor revealed its relationship to the transforming protein v-erb-B from avian erythroblastosis virus (3).

Kawamoto et al.,[23] have described three mAbs, 528 (IgG2a), 455 (IgG1) and 255 (IgG1) to the human EGF receptor. Monoclonal antibodies 528 and 255 competed efficiently for EGF binding and inhibited the mitogenic effect of EGF on HeLa-S cells. Monoclonal antibody 455 bound to the EGF receptor from A431 cells but did not inhibit EGF binding. The 528 mAb was used in an elegant series of experiments which showed that the A431 cell line employed expressed a class of receptors representing only 0.1-0.2% of the total but which possessed high affinity for EGF with an estimated Kd of 7×10^{-11}M. The growth of A431 cells is normally inhibited by EGF, perhaps related in some way to the high numbers of receptors since mutant clones of A431 cells which express more normal numbers grow in response to EGF (24). Kawamoto et al., used a concentration of 528 mAb sufficient to competitively inhibit the binding of EGF to the vast majority of low affinity EGF receptors and showed that EGF occupation of the high affinity sites slightly stimulated mitogenesis.

In another series of experiments Masui et al.,[25] investigated the ability of these mAbs to suppress the growth of tumour cells expressing EGF receptors in vivo. The mAbs inhibited the formation of tumours in athymic mice by A431 and T222 cells but did not affect the growth of the human hepatoma cells Li-7 or HeLa cells. The antibodies were even capable of inhibiting the growth of established A431 cell solid tumours.

POLYCLONAL ANTISERA AND MONOCLONAL ANTIBODIES TO SYNTHETIC SEQUENCES FROM THE EGF RECEPTOR

Although many laboratories have attempted to produce mAbs to the human EGF receptor only a few have been developed which react with the receptor protein. All of these are directed against the extracellular domain of the molecule and three out of five of these are directed against epitopes within or close to the EGF binding site(s).

Since the discovery that the EGF receptor shares sequence homologies in its cytoplasmic domain with members of the src gene transforming protein family (reviewed in 26) it has become highly desirable to generate antibodies against these sequences. One approach now that the complete sequence of the EGF receptor is known (21) is to raise antisera to synthetic peptides. If the polyclonal sera produced react with the native EGF receptor molecule then spleens from mice immunized with these peptides can be used to generate mAbs reactive with the native structure.

There are three basic advantages to this approach. Firstly since no mAbs have ever been produced by conventional immunization that react with the cytoplasmic domain of the EGF receptor then the synthetic peptide strategy may represent the only feasible way of developing such reagents. Secondly, sequences may be selected that represent specially interesting features of the molecule such as the ATP binding site or sites of autophosphorylation. Thirdly, the free peptide is capable of competing for antibody binding providing a very useful assessment of the specificity of binding, particularly in immunohistology and immunofluorescent staining methods.

In collaboration with Dr. J. Schlessinger's group at the Weizmann Institute we have developed polyclonal sera to a sequence from near the C-terminus of the EGF receptor which are capable of recognizing the native molecule. Subsequently we (Gullick, Marsden and Waterfield, unpublished results) have generated two mAbs to this peptide which also react with the native EGF receptor (Table 1). We are currently investigating the usefulness of these reagents, and others raised against a series of peptides, for studies of the v-erb-B transforming protein and to examine the expression of the EGF receptor in primary human malignancies.

SUMMARY

   Antibodies to the EGF receptor have been instrumental in increasing our
knowledge of this complicated molecule. Intitially polyclonal sera were used,
mainly as immunprecipitants. As mAbs have been developed they have been useful
as reagents to assay and purify the receptor and as an aid to examine receptor
biosynthesis. They have also been used to study and influence receptor function
but to date the few mAbs examined have not behaved in very similar ways.
This issue will not be resolved until their binding sites are mapped rigourously.
This can be a complicated and time consuming enterprise (27). An alternative
approach we are pursuing is to generate antibodies to selected synthetic
sequences thereby providing reagents against defined structural features.
This approach should help to answer many outstanding questions concerning
normal and abberrant growth control mechanisms.

REFERENCES

1. Waterfield MD, Scrace GT, Whittle N, Stroobant P, Johnsson A, Wasteson A,
    Westermark B, Heldin C-H, Huang JS, Deuel TF (1983) Nature (London)
    304:35-39

2. Doolittle RF, Hunkapiller MW, Hood LE, Devare SG, Robbins KC, Aaronson SA,
    Antoniades HN (1983) Science 221:275-277

3. Downward J, Yarden Y, Mayes E, Scrace G, Totty N, Stockwell P, Ullrich A,
    Schlessinger J, Waterfield MD (1984) Nature (London) 307:521-527

4. Haigler HT, Carpenter G (1980) Biochem. et Biophysica Acta 598:314-325

5. Horsch M, Schlessinger J, Gootwine E, Webb CG (1983)EMBO J. 2:1937-1941

6. Decker S (1984) Arch. Biochem. Biophys. 228:621-626

7. Stoscheck CM, Carpenter G (1983) Arch. Biochem. Biophys. 227:457-468

8. Das M, Knowles B, Biswas R, Bishayee S (1984) Eur. J. Biochem. 141:429-434

9. Kondo I, Shimizu N (1983) Cytogenet. Cell. Genet. 35:9-14

10. Richert ND, Willingham MC, Pastan I (1983) J. Biol. Chem. 258:8902-8907

11. Schreiber AB, Lax I, Yarden Y, Eshhar Z, Schlessinger J (1981) Proc.
    Natl. Acad. Sci. USA 78:7535-7539

12. Gregoriou M, Rees AR (1984) EMBO J. 3:929-937

13. Parker PJ, Young S, Gullick WJ, Mayes ELV, Bennett P, Waterfield MD
    (1984) J. Biol. Chem. in press

14. Gooi HC, Schlessinger J, Lax I, Yarden Y, Libermann TA, Feizi T (1983)
    Bioscience Reports 3:1045-1052

15. Fredman P, Richert ND, Magnani JL, Willingham MC, Pastan I, Ginsburg V
    (1983) J. Biol. Chem. 258:11206-11210

16. Der Balian GP, Slack J, Clevinger BL, Bazin H, Davie JM (1980)
    J. Ex. Med. 152:209-217

17. Schreiber AB, Libermann TA, Lax I, Yarden Y, Schlessinger J (1983)
    J. Biol. Chem. 258:846-853

18. Hapgood J, Libermann TA, Lax I, Yarden Y, Schreiber AB, Naor Z,
    Schlessinger J (1983) Proc. Natl. Acad. Sci. USA 80:6451-6455

19. Waterfield MD, Mayes ELV, Stroobant P, Bennett P, Young S, Goodfellow PN, Banting GS, Ozanne B (1982) J. Cell. Biochem. 20:149-161

20. Mayes ELV, Waterfield MD (1984) EMBO J. 3:531-537

21. Ullrich A, Coussens L, Hayflick JS, Dull TJ, Gray A, Tam AW, Lee J, Yarden Y, Libermann TA, Schlessinger J, Downward J, Mayes ELV, Whittle N, Waterfield MD, Seeburg P (1984) Nature (London) 309:418-425

22. Gullick WJ, Downward DJH, Marsden JJ, Waterfield MD (1984) Anal. Biochem. in press

23. Kawamoto T, Sato JD, Le A, Polikoff J, Sato GH, Mendelsohn J (1983) Proc. Natl. Acad. Sci. USA 80:1337-1341

24. Gill GN, Buss JE, Lazar CS, Lifshitz A, Cooper JA (1982) J. Cell. Biochem. 19:249-257

25. Masui H, Kawamoto T, Sato JD, Wolf B, Sato G, Mendelsohn J (1984) Cancer Res. 44:1002-1007

26. Hunter T (1984) Scientific American Aug:61-69

27. Gullick WJ, Lindstrom JM (1983) Biochemistry 22:3312-3320

RESUME

   Les anticorps contre le récepteur de l'EGF constituent un instrument nous permettant d'augmenter notre connaissance de cette molécule compliquée. Initialement, des sérums polyclonaux furent utilisés, principalement comme immunoprécipitants. Lors du développement des anticorps monoclonaux, ceux-ci se sont avérés utiles comme réactifs permettant de doser et de purifier le récepteur et comme aide pour étudier la biosynthèse du récepteur. Ils ont aussi été utilisés pour étudier et influencer la fonction du récepteur mais, jusqu'à présent, les quelques anticorps monoclonaux examinés ne se sont pas comportés de manière très similaire. Cette question ne sera pas résolue tant que leurs sites de liaison n'auront pas été déterminés de manière rigoureuse. Ceci pourrait être une entreprise compliquée et qui demande beaucoup de temps (27). Nous poursuivons une approche différente qui consiste à produire des anticorps contre des séquences sélectionnées de synthèse, ce qui fournit des réactifs contre des caractéristiques structurelles définies. Cette approche devrait permettre de répondre à de nombreuses questions éminentes concernant les mécanismes normaux et abérrants de contrôle de la croissance.

Hormones and Cell Regulation, Volume 9
INSERM European Symposium
J.E. Dumont, B. Hamprecht and J. Nunez editors
© 1985 Elsevier Science Publishers B.V. (Biomedical Division)

SITE-SPECIFIC METHYLATIONS INACTIVATE VIRAL PROMOTERS

WALTER DOERFLER, KLAUS-DIETER LANGNER, DAGMAR KNEBEL, HERMANN
LÜBBERT[1], INGE KRUCZEK[2], AND ULRIKE WEYER
Institute of Genetics, University of Cologne, Weyertal 121,
5000 Cologne (Germany)

INTRODUCTION

In comparing genetic coding principles to those known in
linguistics (1) one could equate, metaphorically, the primary
genetic code to the vocabulary of a language. Knowledge of the
vocabulary of a language, however, does not yet provide any
inkling of its grammatic and syntactic intricacies. Similarly,
the primary genetic code lacks a description of the rules that
dominate the expression of the genetic text and its regulation.
It has proved a major task in molecular biology to recognize
regulatory signals in the more complex part of the genetic code
(secondary and tertiary coding principles). Particularly, in
eukaryotes, these regulatory principles appear to be complicated
and have not yet been completely unravelled. There is considerable experimental evidence which was derived from many different
eukaryotic systems that 5-methylcytosine (5-mC) as constituent
of specific nucleotide sequences can play an important role in
the long-term switch off of eukaryotic genes (2-6).

The coding and signal values of the four deoxyribonucleotides
adenosine, cytidine, guanosine, and thymidine in naturally occurring DNAs can be complemented by specific modifications of
some of these nucleotides. The modified nucleotides found in DNA
are 5-mC and N^6-methyladenine (N^6mA).

Present address: [1]California Institute of Technology, Pasadena,
California, U.S.A.
[2]Institute of Biochemistry, University of
Munich, Munich, Germany

In DNA derived from prokaryotes 5-mC and N^6mA are found. In the DNA of higher eukaryotes 5-mC is the predominant modified nucleotide. The biological functions of 5-mC and N^6mA have long been enigmatic. The most clearly established role that DNA modifications have been shown to play is in their generally protective effect against the activity of restriction endonucleases in microorganisms (7). In almost all instances, restriction endonucleases (for review see 8) cannot cleave DNA at nucleotide sequences that have been specifically methylated. Thus, a negative effect on the interaction between restriction endonucleases and DNA is the preponderant one. In contrast, however, the restriction endonuclease pair DpnI and DpnII from _Diplococcus_ _pneumoniae_ is isoschizomeric in that it recognizes and cuts at the nucleotide sequence 5'-GATC-3'. DpnI is dependent in its activity on the presence of N^6mA in the recognition sequence, whereas DpnII is completely inhibited by the modified nucleotide (9). Thus, the directional significance of this functional modifier is _a priori_ not uniquely determined and seems to depend crucially on the nucleotide sequence and the specific type of protein-DNA interaction it is going to influence.

DNA methylations or alterations of specific patterns of DNA methylation have also been implicated to play a role in the repair of DNA, in genetic recombination, in virus latency, in differentiation, in the inactivation of one of the X chromosomes, and possibly in the alterations leading to the transformation of cells to cancer cells. In particular, with the more complex biological processes cited, it is likely that changes in the patterns of DNA methylation directly or indirectly affect the expression of specific genes or of sets of genes. Altered expression patterns of a cell may then cause or be related to more complex changes in the functional phenotype of differentiated cells or cancer cells.

In this review I shall briefly summarize the evidence obtained from the adenovirus system that has led to the notion that in higher eukaryotes, DNA methylations at highly specific sequences in the promoter and 5' region of a gene can lead to gene inactivation. The gene inactivating effect of 5-mC could be mediated by modulating specific protein-DNA interactions at specific sequence motifs of the promoter site of a gene or by structural alterations of the DNA or by a combination of both. Since active demethylations of DNA have as yet not been proven to occur, patterns of DNA methylation can probably only be changed by DNA replication and concomitant inhibition of maintenance DNA methyltransferases. It is therefore plausible to assume that methylations of specific nucleotides in the promoter sequence of a gene will lead to its long-term inactivation. In this context, it is important to realize that a gene can be inactivated also without promoter methylation, e.g. simply due to specific protein-DNA interactions at specific promoter sites. The additional measure of methylation of promoter sites would probably safeguard this inactivation and would render it long-term. Obviously, unmethylated promoter sites do not guarantee gene activity: "An unlocked door is not necessarily open" (10).

The literature on DNA methylation and its possible biological functions has rapidly grown in recent years, and has been extensively reviewed (2-6, 10-15).

SITE-SPECIFIC DNA METHYLATION AND GENE INACTIVATION: A SURVEY OF STUDIES ON THE ADENOVIRUS SYSTEM

Studies on molecular mechanisms in biology have fequently been facilitated by the availability of viral model systems. In this respect, adenoviruses have played an important role in the recognition and elucidation of theretofore unexplained phenomena in eukaryotic molecular biology. The functional significance of DNA methylation, particularly in eukaryotic organisms has long remained enigmatic. Recently, a wealth of information on DNA methylation has become available. Several experimental systems have proved useful in these series of studies; the adenovirus system in connection with the methods of gene technology has again shown its pliability as a spearhead in research in molecular biology.

Virion DNA and free intracellular adenovirus DNA are unmethylated

Adenovirions, which had replicated in the nuclei of human cells growing in tissue culture, can be highly purified. The virions can be freed of cellular DNA contaminants by treatment with DNase. Viral DNA extracted from virions purified in this way has been shown to be devoid of the methylated bases 5-methylcytosine (5-mc) and N^6-methyladenine (N^6mA) (16, 17). Other animal viruses, which replicate and are assembled in the nuclei of their host cells, are likewise deficient in the methylation of their genomes (17-19). In contrast, the iridovirus, frog virus 3, with a complicated pathway of replication involving a nuclear phase but assembly in the cytoplasm, carries > 20 % of 5-mC in its DNA (20).

Human cellular DNA in whose immediate environment adenovirus DNA replicates at an extremely high rate, contains 3.5 - 4.4 % of 5-mC relative to the total C content of its DNA. It is still unknown how adenovirus DNA is capable of escaping the modifying activity of the DNA methyltransferase system of the host. We have been able to show that both adenovirus-infected and uninfected human cells exhibit the same DNA methyltransferase activities (21). Hence, absence of 5-mC in adenovirus DNA is probably not due to a lack or inactivity of the cellular DNA methyltransferase system. Since active enzymatic demethylation of DNA does not appear to be a likely mechanism, the only alternatives remaining to enact changes in the patterns of DNA methylation are DNA replication followed by specific inhibitions of the maintenance DNA methyltransferase. For this reason, DNA methylation is a very stable signal and, if used in the regulation of gene expression, will impose a long-term inactivating effect on gene expression. In this sense, it would be highly inopportune for adenovirus DNA to succumb to this functional control of the host cell. The question remains of how adenovirus DNA can do the trick. Compartmentalization of the DNA methyltransferase system in the host's chromatin and the high rate of viral DNA replication seem to be among the salvaging factors - or else what would adenovirologists do today.

It is interesting to mention in this context that human cell DNA that has become integrated into the DNA of a symmetric adenovirus-host cell DNA recombinant, SYREC for symmetric recombinant, which is incapsidated into virions (22, 23), contains human cell DNA sequences that are unmethylated at their 5'-CCGG-3' sites. The same cellular DNA sequences are strongly methylated at the same sites when investigated within the compartment of the human cellular chromatin (22). Hence, methylation of DNA in mammalian cells does not seem to depend on DNA sequence but rather on the location of DNA in apparently functionally different nuclear compartments.

We have also investigated the state of DNA methylation of free intracellular adenovirus DNA in cells infected productively (e.g. human cells infected with adenovirus type 2) or abortively [e.g. hamster cells infected with adenovirus type 12 (24-26)] with adenoviruses. As will be expected from the aforementioned reasoning, we have not found any evidence for the presence of 5-mC residues in free intracellular adenovirus DNA (27, 28). This statement is based on the results of analyses of intracellular viral DNA using several restriction endonucleases which are sensitive to site specific DNA methylations. Obviously, it is impossible to cover all the CpG dinucleotide combinations in adenovirus DNA with this analytical tool. Hence, we want to conclude with caution that free intracellular adenovirus DNA is devoid of methylated nucleotides as detectable by extensive restriction analyses.

Integrated adenovirus DNA is part of the host genome and ex-hibits specific, functionally highly significant patterns of DNA methylation

We will have to distinguish (survey Table 1) free intracellular adenovirus DNA from adenovirus DNA which has become integrated into host DNA (24, 26, 29, 30), e.g. in adeno-virus-transformed cells or adenovirus-induced tumor cells. These viral DNA sequences - from a structural and from a functional point of view - have been rendered part of the chromatin of the host cell, and within this compartment are subject to the methylation strategies of the host. We have shown - for the first time - that integrated adenovirus DNA exhibits unique and functionally highly significant patterns of DNA methylation, and that inverse correlations exist between the extent of DNA methylation and the degree to which the same viral genes are expressed (27, 31-35).

A particularly clear but, by no means, the only example of such inverse correlations between DNA methylation at specific sites and gene expression (3, 5) is offered by the E2A region of adenovirus type 2 (27). The data are schematically summarized in Table 2. The E2A region, one of the genes that is expressed

TABLE 1

DNA METHYLATION OF THE GENOME OF HUMAN ADENOVIRUSES[a]

Type of Adenovirus DNA	State of Methylation	References
DNA packaged into virus particles	Unmethylated[b]	16, 17
Free viral DNA in infected mammalian cells (productive and abortive systems) both parental and newly synthesized	Unmethylated	27, 28
Integrated adenovirus DNA in established cell lines	Methylated in highly specific patterns: Active genes are hypo-methylated; inactive genes are hypermethylated	27, 31–35
Integrated adenovirus DNA in recently produced tumor cells	Hypomethylated. Gradual increase in DNA methyl-ation with passage of tumor cells. Shift in methylation is non-random	36, 37

[a]The only methylated base detectable is 5-mC (16, 17)

[b]Limit of detectability < 0.04 % 5-mC/C (17)

TABLE 2

INVERSE CORRELATIONS BETWEEN DNA METHYLATION AND E2A EXPRESSION IN CELL LINES HE1, HE2, AND HE3

Cell line[1]	E2A gene and late promoter[2] persisting in integrated form	mRNA[3]	E2A Protein[4]	Methylation[5] of 14 CCGG sites in E2A gene
HE1	+	+	+	-
HE2	+	-	-	+
HE3	+	-	-	+

References
[1]38, [2]39, [3]40, [4]41, [5]27

53

early in adenovirus infection, codes for the 72,000 molecular
weight (72K) DNA binding protein (42) which has now been recognized as a highly interesting protein exerting an astounding
number of different functions. Perhaps for this reason, the cell
is well advised to control the expression of this protein very
tightly once the corresponding gene has been inserted into its
own genome. In this functional analysis we have availed ourselves of a number of adenovirus type 2-transformed hamster cell
lines (38). Cell line HE1 expresses the 72K protein (40, 41),
and all fourteen 5'-CCGG-3' sites in the E2A region of the
integrated adenovirus type 2 DNA are unmethylated (27). In
contrast, cell lines HE2 and HE3 fail to express the 72K protein
(40, 41), and the fourteen 5'-CCGG-3' sites in the E2A region
are all completely methylated (27). We have also investigated
the 5'-GGCC-3' sites in the E2A region of the integrated adenovirus type 2 genomes in the same cell lines and found them
unmethylated irrespective of the state of genetic activity of
the E2A region (43). These data suggested that highly specific
sequences might be involved in gene regulation via DNA methylation.

The type of analysis described here in an exemplary fashion
predicted a role for DNA methylation in the regulation of gene
activity. However, the results did not permit one to distinguish
between DNA methylation being the cause or the consequence of
specific gene inactivations.

Site-specific DNA methylations in the promoter and 5' part of a
gene correlate with its inactive state

Before we attempt to accomplish this distinction, we have to
broach the problem of the significant sites for DNA methylations
to occur in inactivated genes. Results from a variety of
eukaryotic systems (for review 5, 6) support the concept that
the promoter and 5' part of a gene are the functionally decisive
parts of a gene with respect to DNA methylation. This notion is
documented by the results of analyses in which the 5'-CCGG-3'
and 5'-GCGC-3' sites in integrated adenovirus type 12 DNA in
three adenovirus type 12-transformed hamster cell lines have
been investigated for the presence of 5-mC. Methylations of
these sites at the promoter and 5' parts of the integrated
adenovirus genes correlated with the inactive states of the

genes, whereas the extent of DNA methylation in the 3' parts, the main bodies of these genes, seemed to be functionally less important (33). Frequently, but not always, the early viral genes in the integrated adenovirus type 12 genomes were active, the late genes were inactive (44). Correspondingly, the promoters of each group of these genes in integrated adenovirus type 12 DNA were unmethylated or methylated, respectively. In one instance, cell line HA12/7, one of the early regions, E3, was not expressed, although the gene persisted in an integrated form in this cell line. In the promoter and 5' region of the E3 segment of integrated Ad12 DNA all 5'-CCGG-3' sites were methylated (33). It is still unknown what the functional meaning of DNA methylations in the 3' main parts of genes could be, if promoter and 5' methylations sufficed to effect gene inactivation. Do methylations in the 3' part of a gene constitute a safeguard or redundancy, do they have a function unrelated to the control of gene expression or are these methylated sites a consequence of the peculiarities of the enzymatic mechanism of methylation? These and other questions cannot readily be answered at present.

The results quoted in this and the preceding sections were mainly based on restriction enzyme analyses of patterns of DNA methylation. It is important to recall that, using this approach, one would analyze only a subfraction of the total number of CpG dinucleotide pairs in DNA. This dinucleotide is the main, albeit not the only, target of the DNA methyltransferase system. Hence, at present a complete description of the true patterns of DNA methylation including all the 5-mC residues in the sequence of an inactivated promoter is still lacking. It is likely that the recently developed genomic sequencing technique (45) will help to alleviate some of these shortcomings.

It is also interesting to recall that the CpG dinucleotide is underrepresented by a factor of four in comparison to statistical expectation in the DNA of eukaryotes (46, 47). In adenovirus DNA, incidentally the CpG dinucleotide is only slightly underrepresented (48). Superficially, this underrepresentation can be explained by the propensity of 5-mC for deamination and the subsequent conversion of a 5-mCpG dinucleotide to a TpG sequence. More significantly, the distribution of certain se-

quences in DNA is the result of long standing functional selection, and in this context it will be meaningful to recognize
that the CpG sequence that is relatively rare in eukaryotes, is
also very important in long-term gene inactivation. In spite of
the relative underrepresentation of CpG dinucleotides, a
clustering of this sequence has been recognized in the promoter
parts of several eukaryotic genes (3, 21, 49, 50).

Genes in vitro methylated at specific sites are inactive upon
microinjection or transfection into eukaryotic cells

In order to resolve the problem of whether DNA methylation is
the cause or the consequence of gene inactivation, several
laboratories have devised experiments in which cloned viral or
non-viral eukaryotic genes were _in vitro_ methylated using DNA
methyltransferases of known sequence specificities. These
enzymes were frequently of prokaryotic origin. The methylated or
the unmethylated gene was subsequently tested for activity after
microinjection or transfection into eukaryotic cells. The out-
come of many of these experiments demonstrated methylated genes
to be inactive and hence supported the notion that sequence-
specific DNA methylations led to gene inactivation.

In designing _in vitro_ methylation experiments in the adenovirus system, we soon recognized the absolute necessity to base
the plans of these experiments on findings gleaned from investigations on patterns of methylation of adenovirus genes which
were permanently fixed in the genomes of virus-transformed
cells. In planning one of these experiments the data summarized
in Table 2 were considered. The results of analyses on patterns
of DNA methylation of the E2A gene in adenovirus type 2-trans-
formed cells suggested that the fourteen 5'-CCGG-3' sequences in
this gene might be functionally important. We therefore chose to
clone the E2A gene including its promoter and 5' region into a
prokaryotic vector. Subsequently, the E2A region was left unmethylated or was _in vitro_ methylated with the prokaryotic HpaII
DNA methyltransferase which methylates the internal C of the
sequence 5'-CCGG-3'. Complete methylation was ascertained by
diagnostic cleavage with the isoschizomeric restriction endonuclease pair HpaII and MspI (51), by electrophoresis, Southern-
blotting (52) and hybridization to a [^{32}P]-labeled adenovirus
type 2 DNA probe followed by autoradiography. It was essential

to assure complete methylation prior to the performance of microinjection experiments. The methylated or the unmethylated DNA was then microinjected into the nuclei of Xenopus laevis oocytes. The completely methylated E2a gene was transcriptionally inactive, the unmethylated gene was expressed into E2A-specific messenger RNA which was - at least in part - initiated at the same (late) promoter as in adenovirus type 2-infected human cells (53, 54). Unmethylated histone genes coinjected with the methylated E2A gene were expressed (54). E2A gene methylated at the 5'-GGCC-3' sites which were not methylated in transformed cells (see above), continued to be transcribed after microinjection into the nuclei of Xenopus laevis oocytes (43). These data and those from other systems (for review 5, 6) clearly argued in favor of the concept that site-specific DNA methylation was the cause and not the consequence of gene inactivation.

Methylations of one or a few sequences in the promoters of viral genes suffice to inactivate these genes

We have recently refined the studies designed to methylate in vitro viral genes by restricting methylation to very few sequences, notably in the promoters of viral genes.

We have recently described experiments in which partly methylated clones of the E2A gene were constructed (55). In the promoter (5')-methylated construct, three 5'-CCGG-3' sequences at the 5' end of the subclone were methylated. One of these sites is located 215 base pairs (bp) upstream (bp 26,169 of adenovirus type 2 DNA), and two sites are located 5 and 23 bp downstream from the cap site (bp 25,931 and 25,949 of adenovirus type 2 DNA), of the E2A gene. This construct was transcriptionally inactive upon microinjection into nuclei of Xenopus laevis oocytes. In the gene (3')-methylated construct, eleven 5'-CCGG-3' sequences in the main part of the transcribed gene region were methylated in vitro. This construct was actively transcribed in Xenopus laevis oocytes, and at least some of the adenovirus type 2-specific RNA synthesized was initiated at the same sites as in adenovirus type 2-infected human cells in culture (55). Both mock-methylated constructs were transcribed into Ad2-specific RNA in Xenopus laevis oocytes. Mock methylation experiments were carried out similarly to methylation experiments, except that the methyl donor (SAM) was omitted. These results demonstrate that DNA methylations at or close to the promoter and 5' end of the E2A gene cause transcriptional inactivation. Perhaps only one methyl group would be adequate for inactivation.

The E2A promoter of adenovirus type 2 DNA was also inserted into the pSVO CAT construct (56) (see below) and the activity of the unmethylated or the methylated DNA was tested after transfection into human HeLa cells. The results of in vitro promoter methylations were the same in the mammalian system as compared to the amphibian oocyte system: The methylated promoter was inactive the unmethylated promoter was active.

The effect of DNA methylations at specific promoter sites on gene expression was also tested by using a sensitive and quantitative assay system. The plasmid pSVO CAT contains the prokaryotic gene chloramphenicol acetyltransferase (CAT) and a HindIII site in front of it for experimental promoter insertion (56). Upon insertion into pSVO CAT, the E1A and protein IX gene promoters from adenovirus type 12 DNA were capable of mediating CAT expression upon transfection into mouse cells (35). The CAT-promoting activity of the early simian virus 40 promoter in plasmid pSV2 CAT is refractory to methylation by the HpaII or HhaI DNA methyltransferase at 5'-CCGG-3' or 5'-GCGC-3' sequences, respectively, because this promoter lacks such sites. The CAT coding sequence of this plasmid carries four HpaII and no HhaI sites. Methylation of the HpaII or HhaI sites in the coding region does not affect expression. The E1A promoter of adenovirus type 12 DNA comprising the leftmost 525 base pairs of the viral genome carries two 5'-CCGG-3' and three 5'-GCGC-3' sites upstream from the leftmost TATA signal. Methylation of the HpaII or HhaI sites incapacitates this promoter in mouse cells (35). Similar results were obtained with the E1A promoter in human HeLa and hamster BHK21 cells. Again, methylation inactivated this promoter (D. Knebel, U. Weyer, and W. Doerfler, unpublished results). The promoter of protein IX gene of adenovirus type 12 DNA contains one 5'-CCGG-3' and one 5'-GCGC-3' site downstream and two 5'-GCGC-3' sites > 300 base pairs upstream from the TATA motif, and probably outside the promoter. The protein IX promoter is not inactivated by methylation of these sites (35). These data demonstrate that critical 5' methylations in the

promoter region decrease or eliminate transcription; methylations of sites too far upstream and probably sites downstream
from the TATA site do not seem to affect expression.

The complete methylation of CpG sequences in the 5' region of
the γ-globin gene, between nucleotides -760 to +100, eliminates
transcription of this gene (57). There are 11 CpG dinucleotides
in this stretch of 860 nucleotides. It has not yet been elucidated which of these nucleotides are the decisive ones in
shutting off gene expression. The authors have used the in vitro
methylation procedure based on replicating a gene, in this case
the γ-globin gene, in the single-stranded vector M13 in the
presence of 5-methyl dCTP (58). By choosing the appropriate
primer fragment for this in vitro replication reaction, specific
segments of the gene and its flanking sequences can be methylated. In this way, the cytidine residues in a certain segment
will be methylated; this method will therefore not permit
identification of the functionally significant nucleotides. The
cloned γ-globin gene thus hemimethylated in various regions was
then cotransfected into Ltk⁻ mouse cells, an established mouse
cell line, with the thymidine kinase gene of herpes simplex
virus. Cell clones containing both the γ-globin and thymidine
kinase genes were selected. It was found that the CpG methylation pattern introduced in vitro was maintained in both strands
of the gene and was faithfully inherited in this system as
previously reported for other genes (57, 58). Methylation in the
region from -760 in the 5' flanking region to +100 in the γ-
globin gene suppressed globin expression, methylation of C
residues from +100 in the γ-globin gene to +1950 in the 3'
flanking region did not affect globin gene transcription. These
data are in excellent agreement with those described in the
adenovirus system.

The results discussed attest to the methylation sensitivity of
two adenovirus promoters and of the γ-globin promoter. The data
on the viral systems demonstrated that methylation had to occur
at specific sites to elicit gene inactivation. One of the viral
promoters (E2A of adenovirus type 2) was tested in amphibian
oocytes and in mammalian cells; sensitivity toward methylation
was the same in both vertebrate systems. Gene activity and its
shut-down by site-specific promoter methylation were investi-

gated with the viral genes in the extrachromosomal state or with the globin gene integrated into the recipient cell DNA. In conjunction with the bulk of evidence derived from correlative experiments, the data make a strong case for the regulatory function of site-specific promoter methylations in higher eukaryotes.

An insect virus promoter is also sensitive to site-specific DNA methylation in insect cells

It has been documented by many investigators that drosophila DNA is devoid of 5-mC or contains only minute amounts of the modified nucleotide (59). One could, therefore, reason that 5-mC as a regulatory signal may be a late addition in phylogeny to the arsenal of regulatory functions. Although DNA methylation might not play a major role in insect cells, it was nevertheless conceivable that the expression machinery in insect cells was sensitive to site-specific methylations. We investigated this possibility by using a conspicuously strong insect virus promoter in such studies, viz. the p10 gene promoter of Autographa californica Nuclear Polyhedrosis Virus (AcNPV) DNA (60).

In lepidopteran insect cells infected with the baculovirus AcNPV, two major late viral gene products are expressed: The polyhedrin which makes up the mass of the nuclear inclusion bodies, and a 10,000 molecular weight protein (p10) whose function is unknown. The structures of the promoters of these genes have been determined (60-62). These structures conform to those of other eukaryotic promoters, except that the nucleotide sequences of these promoters are rich in adenine-thymine base pairs.

We have used the pSVO CAT construct (56) containing the pro-karyotic gene chloramphenicol acetyl transferase (CAT) to study the function of the p10 gene promoter in insect and mammalian cells. Upon transfection of the pAcp10 CAT construct, which contains the p10 gene promoter of AcNPV DNA in the HindIII site of pSVO CAT, CAT activity was determined. The p10 gene promoter is inactive in human HeLa cells and in uninfected Spodoptera frugiperda (S.f.) insect cells. The same promoter has proved active, however, in AcNPV-infected S.f. cells and exhibits optimal activity when cells are transfected 18 h after infection (63). This finding demonstrates directly that the p10 gene

promoter requires additional viral gene products for its activity in S.f. cells. The nature of these products is unknown at
present.

The p10 gene promoter sequence contains one 5'-CCGG-3' (HpaII)
site 40 bp upstream from the cap site of the gene and two such
sites far downstream in the coding sequence of the gene. In view
of the fact that drosophila DNA has been reported to contain no
5-mC or extremely small amounts of it, we were interested in
determining the effect of site-specific methylation on the
insect virus promoter of the p10 gene. Methylation at the 5'-
CCGG-3' site led to inactivation of the promoter or to a drastic
reduction in its activity (63). These data show that an AcNPV
insect virus promoter can be inactivated by site-specific
methylation even in insect cells.

The data presented are consistent with the idea that insect
cells have the capacity to recognize methylated nucleotides in
specific promoter sequences and respond to this signal in a
similar way as vertebrate systems do. This finding was
surprising in view of the fact that 5-mC is not a nucleotide
modification that abounds in insect cell DNA. We conclude that
5-mC as a regulatory signal, although apparently not extensively
used in insect cells, can nevertheless be recognized by the
transcriptional machinery in insect cells.

Summary of in vitro methylation experiments modifying specific
sites in viral promoters

Table 3 provides such a summary. The data indicate that
promoter sites upstream from the TATA or cap signal are the most
sensitive ones with respect to DNA methylation. As stated
previously, we do not understand yet the biochemical mechanism
by which site-specific promoter methylations lead to gene
inactivation.

DNA methylation as a long-term inactivating signal is suitable
only for certain genes

Among the extensive literature on DNA methylation in
eukaryotes and its biological functions, there are reports that
do apparently not support the main conclusions drawn from the
work discussed in previous sections. Before seeking complicated
explanations for such apparent discrepancies, one should
consider the notion that one mechanism for the shut-off of

TABLE 3

Site-Specific Promoter Methylations and Gene Inactivation

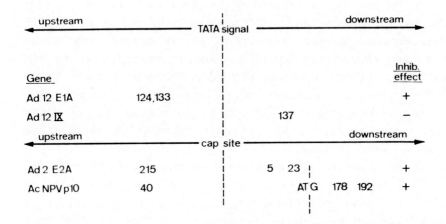

certain genes does not necessarily have to be suitable for the regulation of expression in all genes. Since DNA methylation can serve as a long-term signal in the inactivation of genes, it is not likely that this signal should be selected in the transient inactivation of gene functions. It will be useful in this context to operationally subdivide genes into at least three classes:

1. Genes that become permanently inactivated. In this class of genes, DNA methylation can play an important role.
2. Genes that are intermittently active, but inactive for long periods, or genes that can be reactivated, e.g. by hormone treatment. These genes will probably not be inactivated by site-specific promoter methylations, since it would then be impossible to instantaneously reactivate them.
3. Active genes. The promoters of these genes are obviously unmethylated.

Of course, this simple schematic subdivision of gene classes does not take into account the possibility that the sites of promoter methylation leading to gene inactivation may have to be highly specific. Thus, the occurrence of 5-mC in a promoter sequence may not per se be associated with the inactive state of a gene.

Shifts in the patterns of DNA methylation

It is still a major puzzle to understand how highly specific patterns of DNA methylation are generated. Such elaborate patterns may have functional significance in more than one respect. From the adenovirus system, we should like to describe briefly a set of data which document how complicated the generation of such patterns could be. In adenovirus type 12-induced hamster tumors varying amounts of adenovirus DNA become fixed in the cellular genome by integration (36). These patterns of viral DNA integration have been analyzed in great detail. In these tumor cells only few viral genes are being expressed as messenger RNA. It has, therefore, been surprising to find that the integrated adenovirus DNA is strikingly undermethylated at 5'-CCGG-3' and 5'-GCGC-3' sites (37, 64). These findings demonstrate that the absence of DNA methylation is a necessary but not a sufficient precondition for gene expression.

Upon subcultivation of cells from some of the tumors, the viral genomes were eliminated, apparently in a stepwise process, with segments of the left termini of adenovirus type 12 DNAs persisting the longest. Morphological variants of these tumor cells lost all viral DNA and yet retained the oncogenic phenotype. All 13 independently isolated clones from one revertant line were devoid of adenovirus type 12 DNA. The extent of viral DNA methylation was minimal in adenovirus type 12-induced tumors, although the viral genome was not extensively expressed, if at all. Upon passage in culture, however, the levels of viral DNA methylation increased. It was interesting that establishment of the final methylation patterns of integrated adenovirus type 12 DNA required many cell generations after the fixation of foreign DNA in the host genome. The shift in methylation was nonrandom. The late parts of the inserted viral genomes became methylated more extensively than did early gene segments (37).

Outlook and current research

The evidence accumulated from many different experimental systems provides a rather convincing conceptual framework to postulate that site-specific methylations in the promoter regions of higher eukaryotic genes can affect gene expression. There is evidence that insect cells can also respond to the regulatory signal 5-mC, although in insect DNA itself this modified nucleotide does not seem to abound. We consider 5-mC an interesting and functionally very significant signal which is effective in the regulation of gene expression. Since many biological phenomena can functionally be reduced to the molecular mechanism of gene expression, DNA methylation is likely to play an important role in many complex biological problems, such as differentiation, mutagenesis, oncogenesis, the regulation of the immune response, and others.

Current research on DNA methylation has become a multifacetted approach directed at a number of different problems. Among the most important questions are the following:

How are patterns of methylation controlled in eukaryotes?

What enzymes are involved in the generation and alteration of these patterns?

What functions other than regulation of gene expression does DNA methylation serve in eukaryotes?

We can certainly expect a wealth of new information on DNA methylation and its function within the next few years, and this information will be likely to have important impacts on many different fields in biology.

ACKNOWLEDGMENTS

We are indebted to Petra Böhm for expert editorial work. Research in the authors' laboratory was supported by the Deutsche Forschungsgemeinschaft through SFB74-C1, by the Federal Ministry of Research and Technology (PTB 03 8498 2), and by a donation of the Hoechst Company.

REFERENCES
1. Doerfler W (1982) Medical Hypotheses 9:563-579

2. Razin R, Riggs A (198o) Science 210:604-610

3. Doerfler W (1981) J gen Virol 57:1-20

4. Ehrlich M, Wang RY-H (1981) Science 212:1350-1357

5. Doerfler W (1983) Ann Rev Biochem 52:93-124

6. Doerfler W (1984) Adv Viral Oncol 4:217-247

7. Arber W (1974) Progr. Nucl Acids Res Mol Biol 14:1-37

8. Roberts RJ (1982) Nucleic Acids Res 10:r117-r144

9. Lacks S, Greenberg B (1975) J Biol Chem 250:4060-4066

10. Riggs A, Jones A (1983) Adv Cancer Res 40:1-30

11. Doerfler W (1984) Current Topics Microbiol Immunol 108:79-98

12. Trautner TA (ed) (1984) Methylation of DNA. Springer, Berlin/Heidelberg/New York/Tokyo

13. Taylor JH (1984) DNA Methylation and Cellular Differentiation. Springer, Wien/New York

14. Razin A, Cedar H, Riggs AD (1984) DNA Methylation. Springer, New York

15. Cooper DN (1983) Human Genetics 64:315-333

16. Günthert U, Schweiger M, Stupp M, Doerfler W (1976) Proc Natl Acad Sci USA 73:3923-3927

17. Eick D, Fritz HJ, Doerfler W (1983) Anal Biochem 135:165-171

18. Kaye AM, Winocour E (1967) J Mol Biol 24:475-478

19. Tjia ST, Carstens EB, Doerfler W (1979) Virology 99:399-409

20. Willis DB, Granoff A (1980) Virology 107:250-257

21. Doerfler W, Kruczek I, Eick D, Vardimon L, Kron B (1982) Cold Spring Harbor Symp Quant Biol 47:593-603

22. Deuring R, Klotz G, Doerfler W (1981) Proc Natl Acad Sci USA 78:3142-3146

23. Deuring R, Doerfler W (1983) Gene 26:283-289

24. Doerfler W (1968) Proc Natl Acad Sci USA 60:636-643

25. Doerfler W (1969) Virology 38:587-606

26. Doerfler W (1970) J Virol 6:652-666

27. Vardimon L, Neumann R, Kuhlmann I, Sutter D, Doerfler W (1980) Nucleic Acids Res 8:2461-2473

28. Wienhues U, Doerfler W (1985) in preparation

29. Doerfler W (1982) Current Topics Microbiol Immunol 101:127-194

30. Doerfler W, Gahlmann R, Stabel S, Deuring R, Lichtenberg U, Schulz M, Eick D, Leisten R (1983) Current Topics Microbiol Immunol 109:193-228

31. Sutter D, Doerfler W (1979) Cold Spring Harbor Symp Quant Biol 44:565-568

32. Sutter D, Doerfler W (1980) Proc Natl Acad Sci USA 77:253-256

33. Kruczek I, Doerfler W (1982) EMBO J 1:409-414

34. Sutter D, Westphal M, Doerfler W (1978) Cell 14:569-585

35. Kruczek I, Doerfler W (1983) Proc Natl Acad Sci USA 80:7586-7590

36. Kuhlmann I, Achten S, Rudolph R, Doerfler W (1982) EMBO J 1:79-86

37. Kuhlmann I, Doerfler W (1983) J Virol 47:631-636

38. Cook JL, Lewis AM, Jr (1979) Cancer Res 39:1455-1461

39. Vardimon L, Renz D, Doerfler W (1983) Recent Results Cancer Res 84:90-102

40. Esche H (1982) J Virol 41:1076-1082

41. Johansson K, Persson H, Lewis AM, Pettersson U, Tibbetts C, Philipson L (1978) J Virol 27:628-639

42. van der Vliet PC, Levine AJ (1973) Nature New Biol 246:170-174

43. Vardimon L, Günthert U, Doerfler W (1982) Mol Cell Biol 2:1574-1580

44. Ortin J, Scheidtmann K-H, Greenberg R, Westphal M, Doerfler W, (1976) J Virol 20:355-372

45. Church G, Gilbert W (1984) Proc Natl Acad Sci USA 81:1991-1995

46. Josse J, Kaiser AD, Kornberg A (1961) J Biol Chem 236:864-875

47. Subak-Sharpe H, Burk RR, Crawford LV, Morrison JM, Hay J, Keir HM (1966) Cold Spring Harbor Symp Quant Biol 31:737-748

48. Morrison JM, Keir HM, Subak-Sharpe H, Crawford LV (1967) J gen Virol 1:101-108

49. Felsenfeld G, Nickol J, McGhee J, Behe M (1982) Cold Spring Harbor Symp Quant Biol 47:577-584

50. Tykocinski ML, Max EE (1984) Nucleic Acids Res 12:4385-4396

51. Waalwijk C, Flavell RA (1978) Nucleic Acids Res 5:3231-3236

52. Southern EM (1975) J Mol Biol 98:503-517

53. Vardimon L, Kuhlmann I, Cedar H, Doerfler W (1981) Eur J Cell Biol 25:13-15

54. Vardimon L, Kressmann A, Cedar H, Maechler H, Doerfler W (1982) Proc Natl Acad Sci USA 79:1073-1077

55. Langner K-D, Vardimon L, Renz D, Doerfler W (1984) Proc Natl Acad Sci USA 81:2950-2954

56. Gorman CM, Moffat LF, Howard BH (1982) Mol Cell Biol 2:1044-1051

57. Busslinger M, Hurst J, Flavell RA (1983) Cell 34:197-206

58. Stein R, Gruenbaum Y, Pollak Y, Razin A, Cedar H (1982) Proc Natl Acad Sci USA 79:61-65

59. Achwal CW, Ganguly P, Chandra HS (1984) EMBO J 3:263-266

60. Lübbert H, Doerfler W (1984) J Virol 52:497-506

61. Hooft van Iddekinge BJL, Smith GE, Summers MD (1983) Virology 131:561-565

62. Krebs A, Kuzio JH, Faulkner P, Rohel DZ, Carstens EB personal communication

63. Knebel D, Lübbert H, Doerfler W (1985) submitted for publication

64. Kuhlmann I, Doerfler W (1982) Virology 118:169-180

RESUME

Le nucléotide modifié, 5-méthylcytosine, représente la modification la plus significative dans l'ADN des eucaryotes. Il y a beaucoup d'arguments en faveur d'un rôle très important de cette modification, dans la régulation de l'expression du génôme dans les cellules de vertébrés, dont les mammifères ont été étudiés le plus en détail. Notre équipe a utilisé un système viral, celui de l'adénovirus, pour les études sur la fonction régulatrice, du 5-méthylcytosine, dans les sites très spécifiques des gènes. Il y a plusieurs années que l'on a observé des corrélations inverses entre le degré de méthylation des séquences spécifiques d'ADN dans certains gènes et le niveau de leur expression. Les segments inactifs d'ADN sont fréquemment hypométhylés. Des arguments expérimentaux, basés sur l'étude des systèmes viraux, soulignent le rôle très important des méthylations spécifiques dans les régions promotrices des gènes. Les sites promoteurs, qui sont localisés en amont du signal "TATA" ou du site "CAP" sont particulièrement sensibles à la méthylation. Le signal génétique représenté par la 5-méthylcytosine constitue un déterminant de longue durée parce que les cellules ne peuvent modifier ces méthylations d'ADN que par une réplication d'ADN. C'est-à-dire, s'il faut inactiver un gène de façon permanente, la cellule utilise la méthylation des sites spécifiques. On peut donc conclure que la méthylation peut jouer un grand rôle, en inactivant certains gènes, dans l'explication de phénomènes complexes comme la différenciation, l'oncogénèse, la mutagénèse, l'immunogénétique, etc ...

Hormones and Cell Regulation, Volume 9
INSERM European Symposium
J.E. Dumont, B. Hamprecht and J. Nunez editors
© 1985 Elsevier Science Publishers B.V. (Biomedical Division)

TRANSCRIPTION ENHANCERS

THOMAS GERSTER AND WALTER SCHAFFNER

Institut für Molekularbiologie II der Universität Zürich, Hönggerberg, CH-8093 Zürich, Switzerland

INTRODUCTION

Eukaryotic genes encoding proteins are transcribed by RNA polymerase II. The expression of these genes follows a complicated pathway suggesting several distinct levels at which the abundance of a protein in the cell can be regulated. This pathway can be divided into at least the following steps: (i) accessibility of the gene in the chromatin, (ii) synthesis of the primary transcript, (iii) processing of the primary transcript, (iv) transport of the mRNA out of the nucleus and turnover in the cytoplasm, (v) translation efficiency, (vi) turnover of the protein.

A crucial event in gene expression is transcription by RNA polymerase. Initiation of transcription poses two problems to the cellular machinery: The point of initiation has to be recognized by the polymerase complex and the frequency of initiation events has to be regulated. The start site of transcription is determined by a sequence called the promoter. Work during the last few years has established that the promoter function of RNA polymerase II transcribed genes is usually located within about one hundred base pairs upstream of the transcription initiation site (see e.g. 1,2). The molecular interactions involved in polymerase recognition of promoter sequences are currently being investigated through detailed in vitro studies.

How the initiation frequency is determined is more enigmatic. One obvious mechanism that might determine the accessibility of a gene to RNA polymerase is the modulation of the chromatin state, as clearly shown by the inactivation of genes located on condensed X-chromosomes. However, it is important to note that DNA sequence elements must play a major role in regulating transcription regardless of whether promoter accessibility is determined by a general "opening" of the chromatin structure or by specific factor(s) binding to the gene.

Regulatory sequence elements may be tightly associated with the promoter, e.g. as found in the Drosophila heat shock gene hsp 70 (3), mouse mammary tumor virus (MMTV) (4,5), or the human α_1- and β-interferon genes (6,7). However, regulatory elements can also be located far from the site of transcription initiation. A particularly interesting group of such genetic elements are the transcriptional enhancers that are the topic of this article. Since their discovery some years ago enhancers have been the focus of research in our laboratory. In the course of this work we have been led into the field of tissue-specific gene expression. These studies will thus provide some tools to tackle the basic question of developmental biology: How does a fertilized zygote coordinately differentiate into an organism of highly complex structure?

PROPERTIES OF ENHANCER ELEMENTS

Enhancers were first discovered in papovaviruses, a group of small DNA tumor viruses (8,9,10). In our laboratory we have been studying recombinants containing DNA sequences from the rabbit β-globin gene and the monkey virus SV40. We found that a segment of the SV40 genome dramatically increases transcription from the heterologous β-globin promoter. This SV40 element, called an

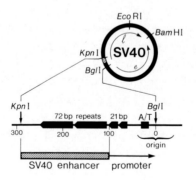

Fig. 1. Map of the 5.2 kb SV40 genome with the KpnI to BglI region expanded to show enhancer (fine hatching) and "early" promoter. Directions of "early" and "late" transcription units are indicated (reprinted from ref.29).

enhancer, contains a tandemly repeated sequence 72 base pairs in length and is located in a region of the viral genome also harboring the origin of replication and the promoters of the "early" and "late" transcription units (Figure 1). As outlined below, the SV40 enhancer exhibits novel features that clearly distinguish it from other upstream promoter components, as was inferred from the work of Benoist and Chambon (11) and of Gruss et al. (12).

Briefly, our studies were usually done with the cloned rabbit β1-globin gene which was introduced into HeLa cells. Two days after transfection using the calcium phosphate coprecipitation technique we extracted cytoplasmic RNA from the entire cell population and analyzed it by the S1 nuclease hybridization assay. Transcripts were barely detectable when β-globin-plasmid recombinants were used. However, at least 200 times more β-globin gene transcripts were found when the β-globin recombinants also contained SV40 DNA and 90% of these transcripts (5000 - 10000 per transfected cell) had the same 5' end as authentic rabbit β-globin mRNA (8). In the latter case, abundant production of β-globin protein was readily detected in a fraction of transfected cells by immunofluorescent staining. Enhancement of β-globin gene expression was dependent upon the 72 bp repeats of SV40 acting in cis but was independent of the closeby viral origin of replication. Surprisingly, the effect of this SV40 enhancer element was independent of its orientation with respect to the β-globin promoter. Furthermore, there was no difference in correct β-globin gene expression whether the enhancer segment was located 1400 bp upstream or 3300 bp downstream of the β-globin gene cap site. An example of such experiments is shown in Figure 2.

As circular molecules have been used in these experiments one could argue that a position downstream of the gene corresponds to a location far upstream. However, this explanation seems unlikely for two reasons (i) cellular enhancers have been found that are located downstream of the promoter (see below) and (ii) it has been observed that plasmid sequences inhibit the transmission of the enhancer effect (8). Therefore, activation of β-globin transcription in constructs like clones 2 and 4 in Figure 2 can only be explained by enhancer activity from a 3' position. The attenuation of the enhancer effect by plasmid sequences could in

part be due to pseudopromoters located therein that might act as
"sinks" for transcriptional activity (13). It has been shown that
from two tandem promoters (or pseudopromoters) in a row the
proximal one with respect to the enhancer becomes preferentially
activated by the enhancer sequence (13-15).

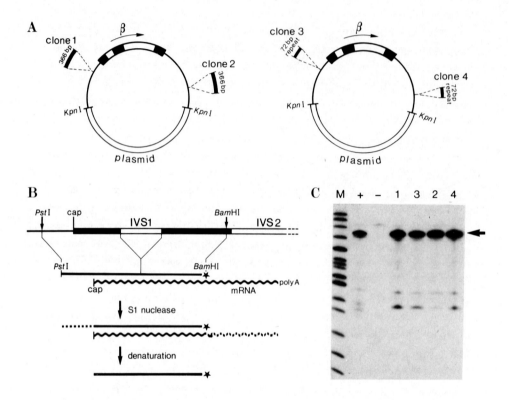

Fig. 2. (A) Recombinants used to assess the effect of distance
(clone 1 vs. clone 2, and clone 3 vs. clone 4). They also allowed
a comparison of the enhancer activity of the 366 bp fragment containing both the origin of replication and the 72 bp repeats with
the 196 bp fragment containing the 72 bp repeats alone (clone 1 vs.
clone 3, and clone 2 vs. clone 4). (B) S1 nuclease mapping scheme
using as a radioactive probe a globin gene clone lacking the first
intervening sequence and end-labelled at the BamHI site. (C) Auto-
radiograph of the S1 nuclease assay. ß-globin gene transcripts
protected a fragment of 354 nucleotides (arrow). (+) Hybridization
to authentic rabbit globin mRNA; (-) hybridization to RNA from
HeLa cells transfected with the original ß-globin clone lacking
enhancer sequences; (1,2,3,4) RNA from cells transfected with the
corresponding clones shown in (A); (M) marker DNA fragments
(reprinted from ref.13).

In experiments analogous to the ones described above, an enhancer sequence was also found in the SV40-related mouse polyoma virus (10,16). This enhancer element revealed a further peculiarity of many enhancers. Whereas the SV40 enhancer seems to be active in a wide variety of tissues and species, the polyoma enhancer shows a host cell preference. The activity of this enhancer is four to five times higher in mouse 3T6 cells than in primate (human HeLa and monkey CV-1) cells (14,17).

ASSOCIATION OF ENHANCERS WITH CELLULAR GENES

Since their discovery in SV40 and polyoma enhancers have been found in the long terminal repeats (LTR) of retroviruses (18,19), in bovine papilloma virus (20), in the early region of adeno virus (21), and in the immediate early region of Herpes simplex virus type I (22). The question arose whether enhancers only play a role in the regulation of viral gene expression or whether they are also involved in regulating transcription of cellular genes. We chose two strategies to investigate this question. One way was the random incorporation of DNA fragments with potential enhancer activity into a non-viable enhancer-lacking SV40 mutant and subsequent screening for resurrection of viral growth ("enhancer trap").

The other approach consisted of testing fragments of characterized genes for enhancer activity. Very promising candidates for enhancer-containing genes were the immunoglobulin genes since their peculiar organization suggested that their transcriptional activation could be accounted for by elements with enhancer-like properties. A functional assay was developed to test this hypothesis and indeed enhancers were found both in the heavy chain and in the kappa light chain loci (see below).

ENHANCER TRAP

For an understanding of the rationale of our "enhancer trap" it seems worthwhile to summarize the life cycle of SV40 (for more details see ref.23): In the initial phase of SV40 infection the enhancer directs transcription of the "early" half of the viral genome which codes for the large T- and the small t-antigens. When enough T-antigen has accumulated it binds to the origin of replication and triggers viral DNA synthesis. Furthermore, T-antigen is also required for expression of the "late" genes

encoding the viral capsid proteins. The infection cycle of SV40
can therefore be blocked at the initial stage by depriving the
virus of its enhancer. If, however, another DNA element with
enhancer activity is ligated to this viral mutant, viability of
the virus is restored. Therefore, we cotransfected cloned
enhancer-less SV40 DNA ("enhancer trap") with a mixture of random
DNA fragments into CV-1 host cells. Cotransfected DNA molecules
are efficiently ligated in tissue culture cells. We assumed that
any fragment with enhancer activity present in the transfected
DNA could be picked up by the enhancer trap. Such a hybrid virus
would be able to replicate and infect neighboring cells allowing
us to isolate SV40 viruses containing foreign enhancer elements
(24; Figure 3).

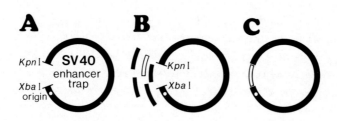

Fig. 3. Schematic diagram of an SV40 enhancer trap experiment.
The viral DNA (A) is mixed with sonicated input DNA (B) and
transfected into monkey CV-1 cells. Fragments with the ability
to substitute for the SV40 enhancer are incorporated in vivo,
giving rise to viable recombinant virus (C) (reprinted from
ref.24).

     To test the sensitivity of this system, random cellular DNA
fragments from man, monkey, and mouse were mixed with decreasing
amounts of a cloned restriction fragment carrying the SV40
enhancer. A dilution of the enhancer in carrier DNA of 10 ppm
still resulted in outgrowth of virus. This corresponds to 150
copies per haploid genome. Lower concentrations of the enhancer
piece were not able to give rise to viable virus. This means that
no more than 150 sequences exist in the genomes of these

mammalian species that exhibit enhancer activity in CV-1 monkey
kidney cells and which also are short enough to fit the size
constraints of SV40 virions. Therefore, the enhancer trap system
may not detect potential enhancers associated with a small number
of genes that are highly active in epithelial cells, i.e. the
CV-1 host cells of SV40. This finding is interesting per se as it
suggests that strong enhancers of the SV40 type do not play a
crucial role in the regulation of most of the genes active in a
given cell type. Furthermore, this result shows that enhancers
have unique sequence features and are not randomly scattered as
"pseudoenhancers" in large amounts throughout DNA of various
sources.

Some of the limitations of detecting cellular enhancers by
the enhancer trap can be overcome by searching for enhancers in
viruses since viral genomes are much smaller. Furthermore, it has
been pointed out by Y. Gluzman that viral enhancers must
generally be stronger than cellular enhancers in order to compete
effectively for factors of the established transcription
complexes in host cells (personal communication). The enhancer
trap has allowed us to find sequences that result in outgrowth of
SV40 in the genomes of Herpes saimiri virus (Schirm, Weber,
Schaffner and Fleckenstein; manuscript submitted) and of both
human and mouse cytomegalovirus (Boshart, Weber, Fleckenstein,
Dorsch and Schaffner, manuscript submitted; Dorsch, Keil,
Schaffner and Koszinowsky, manuscript in preparation) These
sequences have been shown to be true enhancers since they
stimulate transcription of the rabbit β-globin gene in an
orientation-independent manner and also from downstream
positions.

The enhancer trap system has also yielded some important
information on the structure of enhancers. Sequences having
little or no enhancer activity have been rearranged within
transfected cells and have given rise to strong enhancers. In one
such case a subfragment of the polyoma enhancer having only 25%
of the activity was transfected together with the enhancer trap.
The outgrowing virus contained a dimer of the truncated polyoma
enhancer inserted opposite from its normal orientation. This
reconstituted enhancer had the same activity as the wild type
polyoma enhancer and also exhibited the typical host cell
preference of polyoma. In other experiments viruses were obtained

containing no foreign DNA at all. These viruses had duplicated sequences on the "late" side of their genomes thus generating a functional enhancer de novo. The structure of these enhancers is remarkable as they show only limited or no homology to most of the "consensus" sequences for which experimental evidence has suggested some role in enhancer function. These are notably the "core" sequence $(G)TGG^{AAA}_{TTT}(G)$ (25,26), stretches of alternating purine/pyrimidine residues with the potential of forming Z-DNA (27), and the CA-rich sequence motif pointed out by Lusky et al.(20).

IDENTIFICATION OF ENHANCERS IN IMMUNOGLOBULIN GENES

To create a functional immunoglobulin gene one of several hundred gene segments coding for the variable (V) portion of an antibody molecule is translocated in several steps to diversity (D) and joining (J) elements and finally to the locus encoding the constant (C) regions (reviewed in ref.28). The promoters located before each V gene do not become altered during the translocation. However, only the promoter of the functionally rearranged gene becomes transcriptionally active in B-lympho-cytes. Apparently, sequences in the constant region signal the promoter that it is part of a functional immunoglobulin gene. Such transcriptional activation by a DNA sequence located downstream of a promoter is a typical feature of enhancer elements.

To test the hypothesis that immunoglobulin genes contain sequences with enhancer activity, various restriction fragments of an embryonic Cμ heavy chain gene were cloned into a plasmid vector containing the SV40 T-antigen and the rabbit β-globin genes as test genes (Figure 4). Transfection of these constructs into myeloma cells showed that one of the restriction fragments was able to stimulate T-antigen expression as assayed by immuno-fluorescent staining (29). In addition a greatly increased level of β-globin transcripts was observed when this fragment was inserted in both orientations downstream of the gene. This cellular DNA segment therefore behaves like a viral enhancer. The immunoglobulin heavy chain gene enhancer, furthermore, shows a striking tissue specificity: it is active in lymphoid cells but not in 3T6 mouse fibroblasts, human HeLa cells, or mink lung cells. Independently, other researchers have reached the same

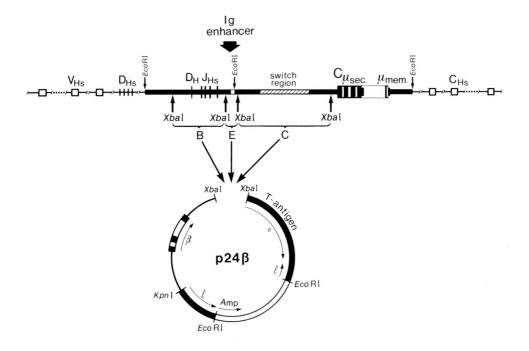

Fig. 4. Detection of an enhancer in immunoglobulin heavy chain
genes. A cloned μ gene in germ line configuration was cleaved with
XbaI and the resulting fragments were inserted in both orientations
between the rabbit β-globin and the SV40 T-antigen genes of vector
p24β. The enhancer activity is contained in the XbaI E fragment
(reprinted from ref.29)

conclusion thus confirming the presence of an enhancer in the
large intron of the Cμ gene (30,31).

In addition to screening the heavy chain genes for enhancer
activity we also searched the light chain loci for similar
elements. Indirect evidence suggested that an enhancer could be
present in the large intron of the C_κ gene as it contains a
segment of 130 bp that is highly conserved between mouse, man and
rabbit (32) and which has a DNaseI hypersensitive site in kappa
light chain producing cells (33). To demonstrate directly that
this sequence is an enhancer it was linked to the rabbit β-globin
gene. Transfection of these recombinants into myeloma cells

resulted in an increased transcription of the globin gene (34; Figure 5).

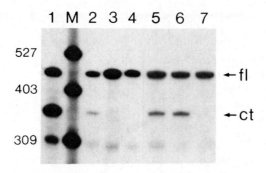

Fig. 5. S1 nuclease hybridization assay of RNAs from myeloma cells transfected with various β-globin recombinants containing immuno-globulin sequences: (1) 1 kb XbaI E fragment of Cμ containing the heavy chain enhancer; (M) DNA size marker fragments; (2) 2.8 kb fragment containing the 3' portion of the large intron and the C_K gene; (3 and 4). The 5' and 3' portions, respectively of a 7.4kb EcoRI fragment containing the λ_1 gene. (5 and 6) Both orientations of the 475 bp fragment containing the kappa light chain enhancer; (7) the 475 bp kappa enhancer fragment cloned into pUC8 cotrans-fected in trans with an enhancer-less β-globin recombinant. S1 mapping scheme as shown in Fig.2. ct: correct 5' terminus; fl: full length probe (reprinted from ref. 34).

Surprisingly, the kappa light chain enhancer is only 5% as active as the heavy chain enhancer. In contrast no enhancer activity has been found to date associated with the lambda light chain genes. In addition to the observation that not all immunoglobulin genes have strong enhancers two groups have isolated myeloma cell variants that maintained high levels of heavy chain gene expression after deletion of the enhancer (35,36). To us this indicates that an enhancer might be dispensable at later stages of B-cell development. Whatever the significance of these findings, it is clear that enhancers - although being important features - are not the only factors which regulate immunoglobulin gene expression.

This point is also illustrated by our recent findings that the heavy chain enhancer is already activated very early in B-lymphocyte ontogeny at a time when only low amounts of mRNA are found in the cytoplasm (Gerster, Picard and Schaffner, manuscript in preparation). The large increase of mRNA observed in plasma cells is due to powerful post-transcriptional mechanisms as indicated by measurements of the transcription rate in isolated nuclei showing that the heavy chain genes are already maximally transcribed in pre-B cells. So it is obvious that both early transcriptional activation of the heavy chain locus and a subsequent increase in stability and/or processing of the primary transcript contribute to the high levels of mRNA found in antibody-secreting plasma cells.

FURTHER EXAMPLES OF ENHANCERS AND RELATED SEQUENCES

The presence of enhancers is not restricted to immunoglobulin genes. Other genes that become highly expressed only in specialized cells but are not subject to developmental chromosome rearrangement have also been found to be associated with enhancers that are strictly tissue specific. Examples are the insulin, chymotrypsin, and α-amylase genes that are highly transcribed in different cell types of the pancreas (37; and T. Edlund, personal communication). Furthermore, a large 2 kb sequence with enhancer activity has been detected upstream of the murine class II major histocompatibility antigen gene E_β^d (38).

Some recent findings suggest that enhancer-like elements may be involved in many other systems as well. DNA sequences 3' to the translation initiation codon of the human adult β-globin gene have been found to be required for induction of transcription in differentiating cells, although it is not known whether the effect of these sequences is position and orientation independent (39). A similar situation is seen in the case of an early embryonic sea urchin histone H2B gene whose transcription is stimulated by parts of the coding sequences even in the reverse orientation (Mous, Stunnenberg, Mächler, Georgiev and Birnstiel, manuscript in preparation). In the closely linked H2A gene an upstream "modulator" element has been described that also stimulates transcription in an orientation-independent manner (40).

WHAT IS THE MECHANISM OF ENHANCER ACTION?

Compared to chromosomal domains which span 30-90 kilo base pairs (41) enhancers are short DNA elements, usually only 100-300 bp in length. On the other hand their length significantly exceeds the short DNA motifs found e.g. in prokaryotic regulatory sequences. Several groups of investigators have tried to define

the important nucleotides in various enhancer segments in order
to deduce common "consensus" sequences (20,21,25-27; N.M. Wilkie,
personal communication). However, as pointed out in the
description of our enhancer trap experiments there exist cases
where such consensus boxes do not appear in DNA segments with
enhancer activity. T. Grundström and P. Chambon (personal
communication) have mutated every nucleotide triplet in the SV40
enhancer. No mutation reduced the activity by more than a factor
of 5, whereas the wild type sequence stimulated transcription
over a 1000 fold. Stafford and Queen have tested deletion mutants
from both sides of the immunoglobulin kappa light chain enhancer.
They found that the activity decreased steadily in proportion to
the number of nucleotides deleted (42). These results suggest
that the enhancer effect is the result of a complex interplay of
various sequence motifs that are distributed throughout the
entire enhancer element and may not be attributed exclusively to
small consensus boxes.

Several mechanisms have been proposed for how such relatively
short DNA segments could stimulate the transcription of remote
genes. For example, an enhancer may, by interaction with a
cellular factor, generate a nucleosome-free "window", or "entry-
site" for RNA polymerase II (43). It is also conceivable that,
starting from an enhancer, an entire chromatin domain is
reorganized into an "unfolded" structure more accessible for
transcription. Also, specific attachment points of the DNA to the
chromosomal scaffold, or matrix (U. Laemmli, personal
communication) could coincide with enhancer sequences. The
finding of cell-type specific enhancers strongly suggest that in
permissive cells specific factors are present that interact with
enhancer sequences and contribute to enhancer action. The
existence of such factors has been further substantiated by in
vivo competition experiments (44). Recently in vitro
transcription systems have been established which might help to
elucidate the mechanism of enhancer function (45; W. Keller,
personal communication; P. Gruss, personal communication).

ACKNOWLEDGEMENTS

We are indebted to our colleagues Julian Banerji, Jean de
Villiers, Laura Olson, Didier Picard, Sandro Rusconi and Frank
Weber for their participation in the work described in this

81

article. We thank Maria Jasin for critical reading of the
manuscript, Silvia Oberholzer for typing it and Fritz Ochsenbein
for preparing the photographs. This work was supported by the
Swiss National Research Foundation, grant No.3.257.82 and the
Kanton of Zürich.

REFERENCES
1. McKnight SL, Kingsbury, R (1982) Science 217:316-324
2. Dierks P, van Ooyen A, Cochran MD, Dobkin C, Reiser J,
    Weissmann C (1983) Cell 32:695-706
3. Pelham HRB (1982) Cell 30:517-528
4. Buetti E, Diggelmann H (1983) EMBO J. 2:1423-1429
5. Zaret KS, Yamamoto KR (1984) Cell 38:29-38
6. Ragg H, Weissmann C (1983) Nature 303:439-442
7. Zinn K, DiMaio D, Maniatis T (1983) Cell 34:865-879
8. Banerji J, Rusconi S, Schaffner W (1981) Cell 27:299-308
9. Moreau P, Hen R, Wasylyk B, Everett R, Gaub MP, Chambon P
    (1981) Nucl.Acids Res.9:6047-6068
10. de Villiers J, Schaffner W (1981) Nucl.Acids Res. 9:6251-6264
11. Benoist C, Chambon P (1981) Nature 290:304-310
12. Gruss P, Dhar R, Khoury G (1981) Proc.Nat.Acad.Sci.USA
    78:943-947
13. Wasylyk B, Wasylyk C, Augereau P, Chambon P (1983) Cell
    32:503-514
14. de Villiers J, Olson L, Banerji J, Schaffner W (1982) Cold
    Spring Harbor Symp.Quant.Biol. 47:911-919
15. Kadesch TR, Berg P (1983) In: Gluzman Y, Shenk T (eds)
    Enhancers and Eukaryotic Gene Expression. Cold Spring Harbor
    Laboratory, Cold Spring Harbor, pp 21-17
16. Tyndall C, La Mantia G, Thacker CM, Favaloro J, Kamen R
    (1981) Nucl.Acids Res. 9:6231-6250
17. de Villiers J, Olson L, Tyndall C, Schaffner W (1982)
    Nucl.Acids Res. 10:7965-7976
18. Laimins LA, Khoury G, Gorman C, Howard B, Gruss P (1982)
    Proc.Natl.Acad.Sci.USA 79:6453-6457
19. Luciw PA, Bishop JM, Varmus HE, Capecchi MR (1983) Cell
    33:705-716
20. Lusky M, Berg L, Weiher H, Botchan M (1983) Mol.Cell.Biol.
    3:1108-1122
21. Hearing P, Shenk T (1983) Cell 33:695-703
22. Lang JC, Spandidos DA, Wilkie NM (1984) EMBO J. 3:389-395
23. Tooze J (ed) (1981) DNA Tumor Viruses (Cold Spring Harbor
    Laboratory)
24. Weber F, de Villiers J, Schaffner W (1984) Cell 36:983-992
25. Weiher H, König M, Gruss P (1983) Science 219:626-631
26. Khoury G, Gruss P (1983) Cell 33:313-314
27. Nordheim A, Rich A (1983) Nature 303:674-679
28. Tonegawa S (1983) Nature 302:575-581
29. Banerji J, Olson L, Schaffner W (1983) Cell 33:729-740
30. Gillies SD, Morrison SL, Oi VT, Tonegawa S (1983) Cell
    33:717-728
31. Neuberger MS (1983) EMBO J. 2:1373-1378
32. Emorine L, Kuehl M, Weir L, Leder P, Max EE (1983) Nature
    304:447-449
33. Chung S, Folsom V, Wooley J (1983) Proc.Nat.Acad.Sci.USA
    80:2427-2431

34. Picard D, Schaffner W (1984) Nature 307:80-82
35. Wabl MR, Burrows PD (1984) Proc.Natl.Acad.Sci.USA 81:2452-2455
36. Klein S, Sablitzky F, Radbruch A (1984) EMBO J. (in press)
37. Walker MD, Edlund T, Boulet AM, Rutter WJ (1983) Nature 306:557-561
38. Gillies SD, Folsom V, Tonegawa S (1984) Nature 310:594-597
39. Wright S, Rosenthal A, Flavell R, Grosveld F (1984) Cell 38:265-273
40. Grosschedl R, Mächler M, Rohrer U, Birnstiel ML (1983) Nucl.Acids Res. 11:8123-8136
41. Paulson JR, Laemmli UK (1977) Cell 12:817-828
42. Queen C, Stafford J (1984) Mol.Cell.Biol. 4:1042-1049
43. Jongstra J, Reudelhuber TL, Oudet P, Benoist C, Chae CB, Jeltsch JM, Mathis DJ, Chambon P (1984) Nature 307:708-714
44. Schöler HR, Gruss P (1984) Cell 36:403-411
45. Sassone-Corsi P, Dougherty JP, Wasylyk B, Chambon P (1984) Proc.Natl.Acad.Sci.USA 81:308-312

RESUME

Les activateurs ("enhancers") de transcription sont des séquences de DNA agissant en cis qui stimulent fortement la transcription devant le promoteur de gènes proches. Ces activateurs peuvent agir sur des grandes distances couvrant plusieurs milliers de paires de base dans les deux orientations, et même à partir de positions en aval du site d'initiation de transcription du gène. Les activateurs ont été découverts dans les papovavirus. Un dénommé "piège à activateur" a été construit pour rechercher les activateurs dans du DNA de différentes sources telles que d'autres virus et des génomes de mammifères. Cette approche a conduit à la détection d'activateurs forts dans les régions très précoces de différents types de virus, notamment du groupe du virus de l'herpès.
Par une approche différente, des activateurs de transcription ont été trouvés dans les locus μ et K des immunoglobulines. Ces activateurs montrent une spécificité de tissu frappante, c-à-d qu'ils sont uniquement actifs dans les cellules lymphoides et sont donc les premiers éléments de DNA jouant un rôle de régulation pour lesquels on a pu montrer une spécificité pour un type cellulaire donné. La présence de ces activateurs contribue à expliquer comment des gènes d'immunoglobulines réarrangés fonctionnellement sont exprimés pendant l'ontogénèse du lymphocyte B.

CYTOSKELETON IN CELL FUNCTION
CYTOSQUELETTE ET FONCTION CELLULAIRE

Hormones and Cell Regulation, Volume 9
INSERM European Symposium
J.E. Dumont, B. Hamprecht and J. Nunez editors
© 1985 Elsevier Science Publishers B.V. (Biomedical Division)

MICROTUBULES IN CELL ORGANIZATION AND MOTILITY

M. DE BRABANDER, J. DE MEY, G. GEUENS, R. NUYDENS, F. AERTS, R. WILLEBRORDS and
M. MOEREMANS

Division of Cellular Biology and Chemotherapy, Department of Life Sciences,
Janssen Pharmaceutica Research Laboratories, B-2340 Beerse, Belgium

Since the pioneering work of Keith Porter (27) microtubule systems have con-
vincingly been shown to exert a central influence on almost all aspects of cellular organization and motility, in virtually all eukaryotic cells.

Weisenberg's (34) early finding that microtubules can be reconstructed from
their subunits in the test tube has led molecular biology to learn us a lot
about microtubule self-assembly in vitro. Borisy and Taylor's (5) experiments,
showing that the long known mitotic inhibitor colchicine binds specifically to
tubulin and thereby inhibits its polymerization provided the microtubule freaks
with an excellent specific tool to probe microtubule assembly and function both
in vitro and in vivo. Synthetic inhibitors, such as nocodazole, were introduced
later (9). It not only widened the arsenal but also allowed to do reversibility
experiments in a proper way. More recently we introduced a new synthetic microtubule inhibitor, tubulozole-C, whose transisomer, tubulozole-T, is in contrast completely inactive (Geuens et al., in press). These compounds should be
valuable tools whenever questions about specificity arise. Rather recently, S.
Horwitz's group at Albert Einstein College of Medicine discovered the property
of a plant diterpenoid, taxol, to bind specifically to microtubules and to inhibit the dissociation of subunits in a manner quite comparable to endogenous
microtubule associated proteins (MAP's) (28). Notwithstanding the great deal
of work of many excellent scientists, our knowledge of the molecular mechanisms
governing microtubule assembly and exerting their various functions in living
cells, has been progressing at a slow pace. It is my belief, however, that we
are standing at the verge of a new era mainly due to the introduction of several
new techniques which aim at the study of the molecular biology of the cell in
the living state. It is a privilege to have been invited at this turning point,
to summarize the state of the art in this field and to be able to show some of
the first exciting new developments arising in several laboratories including
our own.

MICROTUBULE ASSEMBLY

Mechanisms and cofactors governing microtubule assembly in the test tube are
now fairly well characterized. The essentials are the following. Tubulin
dimers associate spontaneously under physiological conditions, whenever a given

86

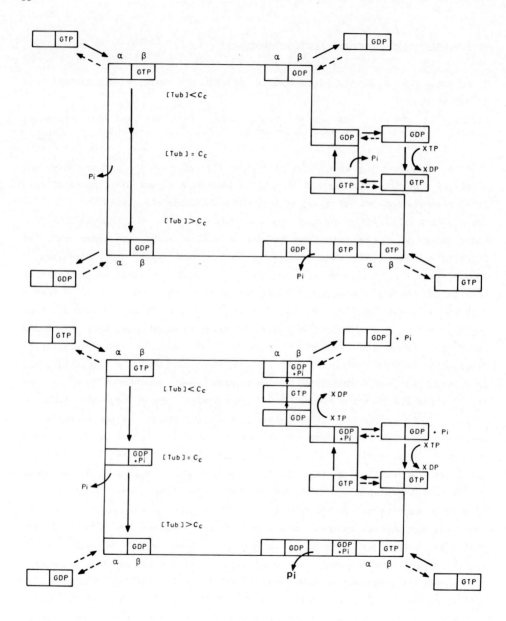

Fig. 1. Possible interactions of GTP binding and hydrolysis with MT assembly
and disassembly. Only the exchangeable nucleotide binding site, located on the
β subunit is pictured. Predominant reactions are denoted by arrows and reactions unlikely to occur in the living cell by dotted arrows. The right hand
portion of each figure depicts 3 situations: steady-state [Tub] = C_c, elongation [Tub] > C_c, shortening [Tub] < C_c. In the classical scheme (upper)
it is believed that GTP subunits bind to the MT ends which is followed by hy-

critical concentration (C_c) of subunits is exceeded. The absolute level of subunits (C_c) depends on many cofactors. It is shifted upwards whenever the association rate of subunits is slowed down or whenever the dissociation rate is increased. Inversely the C_c is lowered when the subunit dissociation rate is decreased or the association rate is increased.

Millimolar calcium, low temperature, microtubule inhibitors (colchicine, vinca alkaloids, nocodazole, tubulozole-C etc...), sulfhydryl scavengers and some ill-characterized endogenous inhibitors all increase the C_c. Guanine nucleotides (GTP and non-hydrolysable analogs even more), microtubule-associated proteins (MAP_1, MAP_2, Tau), sulfhydryl protectants and taxol all decrease the C_c.

At equilibrium the overall dissociation rate equals the association rate. The equilibrium C_c is lower than the C_c necessary to start assembly which is called nucleation. This suggests that for nucleation an activation energy is needed while elongation can proceed at a lower free tubulin concentration. The equilibrium state is only apparent, in fact it is a steady-state because microtubules are polarized structures. The subunits are dimers (α β) which are all arranged in the same direction. The ß-subunit contains an exchangeable guanine nucleotide binding site. Maybe because at one end ß-subunits are free while they are buried at the other end the critical concentration for one end (C_c^+) is lower than the C_c^- of the other end. Therefore at steady-state, in the presence of hydrolyzable GTP, a net incorporation of subunits is seen at the +

drolysis. A transient cap of GTP subunits may exist on rapidly assembling MT. At steady-state or during disassembly the terminal subunits are believed to carry GDP and only GDP carrying subunits are released. The kinetics of association-dissociation may be different at opposite ends because the ß subunits and thus the GTP binding site is exposed differently. Consequently the rate of GTP hydrolysis may differ. It is particularly difficult to explain with this model why in vivo in energy-depleted cells disassembly is blocked while assembly continues (De Brabander et al., 1981b), and why in vitro subunits dissociate much faster soon after they have been added to the polymer than after a certain time (see Carlier 1982 for review).

The lower scheme can to the best of our knowledge explain all existing data. The major additional assumption which is made is that only subunits which have recently hydrolyzed their GTP, and thus may contain GDP and phosphate bound, are released quickly while GTP and GDP subunits are released very slowly only. As a consequence a MT cannot depolymerize at all unless a nucleotide triphosphate is available to regenerate GTP onto the terminal subunits containing GDP. After this regeneration the GTP is hydrolyzed to GDP + Pi and the subunits can be detached. The energy of hydrolysis is thus used to make the polymerization an easily reversible process. Moreover, such a scheme provides the possibility to create very different kinetics at opposite ends of the MT. Because at one end the ß subunit of the terminal is burried the exchange or transphosphorylation may occur much slower. In the extreme case, the left end of the MT in this picture would thus only be able to add subunits and not to loose subunits. Probably, however, the difference is only quantitative.

end and a net loss of subunits at the – end. It looks as if subunits move from the + end to the – end in the tubule, a phenomenon rather misleadingly called treadmilling (24). Somewhat modified from the in vitro data of mainly Carlier and Pantalani (for review see 6) and bringing into consideration the data from De Brabander et al. (11) on the role of the triphosphate pool in living cells one may construct a scheme explained in Figure 1 concerning the role of GTP in microtubule assembly in vivo (see Figure 1). The essential data obtained with living cells suggesting this scheme are that: depletion of ATP and GTP strongly inhibits disassembly but does not hamper assembly per se. However, in energy-depleted cells random assembly prevails instead of organized assembly (11). Further yet unpublished data (De Brabander et al., in press) show that cells in which a disordered network of free microtubules is induced (e.g. by taxol, see further) will grow organized microtubules from the centrosomes if left undisturbed. In the absence of triphosphates, however, the free microtubules are not dismantled. These data provide experimental evidence which largely confirms the hypothesis put forth first by Kirschner (23): cells use the energy of GTP hydrolysis to maintain a disparity in the critical concentration of the two mi-

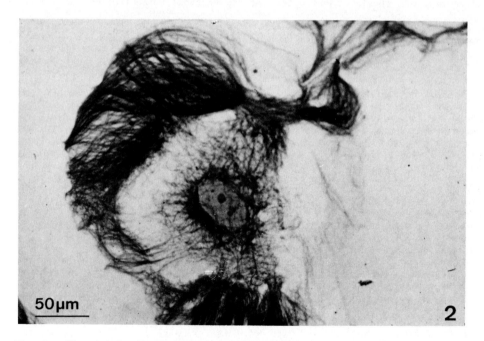

Fig. 2. Microtubule disassembly in a hamster fibroblast treated with nocodazole (0.4 µg/ml) for 4 h is unexpectedly seen to occur predominantly in a centrifugal way. Whole cell stained with antitubulin antibody and the PAP procedure.

crotubule ends. At steady-state organized microtubules that have their minus
end capped, by association with a microtubule organizing centre (MTOC) are more
stable than free microtubules, having both ends free because the tubulin concentration will drop to the C_c^+ level. At this level free minus ends loose
subunits and free microtubules must disappear. Although it was a very stimulating hypothesis, it appears now that certain aspects of it are incorrect.
Mainly the predictions that MTOC-associated microtubules are all attached with
their minus end, and that this end is capped for both tubulin association and
dissociation events have proven incorrect. Microtubules are associated with the
kinetochores and the midbody, two important MTOC's during mitosis, by their plus
end (25). Careful morphological observation shows that most microtubules associated with the MTOC's (centrosomes, kinetochores, midbody) do not have a capped
end (e.g. 13). Also functionally the prediction is incorrect. Indeed the model
states that microtubule disassembly should be entirely from the peripheral ends
towards the centrosome. Figure 2 (unpublished data) shows that in many cells
disassembly may proceed predominantly from the center towards the periphery.
The Kirschner model also predicts that very few if any free microtubules should
occur. Recently, however, he showed that many undisturbed cells do have a substantial complement of free MT superimposed on the organized network (20).
Finally the model cannot explain the preferential nucleation of new microtubules
in the vicinity of MTOC's such as centrosomes and kinetochores, an observation
made by several authors in different cells (13, 36).

De Brabander et al. (8, 12) have proposed a model for the MTOC that does not
suffer from the objections summarized above. It incorporates the ideas of
Kirschner but adds some essential new aspects. It is depicted schematically in
Figure 3. Kirschner too has realized these objections and more recently has
suggested that a "cap" may be "leaky" meaning that subunit addition or loss can
occur (18). One wonders if it is not confusing to still use the term "capping"
in this case. The key observations leading to our model were those of De Brabander et al. (13) and Witt et al. (36) showing that new microtubule assembly
occurs in the vicinity of the MTOC's and not as an outgrowth from seeds or tem-
plates in the MTOC's. The essential elements of the model are the following:
MTOC's induce microtubule assembly by decreasing locally the critical tubulin
concentration. Mitchison and Kirschner, recently corroborated our in vivo data
and hypothesis derived from them by showing that in vitro too centrosomes can
nucleate microtubules far below the steady-state concentration for free microtubules (personal communication, manuscript in press). Anchorage of the microtubules to the MTOC may be a separable event and can occur either through the
terminal subunits (producing capping) or through lateral association of the

terminal portion of the microtubule with some binding proteins. In the latter case, the terminal subunits are free and can either dissociate or bind new subunits. Because, in the vicinity of the MTOC the C_c is lower than elsewhere in the cytoplasm, associated microtubule ends, even when uncapped, are more stable than peripheral microtubule ends. It may even be that associated minus ends are more stable than peripheral plus ends. Indeed, estimates of the disparity between C_c^+ and C_c^- are around a factor of 2 only (see e.g. 4). The capacity to lower the critical tubulin concentration regionally and the property to bind microtubules may be performed by the same or by different molecules or structures.

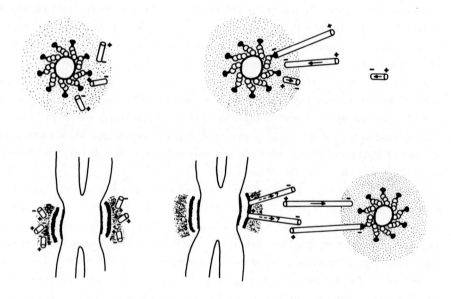

Fig. 3. Some aspects of our model for MTOC activity. The centrosome and kinetochore are surrounded by material (MAP-like?) that lowers the C_c. The left schemes depict the situation after release from nocodazole-induced disassembly. New MT are nucleated preferentially in the vicinity of the MTOC's. At steady-state (right panel) some MT have anchored to centriolar satellites other MT merely maintain their minus end in the assembly promoting environment. Because the free + ends incorporate subunits until the tubulin concentration has dropped to C_c^+, free MT or MT growing from the MTOC with their minus end free must disassemble gradually.
During mitosis (right lower panel) kinetochore MT have become anchored with their + end in the kinetochore plate. The peripheral – ends can only grow or persist if one assumes a stabilizing influence arising from the vicinity of the centrosome or from the interaction with centrosome-derived MT.

This model gained strong support by subsequent observations by De Brabander et al. (12) that taxol, a compound which lowers the critical tubulin concentration _in vitro_ and _in vivo_, not only induces assembly of free microtubules but also, paradoxically in view of the earlier MTOC models, leads to disassembly of pre-existing organized microtubules. Figure 4 shows unpublished quantitative data which confirm our conclusion derived from whole cell immunocytochemistry. Our model can explain these data. If taxol lowers the C_c in the cytoplasm to a level equal to or lower than the C_c prevailing normally around the MTOC then free MT should become as stable as organized MT and be able to detract subunits from them.

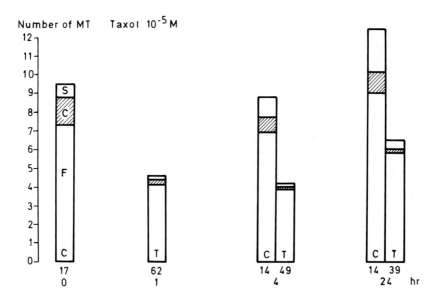

Fig. 4. Number of MT counted in 3.3 μm^2 field surrounding the centrosome, in control (C) cells and taxol (10^{-5} M)-treated (T) cells. The MT are classified as free (F), associated with the cloud (c) or anchored to the satellites (S). The number of sections counted is given below the bars. Taxol paradoxically induces disassembly of MTOC-associated MT while the total amount of polymer in the cytoplasm is increased.

Because our model obviously suggests that the MTOC is surrounded by a factor(s) (protein or other) that binds microtubules and decreases the C_c, e.g. by inhibiting the dissociation rate of subunits, one thinks immediately of a MAP that could fulfil these functions. Unfortunately, none of the known MAP's is con-

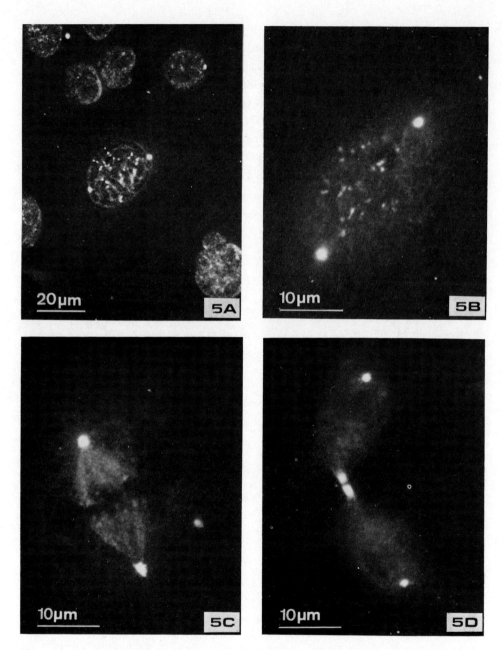

Fig. 5. Immunofluorescent localization of the monoclonal antibody JA2, that reacts with brain MAP1, in PTK$_2$ cells at different stages of mitosis. Note the staining of the centrosome throughout the cell cycle, and an increase of intensity during mitosis. The diffuse spots seen in interphase nuclei (a) become clearly double spots located at the kinetochores at late prophase to early prometaphase (b). Kinetochore staining persists throughout metaphase (c).

centrated in the MTOC's. Ultrastructural evidence points rather to an exclusion (see e.g. 14). Not giving up hope we tried to fish out unknown MAP's by making monoclonals against isolated spindles, centrosomes, etc. and brain microtubules. Recently, our efforts were rewarded as is shown in compound Figure 5 (De Mey et al., in press). We found a monoclonal antibody that reacts specifically with all known MTOC's in various tissue cultured cells: the centrosomes, kinetochores and midbody. Moreover, the staining intensity depends on the cell cycle stage and can be clearly correlated with functional changes in the MTOC. In all cells a faint staining of mitotic microtubules is also seen. In some cells (e.g. neuroblastoma) microtubules are heavily stained also during interphase.

Most interestingly the monoclonal antibody cross-reacts with MAP 1. Our antibody is not unique. Davis et al. (7) found a monoclonal that stained the nucleus and was later found by Vandre et al. (32) to also stain MTOC's exactly like our JA 2 antibody. They have now detected (personal communication) that it also cross-reacts with MAP 1 (A, B and C). We believe that both antibodies recognize a protein that has antigenic sites in common with MAP 1 and that is functionally important in inducing microtubule assembly at the MTOC, in stabilizing associated microtubule ends and in holding the microtubules more or less firmly fixed. Figure 6 depicts schematically our thinking.

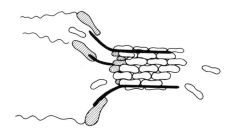

Fig. 6. Schematic representation of how a MAP-like molecule (dark lines) having an affinity for the MT wall and a binding site for structural components of the MTOC may bind microtubules, without arresting assembly and while slowing down dissociation of subunits by strengthening longitudinal bonds.

Kinetochore fibers are also weakly stained at this stage. At telophase (d), the central ends of the midbody MT are stained with the exception of the central plate itself.

The dynamics of microtubule assembly-disassembly in the living cell are also slowly being revealed. Several sets of recently published or yet unpublished data are convergent in showing that microtubule assembly and disassembly in the living cell can occur much faster than suspected from test tube data. Salmon et al. (in press) injected colchicine in sea urchin eggs or treated PTK_2 cells with nocodazole and measured birefringence decay to derive rate constants for microtubule disassembly. Subunit exchange rates of 400-600 dimers/sec were found, which is 200-300 times greater than in vitro. We arrived at the same conclusion using nocodazole treatment and time lapse fixation (De Brabander et al., submitted). Moreover, during interphase microtubules had a 10-100 times slower rate of disassembly than during mitosis. Both during interphase and during mitosis certain subsets of microtubules (e.g. aster microtubules) are much more labile than others (e.g. kinetochore fiber microtubules). Salmon (in press) also injected fluorescently-labelled tubulin in cells and observed recovery of fluorescence after photobleaching as an indication of assembly rates. Again assembly was much faster during mitosis ($\pm$ 1 min) than during interphase (15-30 min). Taking into account also the alterations that occur during mitosis in taxol-treated cells (12; De Brabander et al., submitted) we arrive at the following conclusions. During mitosis the threshold for microtubule assembly (C_c) rises considerably by the appearance of an inhibitor of association maybe released from the nucleus when its membrane opens at prometaphase. This phenomenon may be useful because it enhances the security of the system: MTOC-associated assembly would be strongly favored. Moreover, it increases the turnover rate which is needed during this dynamic stage of the cell cycle.

Microtubules may be heterogenous in many ways. Their association with organizing centres may be peculiar or different kinds of MAP's may be associated with them. The association of calmodulin with certain subsets of mitotic microtubules is a good example (15). Recently it has been shown that certain monoclonal antibodies decorate only a subfraction of microtubules in cultured cells (31). Bulinski et al. (in press) have found, using peptide specific antibodies, that microtubules are heterogenous with respect to the relative content of tyrosylated tubulin. It will be interesting to try to find out possible functional consequences.

CELL ORGANIZATION AND MOTILITY

Microtubules are involved in determining cell shape, polarity and direction of migration (for review see 10, 16). As we have debated before (16) this cannot be explained by invoking a true cytoskeletal role of microtubules. Rather, more and more evidence accumulates, showing that microtubules influence these

processes in an indirect way (e.g. as signal transducers) by determining where
and when the actomyosin system is active to form protrusions or inactive (e.g.
to form stress fibers). These ideas have first been expressed by Vasiliev et
al. (33).

Microtubules also determine intracellular organelle topography. Clearly this
is achieved in a dynamic way. Virtually all organelles move in a typical fashion
along paths defined by the organized microtubule system. Some extreme examples
are axonal transport, pigment migration in chromatophores and chromosome transport during mitosis. Remarkable progress is now being achieved in this area by
the advent of several exciting new techniques. Allen (2) and Inoué (19) introduced high resolution video microscopy which visualizes details in living cells
down to the theoretical limit of resolution of light microscopy (~ 200 nm, and
maybe even somewhat smaller). Microinjection of antibodies, fluorescent analogs
of endogenous proteins (30) and synthetic particles have become a routine procedure in many laboratories.

Allen's group (17) has shown that saltatory motion of organelles occurs along
single microtubules in both directions suggesting that the polarity of the microtubule is probably of little importance. These data, and experiments showing
that latex beads covered with myosin beads travel along actin cables in a cell
free system (lysed algae, 29) have strengthened the believe that the force pro-
ducers which move organelles are embedded in the membrane of the organelles that
crawl autonomously along "cytoskeletal" structures. An opposite view, however,
is supported by recent experiments of Adams and Bray (1) and Beckerle (3) who
showed that polystyrene beads (0.26-0.5 µm in diameter) and other negatively
charged particles, microinjected into axons or chromatophores moved in a typical
micotubule- and ATP-dependent saltatory fashion through the axo/cytoplasm. We
recently (De Brabander et al., submitted) developed a new approach which enables
one to accurately visualize and follow submicroscopic particles having sizes
(20-40 nm) corresponding to the smallest subcellular organelles such as coated
vesicles or even smaller than some proteins (5 nm).

The particles are colloidal gold (5, 10, 20, 30, 40, 80 nm or larger) and lend
themselves easily to coupling with various kinds of proteins (e.g. antibodies).
They are visualized by reflection of polarized light through a crossed analyzer
or by bright field transillumination using a high resolution video camera capable of electronic contrast enhancement (manually or automatically). We would
like to call this approach: nanoparticle video ultramicroscopy or shortly Nanovid ultramicroscopy. The small size of the particles should enable one to approach the molecular biology of the cell in the living state at a heretofore
unknown resolution.

96

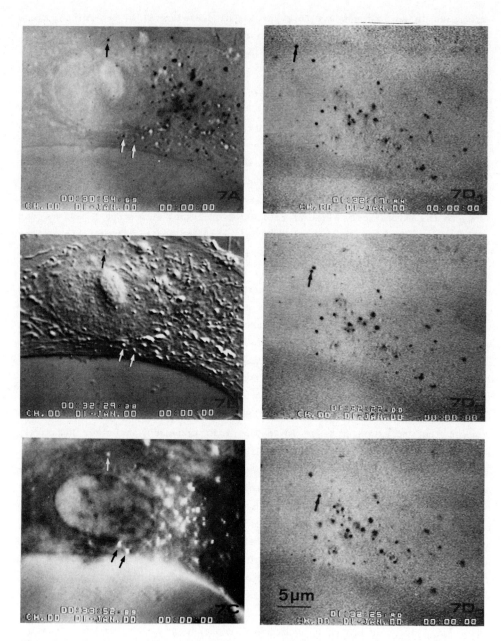

Fig. 7 A, B, C. A cell was injected with colloidal gold (~40 nm coated with albumin and stabilized with PEG) at time -30 min and sequentially viewed with transmitted light (A) DIC (B) and epipolarization (C). A few particles that are coincident in the 3 pictures are denoted by arrows.
D: The rapid linear movement of a particle (arrow) is followed in time. Other particles move also or disappear by leaving the focal plane.

Our first data show that these particles too move through the cytoplasm exactly as endogenous organelles do. Figure 7 shows representative pictures taken from a video tape recording. Figure 8 provides some relevant kinetic data. The movement is entirely microtubule- and ATP-dependent. Interestingly, the movement of both endogenous organelles and gold particles is slightly but signifi-

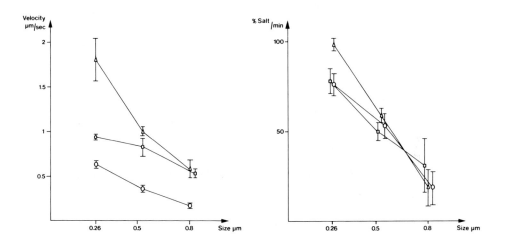

Fig. 8. Quantitative data on the motion of intracellular organelles and injected gold particles (~40 nm).
A. Velocity of moving organelles and gold particles. Granules were followed from the time the linear movement started until it stopped. The apparent size (diameter) of the particles was measured on the monitor after calibration. For each size category at least 10 measurements were done. Mean values and standard deviations are given. Gold particles (o) intracellular organelles in cells of the same preparation (Δ) and in cells of another preparation (□) move at approximately the same speed. Smaller particles consistently move faster than larger ones. We do not know yet whether the small difference between the speed of the gold particles and that of endogenous organelles is significant or only due to slight variations in experimental variables (temperature, illumination, etc...) which have a great impact on the velocity.
B. Frequency of saltation of gold particles (o) and endogenous organelles in the same preparation (Δ) or in another preparation (□) was determined as follows. In a 4 cm^2 area on the screen granules were marked on a still image, and classified according to size. The number of particles that started to move in a typical smooth linear way within a time span of one minute was determined after starting the tape. Data are expressed as percentages and the standard deviation is given. The number of measurements was 30-50 for each size class except for the larger ones (10-20 determinations). The frequency of saltation which is much less influenced by small experimental variances is virtually identical in the 3 samples but also consistently dependent on the size of the granules.

cantly influenced by their size. Both velocity and frequency of saltation correlate negatively with the diameter. Whole mount electron microscopy (Fig. 9)
makes clear that the injected gold granules are not wrapped by a membrane, unlike gold fed extracellularly and captured by pinocytosis. Most granules are
individual and often associated with microtubules. Particles not engaged in
saltatory motion, probably detached from a microtubule show slow oscillatory
motions through the cytoplasm at a speed $\pm$ 10-100 times slower than granules in
the extracellular aqueous milieu, supporting the idea that the ground cytoplasm
is indeed a structured gel (37).

Altogether the data suggest that cells have a microtubule-based transport system
capable of moving almost any particle of suitable charge (1). Adsorbtion of
force producing proteins onto the granules cannot be excluded yet. However, it
would be rather astonishing that this would occur to particles of such different
character in such an efficient way. Most likely the force is produced by the
microtubules or some proteins associated with them. In order to see if microtubule treadmilling could be involved we injected gold coupled to a monoclonal
antibody to tubulin (generous gift of J.V. Kilmartin, 22). Most of these markers
assumed an entirely fixed position, often forming linear arrays and not moving
for over 2 h. This does indeed seem to make treadmilling an unlikely explanation. Consequently, we feel that some form of a microtubule-associated micro-
stream (35) is the most likely possibility left.

An interesting aspect of the inherent capacity of microtubules to organize the
cytoplasm is being studied by McNiven et al. (26). Melanophores, when stimulated, transport pigment centripetally towards the centrosome along parallel
microtubule arrays. In arms cut off the cell body, initially microtubules retain their original orientation and polarity and pigment is transported towards
the cut end. After a while, however, microtubules reorganize to form a radial
pattern and pigment is now transported towards the centre of the cell fragment
although it contains no centrosome.

Mitosis is one of the most exciting but also one of the most complex microtubule-based cellular activities. Actually mitosis and meiosis do not exist
without microtubules. The genetic diversity and evolutionary advantage of the
eukaryotes is thus dependent entirely on microtubules. Both the assembly of the
mitotic spindle and the way in which it separates in a virtually failsafe process the chromatids are still largely unknown and few are the young people that
dare to join the older students of mitosis. Molecular biology and genetic en-
gineering seem safer ways to scientific proliferation.

A few years ago we have revived the idea that the assembly of a functional
spindle is not merely achieved by the activity of the two poles but essentially

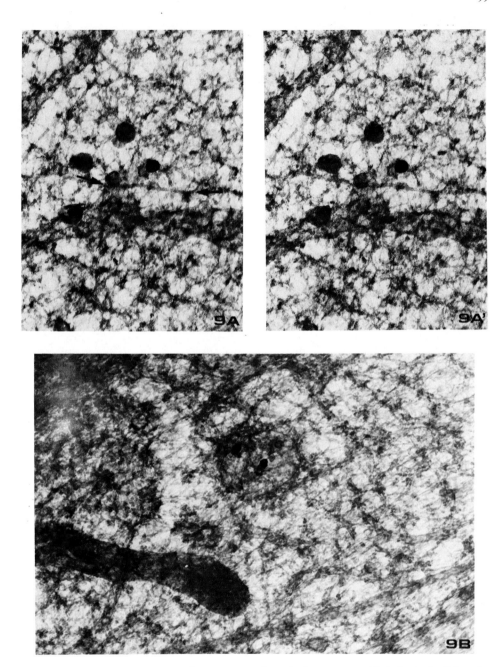

Fig. 9 A. Stereo electron micrographs of a whole mount cell injected with colloidal gold (~40 nm). A single particle is seen enmeshed in the trabecular
network in the vicinity of a microtubule (arrow) and of a vesicle.
    B. Two gold particles located in a large vesicle in a cell that was fed
colloidal gold (40 nm) extracellularly.

dependent on an interaction between the chromosomes (probably the centromeres) and the poles (8, 13). Recently, Karsenti et al. (21) have provided additional evidence supporting this conclusion by showing that in sea urchin eggs the MTOC activity of the centrosome is strongly influenced by the presence in its neighbourhood of nuclei or chromatin. Our finding that taxol paradoxically inhibits nucleation of microtubules by the kinetochore is consistent with the evidence that the kinetochore microtubules are attached with their + end to this MTOC (25). This would make them kinetically the most unfavored class of microtubules. Teleologically this may be useful if one assumes that these microtubules can only grow and construct a kinetochore fiber if they receive a stabilizing influence from outside: the centrosome or centrosome-derived microtubules. This would ensure that kinetochore fibers interact with pole-derived microtubules. We have now further evidence that intermingling of kinetochore and pole microtubules is indeed functionally important. In taxol-treated cells kinetochores do not nucleate microtubules but become attached to pole microtubules directly often in a bipolar fashion. This does, however, not result in chromosome movement or separation unlike what happens during normal mitosis (De Brabander et al., submitted).

Our yet unpublished experiments with taxol have also provided new information on the midbody. Figure 10 shows that in taxol-treated cells multiple midbodies are formed at the aberrant telophase stage despite the absence of a spindle and anything that resembles anaphase. It looks thus that the midbody material is able to construct these structures rather autonomously, and that it becomes active at the right time and disappears again in a programmed way. Our experiments have also provided more information on the timing of the mitotic clock. Table I summarizes our data. Cells treated with nocodazole are arrested at abortive metaphase for about 6 h instead of the usual 30-40 min. In the presence of taxol mitosis is essentially arrested in an identical way, morphologically spoken. However, the arrest takes only 90-120 min and taxol induces telophase prematurely in nocodazole-pretreated cells. The presence of microtubules even without forming a functional spindle influences thus the timing of other mitotic events such as chromatin decondensation, nuclear envelope formation, etc. Maybe this is achieved by the microtubules forming a substrate for the concentrated deposition of the midbody material.

The removal of this material from the cytoplasm may be the trigger that switches on other programmed telophase events such as furrowing, chromatin decondensation, nuclear envelope formation and cell spreading. In any case the data provide yet another striking example of how the microtubule system exerts a central role in both the spatial and temporal coordination of the cell's life cycle.

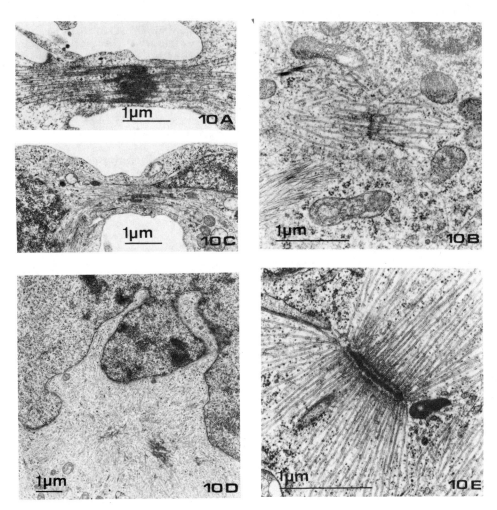

Fig. 10. Ultrastructural aspect of a normal midbody (a) and midbody-like structures in cells treated with taxol; (b): premidbody-like structure in a cell at early telophase (taxol 10^{-4} M); (c-d): midbody-like structures in the central cytoplasm and in a constriction of the same spreading cell (taxol 10^{-4} M); (e): late bipolar remnant in a multinucleate interphase cell (taxol 10^{-4} M) consisting of a dense layer from which MT radiate.

CONCLUSION

I hope this short overview of recent developments in the field of microtubule research has served a useful purpose. It is clear that as far as the living cell is concerned very little is really known. Fortunately, we can gradually leave the purely descriptive approach and start to probe the molecular machinery with finer tools.

TABLE I

Arrest times (mean values and standard deviations) in abortive metaphase of cells treated with taxol or nocodazole. Cell rounding after initiation of prometaphase was arbitrarily taken as the starting point of metaphase. The end of this stage was defined as the initiation of furrowing. Duration of the various stages in untreated PTK_2 cells is: prophase 30–60 min, prometaphase 11 min $\pm$ 3.4, metaphase 14 $\pm$ 4.9, anaphase 7.6 $\pm$ 1.7, furrowing 4.8 $\pm$ 1.3. These data agree very well with those published by others.

	Arrest time in abortive metaphase (min)			
	Pretreated	n	mean $\pm$ S.D.	
Taxol 10 x 10^{-4} M	2–7 h	10	97.2 $\pm$ 17.5	
Taxol 2.5 x 10^{-5} M	2 h	7	98.9 $\pm$ 23.8	
Taxol 10^{-5} M	2–6 h	9	158.2*$\pm$ 103*	
Taxol 2.5 x 10^{-6} M	2 h	1	142	
	4 h 32 min	1	96	
	4 h 33 min	1	92**	
Taxol 1.25 x 10^{-6} M	2 h	1	102	
	2 h	1	157	
Taxol 0.6 x 10^{-6} M	2 h	1	94	
Taxol 0.3 x 10^{-6} M	2 h	1	61	
	2 h	1	66**	
	5 h 17 min	1	70**	
	7 h	1	60**	
Nocodazole 2.5 x 10^{-6} M	2 h	5	384 $\pm$ 132	
Nocodazole 2.5 x 10^{-6} M followed by Taxol 10^{-4} M + nocodazole	2 h	6	82 $\pm$ 14	time in meta before perfusion
		6	155 $\pm$ 52	total time in meta
Nocodazole 2.5 x 10^{-6} M + Taxol 10^{-4} M	0–3 h	7	145 $\pm$ 37	
Taxol 10^{-5} M at prometaphase		6	231 $\pm$ 116	
Taxol 10^{-5} M at metaphase without anaphase		9	158 $\pm$ 186	
with anaphase		1	19	
		1	170	
		1	27	
Nocodazole 2.5 x 10^{-6} M at metaphase		6	409 $\pm$ 98	

* Three very high values, maybe due to unknown variables, appear to shift this value upwards. The median value is: 108 min.
** Normal development through meta-anaphase but prolonged prometaphase.

ACKNOWLEDGEMENTS

We are indebted to L. Leijssen and G. Jacobs for photography and B. Wouters for typing the manuscript. Supported by a grant from the I.W.O.N.L. Brussels, Belgium.

REFERENCES

1. Adams RJ, Bray D (1983) Nature (London) 303:718-720

2. Allen RD, Allen NS, Travis JL (1981) Cell Motility 1:291-302

3. Beckerle MC (1984) J Cell Biol 98:2126-2132

4. Bergen LG, Borisy GG (1980) J Cell Biol 84:141-150

5. Borisy GG, Taylor EW (1967) J Cell Biol 34:535-548

6. Carlier MF (1982) Mol Cell Biochem 47:97-113

7. Davis FM, Tsao TH, Fowler SK, Rao PN (1983) Proc Natl Acad Sci USA 80:2926-2930

8. De Brabander M (1982) Cell Biol Int Rep 6:901-915

9. De Brabander MJ, Van de Veire RML, Aerts FEM, Borgers M, Janssen PAJ (1976) Cancer Res 36:905-916

10. De Brabander M, De Mey J, Van de Veire R, Aerts F, Geuens G (1977) Cell Biol Int Rep 1:453-459

11. De Brabander M, Geuens G, Nuydens R, Willebrords R, De Mey J (1981) Cell Biol Int Rep 5:913-920

12. De Brabander M, Geuens G, Nuydens R, Willebrords R, De Mey J (1981) Proc Natl Acad Sci USA 78:6508-6512

13. De Brabander M, Geuens G, De Mey J, Joniau M (1981) Cell Motility 1:469-483

14. De Brabander M, Bulinski JC, Geuens G, De Mey J, Borisy GG (1981) J Cell Biol 91:438-445

15. De Mey J, Moeremans M, Geuens G, Nuydens R, Van Belle H, De Brabander M (1980) In: De Brabander M, De Mey J (eds) Microtubules and Microtubule Inhibitors, 1980. Elsevier/North-Holland, Amsterdam, pp 227-240

16. Dustin P (1978) Microtubules. Springer-Verlag, Berlin, Heidelberg, New York

17. Hayden JH, Allen RD, Goldman RD (1983) Cell Motility 3:1-20

18. Hill TL, Kirschner MW (1982) Int Rev Cytol 78:1-125

19. Inoué S (1981) J Cell Biol 89:346-356

20. Karsenti E, Kobayashi S, Mitchison T, Kirschner M (1984) J Cell Biol 98:1763-1776

21. Karsenti E, Newport J, Hubble R, Kirschner M (1984) J Cell Biol 98:1730-1745

22. Kilmartin JV, Wright B, Milstein C (1982) J Cell Biol 93:576-582

23. Kirschner MW (1980) J Cell Biol 86:330-335

24. Margolis RL, Wilson L, Kiefer BJ (1978) Nature (London) 272:450-452

25. McIntosh JR, Euteneuer U, Neighbors B (1980) In: De Brabander M, De Mey J (eds) Microtubules and Microtubule Inhibitors, 1980. Elsevier/North-Holland, Amsterdam, pp 357-372

26. McNiven MD, Wang M, Porter KR (1984) Cell 37:753-765

27. Porter KR (1966) Ciba Foundation Symposium on Principles of Biomolecular Organization. Ciba Foundation, Churchill, London

28. Schiff PB, Fant J, Horwitz SB (1979) Nature (London) 277:665-667

29. Sheetz MP, Spudich JA (1983) Nature (London) 303:31-35

30. Taylor DL, Wang YL (1980) Nature (London) 284:405-410

31. Thompson WC, Asai DJ, Carney DH (1984) J Cell Biol 98:1017-1025

32. Vandre DD, Davis FM, Rao PN, Borisy GG (1984) Proc Natl Acad Sci USA, in press

33. Vasiliev JM, Gelfand IM (1976) In: Goldman R, Pollard T, Rosenbaum J (eds) Cell Motility. Cold Spring Harbor, pp 279-304

34. Weisenberg RC (1972) Science 177:1104-1105

35. Weiss DG, Gross GW (1982) In: Weiss DG (ed) Axoplasmic Transport. Springer Verlag, Berlin, pp 362-383

36. Witt PL, Ris H, Borisy GG (1980) Chromosoma 81:483-505

37. Wolosewick JJ, Porter KR (1979) J Cell Biol 82:114-128

RESUME

Une revue est présentée de données récentes, sur l'assemblage et les fonctions de microtubules. Concernant l'assemblage, nous proposons que l'hydrolyse de GTP sur des unités de tubuline terminales accroit la vitesse de dissociation. Cette énergie est ainsi utilisée afin d'assurer la reversibilité de la polymerisation. La réaction implique l'échange de GTP pour GDP sur les sousunités β qui sont exposées différemment aux extrémités opposées des microtubules. La vitesse de dissociation peut ainsi être largement différente aux deux extrémités.

De nouvelles expériences confirment que les centres organisateurs diminuent localement la concentration critique. Un nouvel anticorp monoclonal suggère qu'une proteine apparantée à la MAP1 pourrait être responsable pour cette activité.

Afin d'étudier la motilité subcellulaire liée aux microtubules nous avons mis au point une nouvelle technique appélée ultramicroscopie Nanovid. Des particules d'or colloidal de taille submicroscopique (5-40 nm) sont micro-injectés dans les cellules. Les particules, invisibles à l'oeil nu au microscope sont repérables à l'aide d'une caméra video à haute résolution et d'un circuit d'amplification de contraste. Les particules sont transportées dans les cellules le long des microtubules de la même facon que des organites endogènes. Des particules couplées à des anticorps antitubuline se fixent sur les microtubules et ne migrent pas. Les données suggèrent que la force motrice est associée aux microtubules et que le flux de sousunités tubuline n'est pas impliqué. La possibilité de suivre des particules individuelles d'une taille comparable à celle d'anticorps et d'autres proteines qui peuvent y être couplés promet d'être très utile dans l'étude de la biologie moléculaire de la cellule vivante.

Hormones and Cell Regulation, Volume 9
INSERM European Symposium
J.E. Dumont, B. Hamprecht and J. Nunez editors
© 1985 Elsevier Science Publishers B.V. (Biomedical Division)

SLIDING OF STOP PROTEINS ON MICROTUBULES : A PHENOMENON OF POTENTIALLY GENERAL
SIGNIFICANCE IN CYTOSKELETON DEPENDENT CELLULAR FUNCTIONS.

Didier JOB [1], Michel PABION [1], Robert L. MARGOLIS [2].

[1] Laboratoire de Biochimie Endocrinienne U244, Institut National de la Santé et
de la Recherche Médicale U.S.M.G. 38402 Saint Martin d'Hères Cédex, France
[2] The Fred Hutchinson Cancer Research Center, Seattle, Washington 98104, USA.

I. INTRODUCTION

Microtubules are central to the organization of intracellular space and cell
motility (1-3) in eucaryotes.

This is particularly obvious during mitosis : the crucial role of
microtubules in the spindle for chromosomal migration and formation of the
actin ring has long been recognized (3). Microtubules are also involved in
other cell motility phenomena such as the transport of subcellular organelles or
changes in cell shape (1,3).

Furthermore the assembly state of cytoplasmic microtubules is apparently of
importance to the overall metabolic state of the cell : disruption of cytoplasmic microtubules can induce quiescent Go cells to enter a mitotic cycle
(4,5) and can induce the formation of gap junctions in developing tissue (6).

The molecular mecanisms which underly this wide variety of functions are
still obscure. Nevertheless one common feature is the remarkable plasticity of
the microtubular system whose assembly state seems to be under specific
regulation both in time and in space.

One of the possible ways for the cell to perform this control is to establish
reversible transitions from stable to labile states of cytoplasmic microtubules.
That this type of control exists in vivo is now well documented : thus during
mitosis those microtubules that attach to kinetochores are cold stable while
other polymers are labile and in constant equilibrium with subunits (3). A
striking transition between stable and labile state also occurs during induction
of neurites growth in PC12 pheochromocytoma cells (7) and in the cerebellum
Purkinje cells of thyroid hormones deprived rats (8).

In addition to their plasticity microtubules are also remarkable by their
ability to interact with many different enzymes and cell organelles whose
intracellular movements they seem to direct (1-3).

In the following report we shall discuss the mecanisms and regulation of
microtubules stabilization as observed in vitro. We shall show that the detailed
study of the interaction of some stabilizing proteins with microtubules provided

a model system that led us to propose what we believe might serve as a unifying working hypothesis to explain how cytoskeletal structures could organize the intracellular distributions of enzymes and organelles.

II. BRAIN STABLE MICROTUBULES

We now know that brain tissue contains at least two distinct classes of stable microtubules. In the following sections we shall distinguish the "cold stable microtubules" and the "superstable microtubules".

The cold stable microtubules resist to temperatures down to 0 - 5°C but are immediatly disassembled by freezing or sonication (9). The superstable microtubules resist freezing and sonication (10).

II.1. **Cold stable microtubules** : the existence of a sizable subpopulation of cold stable microtubules in brain has long been recognized (11). Nevertheless, they have only recently been systematically investigated. The major reason is certainly that they proved to be difficult to purify from large animals brains.

The first comprehensive study of these polymers has been performed by WEBB and WILSON (12) on mice brain. They established some fondamental parameters of the system : they showed that cold stable microtubules could be solubilized by millimolar Ca^{++} at 0°C, that they were made of normal tubulin and that in fact cold stability could be reconstituted by addition, to cold labile microtubules, of a fraction obtained by chromatography of cold stable preparation on DEAE. They believed that the cold stabilizing factor was a small molecular weight-compound, acting stoechiometrically by incorporation into the microtubules structure during polymerisation.

Subsequently Margolis and Rauch (13) showed that in rat brain crude extracts the transition from lability to stabilility was a time dependent process that correlated with the endogenous nucleotide triphosphate concentrations and that this transition could be reversed by additon of ATP. They suggested that a phosphoprotein that they found uniquely phosphorylated in cold labile polymers was responsible for a reversible shift from lability to stability ; Although we now know that this "switch protein" has no crucial role in the cold stabiliza-tion process the concept of a substoechiometric protein being responsible for cold stability has proved to be correct.

A key to subsequent steps was the finding that at 30°C, cold stable micro-tubules were calcium resistant but rapidly disassembled when exposed to a mixture of calcium and calmodulin (14). It appeared quite clear from dose effect curves that the dissolution of cold stable polymers was likely to be a two steps event. First, the calmodulin calcium complex bound to the stabilizing factor and then the polymer was disassembled by free calcium. Because the needed concentration of calmodulin was very much substoechiometric to tubulin concentration, we concluded that cold stability was likely to be due to substoechiometric "blocks" probably proteins, interacting with calmodulin-calcium and adding to otherwise usual cold labile microtubules.

A more direct demonstration followed : thus we showed that mechanical shearing of cold stable microtubules generated cold labile fragments (9). This was a useful observation for two reasons : first it was a direct demonstration that the cold stabilizing factor was substoechiometric to tubulin and only present at few locations on microtubules. Second, it allowed us, later on, to distinguish between cold stable and superstable microtubules because only the first subpopulation is sensitive to fragmentation while both of them are solubilized by calcium and calmodulin (10,14).

Recycling of calcium or fragmentation solubilized microtubules led to similar results showing in both cases that initially cold stable polymers reassembled into an only partially stable preparation. This fact and other observations led us to model the generation of cold stability as being due to the random addition of blocks along labile microtubules (9).

Chromatography of solubilized cold stable preparation on calmodulin affinity columns confirmed that blocks specifically bound to calmodulin.

Comparison of filtrates from calmodulin columns run in the presence of EGTA ("inactive columns") or Ca^{++} ("active columns") showed that a family of proteins (STOP proteins) was apparently related to cold stabilization. Since that time, small quantities of active STOPs have been extensively purified in one of us laboratory (15). It appears that the active moiety is a 145 000 K multimeric protein probably which represent what we initially called STOP 1 (9).

At this point, the whole system already looked as a convenient and economical way for cells to regulate microtubules stability by a rather complex set of interactions between tubulin, blocks, Ca^{++} and calmodulin.

Protein phosphorylation soon appeared to introduce another level of regula-tion (16). Thus we found that addition of ATP, but not AMP PNP, to purified cold stable microtubules made them cold labile. The reaction was greatly enhanced by the addition of Ca^{++} - calmodulin at a concentration too low to trigger any disassembly by itself. By manipulation of the homogenisation conditions or more recently, by examination of various animal species we could obtain extracts that were sensitive to both or only one of the two ATP dependent reactions i.e. the calmodulin dependent or independent ones.

In all cases, there was a perfect correlation between the functionnal effect (i.e. disassembly of microtubules) and the presence in the active fraction containing the stabilizing activity, (DEAE filtrate, see above) of phosphoryla-ted substrates : gentle homogenization of rat brains yielded extracts that responded slowly to ATP alone but more rapidly to ATP + Ca^{++}-calmodulin and exhibited Ca^{++}-calmodulin dependent and independent phosphorylation of few peptides in the DEAE eluate ; gently homogenized extracts responded to ATP only in the presence of Ca^{++} calmodulin and correspondingly the peptides phosphory-lation was totally Ca^{++} calmodulin dependent ; beef extracts are responsive to ATP alone but Ca^{++} calmodulin has no enhancing effect (17). Accordingly the Ca^{++} calmodulin dependent phosphorylation of peptides is missing.

Taken together, these observations led us to the conclusion that cold stability could be regulated by calmodulin dependent or independent protein kinases. We now believe that the calmodulin dependent kinase is homologous to the "synaptin kinase" that has recently been purified by several groups. It is uncertain wether the phosphorylated species are mere witnesses of the presence of protein kinases or if they have a role in the stabilization process.

At this point, we spent a lot of time designing procedures suitable for the isolation of cold stable microtubules from larger animals than rodents. This was not easy because even in pure inbred rats yields were highly dependent on the homogenization procedure and on other apparently seasonnal, unpredictible factors. The final answer was nevertheless very simple. Making use at first of sheep brain crude extracts (18) in which assembly can be monitored by O.D. measurements (as in rat but contrarily to beef extracts which have a high baseline turbidity) we developed a method that was subsequently applied with minor modifications to beef (10). Pragmatically, two factors are important to obtain suitable preparations : freshness of the brain and dilution of the extract during assembly.

In fact, we are now able to isolate STOP proteins by direct chromatography of crude extracts on ion exchangers (19). But obtaining large quantities of the cold stable microtubules themselves was nevertheless a perequisite for the study of superstable microtubules.

II. 2.Superstable microtubules (10)

These are microtubule fragments that resist freezing and sonication. They are present in cold stable preparations in which they can be assayed as "seeds" (for details, see 10).

They can be disrupted by Ca^{++} at 0°C but are much more rapidly disassembled in the presence of calmodulin.

They can be reformed by recycling of Ca^{++} on calmodulin Ca^{++} treated preparation. They then appear in an explosive manner after the beginning of polymerisation.

They are probably made of "normal" microtubules to which stabilizing peptides add in a cooperative way reaching probably high concentrations along the polymer.

It has been somewhat desappointing to us that in our initial study we could only see an enrichment in few protein bands in superstable fragments. Using other gel systems we now know that at least two major peptides are specific of superstable polymers : a special high molecular weight MAP and a 145.000 K peptide. These proteins are now under extensive analysis in our labs.

III. A KINETIC STUDY OF THE INTERACTION OF STOP PROTEINS WITH MICROTUBULES

In the following section, we shall briefly present results which will be described extensively elsewhere (20).

In this work we have made use of a previously developed filter assay to quantify radioactive regions of microtubules (21). Microtubules are assembled in the presence of ^3H GTP. Polymers are trapped on glass fiber filters and the corresponding radioactivity is counted. The only modification of the original method in the present study has been the inclusion of PLN in the stabilizing buffer to avoid reassembly of cold disassembled polymers during the assay. Figure 1 shows that this filter method gives results that are equivalent to the usual OD_{350} determination of polymerized tubulin concentration.

This method made possible to address some basic questions about the behaviour of STOP proteins :

a/ are STOPs incorporated into microtubules structure or do they bind on the surface of the polymers ?

b/ what is the kinetic of cold stabilization ?

c/ can STOPs exchange between polymers ?

d/ can STOPs move along a given polymer ?

The answer to question a/ was that STOPs bind to microtubules surface : figure 2 shows that when soluble tubulin is assembled in the presence of a given concentration of STOPs, the resulting amount of cold stability is equivalent to the one that is observed when the same concentration of STOPs is added to preassembled microtubules.

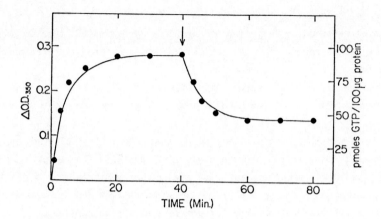

Fig. 1. Comparison of spectrophotometric and filter assay of microtubule assembly state.
Assembly of cold stable microtubules was monitored by turbidity change at a wavelength of 350 nm (solid line). The arrow indicates the point at which the samples were rapidly cooled to 7°C. A portion of the same sample was incubated with (^{3}H)GTP and aliquots were filter assayed at time points as described under Materials and Methods. Radioactivity incorporation into microtubules and retention in cold stable regions is shown (solid circles), expressed as specific pmoles of GTP incorporated. Assembly, at 30°C, was conducted with 2 mg/ml protein, 50 μM GTP. Acetyl phosphate and acetate kinase were also present.

Furthermore, we find that the kinetic of STOP binding to preassembled polymers is very rapid (figure 3). The same figure shows that although STOPs have affinity for free subunits which compete with polymers for STOPs binding, all of these finally bind to microtubules.

This fits with our initial observation that centrifugation of the cold stable polymers fraction in a recycled preparation leaves a totally cold labile supernatant (9).

This competition of free subunits for STOPs binding probably accounts for the fact that as shown in figure 4, percents of cold stability are almost constant during assembly of recycled preparations, a result that would be otherwise inconsistant with the observed rapid kinetics shown in figure 3.

Figure 5 shows that despite of the fact that STOPs are most likely to be surface binding proteins, they do not exchange between microtubules. Thus in a mixture of cold stable and cold labile polymers both species remain indefinitely distinct. If there was any sizable exchange of STOPs initially cold stable or cold labile polymers would become undistinguishable upon incubation of the mixture.

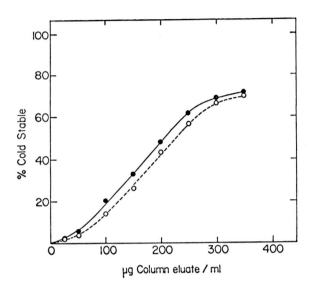

Fig. 2. Comparison of cold stability induced by STOP protein addition prior to or after microtubule assembly.
Recycled (3X) cold labile microtubule protein (1 mg/ml) was assembled to steady state with (^{3}H)GTP and filter assayed for total label incorportion and for cold stable level at each data point. The DEAE column flow through eluate with cold stabilizing activity was added in different concentrations, as indicated, either prior to assembly (●———●) or after cold labile microtubules had attained steady state by preassembly at 30°C for fifty minutes (o – – –o). For samples where the active fraction was added prior to assembly, microtubules were assembled at 30°C for 50 min, total label incorportion was determined, then an aliquot was chilled to 7°C for 40 min after which the cold stable level was assayed. For the steady state addition experiments, microtubules assembled for 50 min were mixed with the active fraction, incubated a further 20 min at 30°C, and assayed for total incorporation. Aliquots were then cooled to 7°C for 40 min, then assayed for cold stability.

Finally figure 6 and figure 7 show that STOP proteins can move along a given polymer. To prove this, we made use of recycled cold stable microtubules that were labelled either in their cold stable or cold labile parts only. Then the evolution of the situation with time was monitored by assaying cold stability in initially cold stable or cold labile portions of the polymers. It was evident that initially cold stable region became progressively labile and initially cold labile regions became increasingly stable during incubation. This must be due to an apparent sliding of STOPs along microtubules surface.

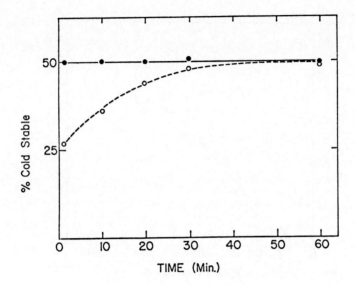

Fig. 3. Kinetics of the binding of STOPs onto preformed microtubules : the influence of free tubulin.
Cold labile (3X purified) microtubules were assembled for 40 min at 30°C with (^{3}H)GTP under the standard conditions for filter assay, and total assembly and blank values were obtained. At time 0, 200 µg/ml (final concentration) of DEAE column derived active fraction as added with 10 µM PLN. Aliquots were cooled to 7°C at the indicated time points and cold stable levels assayed (●———●). The identical experiments was performed, except for the addition of unpolymerized tubulin at time 0, to a final concentration of 4 mg/ml (o— — — — —o). Final dilutions of microtubules and of STOP protein were the same in both experiments.

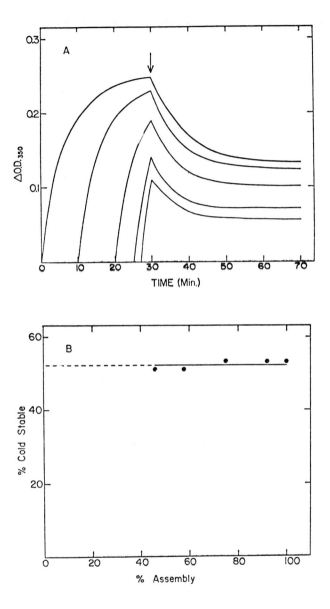

Fig. 4. Cold stability level as a function of the extent of assembly.
(A) Disassembled cold stable microtubule protein was preincubated at 30°C, and assembly initiated by addition of 1.0 mM GTP at the indicated time points. At the arrow the samples were rapidly cooled to 7°C. Both assembly and disassembly were monitored continuously by turbidity measurement.
(B) The measured cold stability is plotted as a function of the extent of assembly attained prior to chilling.

114

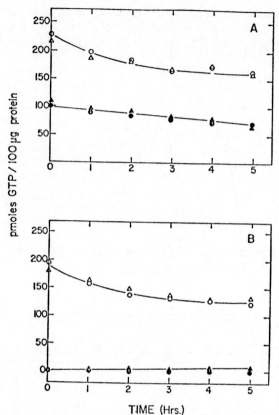

TIME (Hrs.)

Fig. 5. Assay of the exchange of STOPs between microtubules.
(A) Competition of cold labile microtubules for STOPs on cold stable
microtubules. Control : cold stable microtubules (2 mg/ml) were assembled for 50
min at 30°C in the standard conditions for filter assay (including 50 μM ^{3}H-
GTP).   At steady state (time 0) 10 μM PLN and 1 mM GTP were added, and total
(o———o) and cold stable (●———●) microtubule levels were filter assayed as
a function of time. Competition : at steady state, preassembled cold labile
microtubules, also at 2 mg/ml, were added with PLN. Final PLN, GTP, GTP
regenerating system and protein concentrations were the same as for the control.
Three volumes of cold labile microtubules were added for one volume of cold
stable microtubules. Total incorporation (△———△) and cold stable levels
(▲———▲) were assayed as a function of time. Aliquots of 50 μl of control
samples and 200 μl of competition samples were loaded onto filters, to keep the
quantity of cold stable microtubule protein constant. One can expect a maximum
four fold drop in cold stability of labelled mirotubules, relative to controls,
if competition were effective.
(B) The reverse experiment, induction of cold stability by exchange between
microtubules. Control : cold labile (3X cycled) microtubules (2 mg/ml) were
assembled for 50 min at 30°C in the standard conditions for filter assay
(including 50 μM ^{3}H-GTP). At steady state (time 0) 10 μM PLN and 1.0 mM GTP were
added. Total (o- ———o) and cold stable (●———●) levels were filter assayed at
the indicated time points (50 μl/filter). Competition : at steady state,
preassembled cold stable microtubules at 2 mg/ml were added, 3 volumes to one,
along with 10 M PLN. Final concentrations of all materials were the same as for
controls. Aliquots (200 μl/filter) were assayed at time points for total label
(△———△) and cold stable levels (▲———▲).

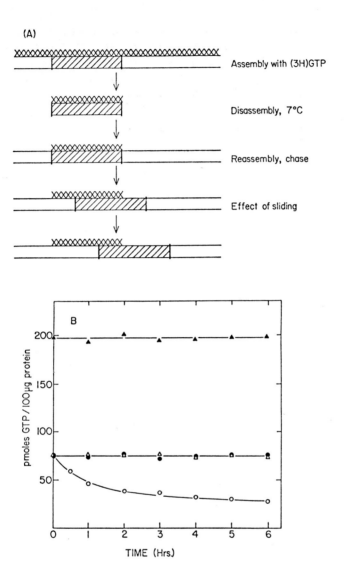

Fig. 6. Assay of the movement of STOP proteins on microtubules.
(A) Diagram of the experimental protocol and expected result, if sliding occurs.
Cold stable microtubules are assembled with (^{3}H)GTP, disassembled at cold
temperature to leave only residual cold stable regions, and reassembled in 20
fold excess GTP so that only cold stable regions are labeled. If STOP proteins
move relative to labeled subunits, the subunits that come to lie in regions of
the polymer external to STOPs will become cold labile. Diagonal lines indicate
the cold stable region of the polymer, delimited by heavy vertical lines
representing STOPs ; XXX indicates labeled subunits.

(B) Sliding experiment. Cold stable microtubules (2 mg/ml) were assembled in (³H)GTP under standard conditions for filter assay. After 50 min at 30°C, the protein was exposed to 7°C for 40 min and separated into two parts. The first sample, the control, was reassembled under the same conditions (the polymer remained fully labeled) and filter assayed at the indicated time points (time zero being the time of rewarming) for total label incorporation (▲———▲), and for cold stability (△———△) after cooling aliquots to 7°C for 20 min. The second part, the experimental, was reassembled at the indicated time 0 in the presence of 2 mM unlabeled GTP. At the indicated time points, the total label at 30°C (corresponding to the initial cold stable part) was assayed (●———●), and the residual cold stability of the labeled region was assayed (o———o), after cooling to 7°C for 20 min.

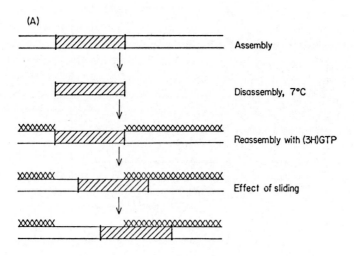

Fig. 7. The reciprocal experiment, demonstrating movement of STOP proteins into cold labile regions of microtubules.
(A) Diagram of the experimental protocol and expected result, if sliding occurs. Cold stable microtubules are assembled in the absence of label, disassembled to residual cold stable regions at 7°C, 20 min and reassembled in (³H)GTP to label, specifically, the labile regions of the polymer. If STOPs move relative to labeled subunits, the subunits that come to lie in regions of the polymer between STOP proteins will become cold stable. Diagonal lines indicate the cold stable region of the polymer, delimited by heavy vertical lines which represent STOPs ; XXX indicates labeled subunits.

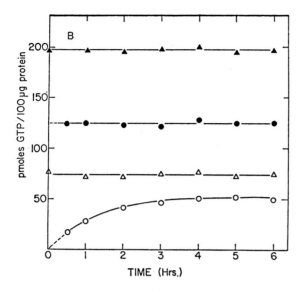

(B) Sliding experiment. Cold stable microtubules were assembled under standard conditions for filter assay but omitting (^{3}H)GTP. They were then cooled to 7°C for 40 min, and rewarmed (indicated time 0)for a second cycle of assembly in the presence of (^{3}H)GTP, to specifically label cold labile regions of polymers. Time points were then taken to assay for total label in the polymers, corresponding to the initial cold labile regions (●——●), and for label in cold stable regions (o———o), determined by chilling aliquots to 7°C for 20 min prior to filter assay. A control was conducted for this experiment by assaying fully labelled microtubules for total label with time and for label in cold stable regions. The procedure was identical to that for the control shown in figure 6B. Total label incorporation (▲——▲); cold stable region (△—△).

IV. ON THE POSSIBLE SIGNIFICANCE OF STOPs PROTEIN SLIDING

In general, very little seems to be known on the nature and kinetics of binding of microtubules associated proteins to these polymers. This is undoubtedly due to the fact that there are no obvious way of conveniently assaying the presence or absence of a given protein on a particular microbutule.

STOPs protein provided a useful model system because their binding to microtubules can be assayed by cold stability levels measurements.

Their behaviour can at first, look rather extraordinary when compared to usual enzymatic or receptors binding systems. Thus their ability to bind to microtubule surface in an apparently irreversible way and then slide along the polymer without equilibrium release might be regarded as a very special and specific property. There is a possibility that this is indeed the case. But we believe that this type of behavior will be found to be shared by many cytoskeleton associated proteins because it should be a general property of all binding systems in which one element is represented by a polymer in solution.

Thus the translocation of STOPs on microtubules is reminiscent of the sliding of many proteins on polynucleotide substrates. As examples DNA helicase (22) T4 DNA polymerase (23) DNA methylase (24) Rec BC (25) restriction endonuclease (26) and the eucaryote 40S ribosomal subunit (27) all exhibit processivity on their bound polynucleotide strand without equilibrium release.

Theoretical considerations as well as extensive experimental investigation of some model systems (28) show that the requirements for sliding of a ligand on a polymer to occur, are minimal : all is needed is a non specific electrostatic interaction between the ligand and the polymer. Then, because a polymer actually is a very high local concentration of a particular protein, the apparent affinity of any ligand for it will be higher than for the free subunits. Thus the ligand will concentrate on the polymer.

Then it will tend to diffuse along a given polymer rather than exchange between polymers. This is due to the fact that sliding makes the ligand to move on an isopotential surface contrarily to dissociation which requires net charge displacement and is thermodynamically unfavorable (28). Concentration and sliding of ligands on cytoskeletal structures might have profound physiological significance.

Thus it is possible to speculate extensively on the mechanical properties of a system made of treadmilling microtubules embedded in a "sleeve" of associated proteins possibly attached on other fixed structures but sliding along the polymer beneath them.

More generally, cytoskeleton could function as an activatory surface for the cells. Those ligands that slide along the cytoskeleton diffuse in one dimension (if we consider cytoskeletal filaments as linear structures) instead of three. This could dramatically improve the efficiency of multi enzymatic systems or accelerate intracellular movements of organelles. A rapidly increasing number of cytosolic enzymes is found to interact with cytoskeleton. Furthermore enzymes of a given metabolic pathway tend to form multienzymatic active complexes. The rate of formation of these active complexes should be enormously increased if all of the constitutive enzymes diffuse along the cytoskeleton. Even in free solution the reduction of dimentionnality of the diffusion process would result in a much higher efficiency (28).

This might be even much more spectacular in cells where an intricate membranes network probably makes non directed diffusion problematic.

STOPs obviously are a poor example of such an increase in diffusionnal efficiency. Maximum rate of sliding should approach a limit value of aproximatively 1.0 μm/s over a non specific binding surface (28) . This is several orders of magnitude above the rate of sliding observed in the present study for STOPs.

We have some evidence that these proteins have high affinity binding sites on each tubulin dimer. These would retard sliding motility enormously.

Increase in diffusionnal efficiency could also be important to intracellular organelles movements. Because this is directionnal one could think that some kind of energy spending directionnal mechanism has to be superimposed to passive diffusion.

Although this could exist the concomitant existence of an effective diffusion system and of localized high affinity sites in cells would generate by itself a rapid apparent directionnal migration of organelles (29).

It is probably of interest that since as outlined above the requirements for polymers to direct the diffusion of a ligand are minimal, specific linkers between cytoskeleton and organelles need not to exist for the cytoskeleton to organize organelles movements.

Finally the ability of labile cytoskeletal structures as microtubules to concentrate and to compartimentalize enzymes offers a relatively straight forward explanation for the substantial effect of their assembly state on the metabolic state of the cell.

In summary although the physiological role of the sliding of STOPs cannot be assessed at present we believe that they should serve as a useful experimental system to help unravel the mechanism of cytoskeleton dependent cellular functions.

Acknowledgments

We are very much endebted to Pr E.M. Chambaz for continuous support during this work.

In several of the reviewed studies we have been directly collaborating with Pr E.H. Fischer. His constant help and disponibility are gratefully acknowledged.

Mr Charles Rauch and Mrs Fabienne Pirollet have been directly involved in many of the presented experiments.

Finally we wish to aknowledge the fine technical assistance of Françoise Gilles.

This work was supported by grants from the National Institutes of Health, from the Amyotrophic Lateral Sclerosis Society of America, from the Institut National de la Santé et de la Recherche Médicale and from the Ministry of Research and Industry, France.

120

REFERENCES

1. SNYDER JA, Mc INTOSH JR (1976) Ann Rev Biochem 45:699-720

2. DUSTIN P (1978) Microtubules (Springer New York)

3. MARGOLIS RL, WILSON L (1981) Nature (London) 292:705-711

4. CROSSIN KL, CARNEY DH (1981) Cell 23:61-71

5. FRIEDKIN M, LEGG A, ROZENGURT E (1979) Proc Nat Acad Sci USA 76:3909-3912.

6. TADVALKAR G, PINTO DA SILVA P (1983) J Cell Biol 96:1279-1287

7. BLACK MH, COCHRAN JM, KURDYLA JT (1984) Brain Res 255:255-263

8. FAIVRE C, LEGRAND C, RABIE A (1984) Int J Dev Neurosci, in press

9. JOB D, RAUCH CT, FISCHER EH, MARGOLIS RL (1982) Biochemistry 21:509-515

10. JOB D, MARGOLIS RL (1984) Biochemistry 23:3025-3031

11. GRISHAM LM (1976) Dissertation (Standford Univ Stanford CA)

12. WEBB BC, WILSON L (1980) Biochemistry 19:1993-2000

13. MARGOLIS RL, RAUCH CT (1981) Biochemistry 20:4451-4458

14. JOB D, FISCHER EH, MARGOLIS RL (1981) Proc Natl Acad Sci USA 78:4679-4682.

15. MARGOLIS RL, RAUCH CT, manuscript in preparation

16. JOB D, RAUCH CT, FISCHER EH, MARGOLIS RL (1983) Proc Natl Acad Sci USA 80:3894-3898.

17. JOB D, MARGOLIS RL, unpublished observations

18. PIROLLET F, JOB D, FISCHER EH, MARGOLIS RL (1983) Proc Natl Acad Sci USA 80:1560-1564.

19. PIROLLET F, JOB D, MARGOLIS RL, unpublished observations

20. PABION M, JOB D, MARGOLIS RL (1985) Biochemistry, in press

21. WILSON L, SNYDER KB, THOMPSON WC, MARGOLIS RL (1982) in Methods in Cell Biology vol 24A (Wilson Ed) 159-162 (Academic New York)

22. YARRANTON GT, DAS RH, GEFTER ML (1979) J Biol Chem 254:12002-12006

23. WINTER RB, BERG CG, VON HIPPEL PH (1981) Biochemistry 20:6961-6977

24. DRAHOVSKY D, MORRIS NR (1971) J Mol Biol 57:475-489

25. ROSAMUND J, ENDLICH B, LINN S (1979) J Mol Biol 129:619-635

26. ROSAMUND J, TELANDER RM, LINN S (1979) J Biol Chem 254:8646-8652

27. KUZAK H (1980) Cell 22:459-467

28. BERG OG, WINTER RB, VON HIPPEL PH (1982) Trends in Biochem Sci 7:52-55

29. SHERIDAN PJ, BUCHANAN JM, ANSELMO VC, MARTIN PM (1982) Nature (London) 282:575-582

RESUME

Les microtubules peuvent être rendus résistants à la dépolymérisation induite par le froid par de faibles quantités de "STOP protéines". Cette stabilisation peut être réversée par le complexe Ca^{++}-calmoduline et par des réactions de phosphorylation dépendantes ou non de la calmoduline. Une étude cinétique de l'interaction entre STOPs et microtubules a été effectuée. Elle a montré que : 1/ les STOPs agissent aussi bien lorsqu'on les ajoute à de la tubuline désassemblée que lorsqu'on les ajoute à des microtubules préformés ; 2/ qu'elles se lient très rapidement aux microtubules préformés ; 3/ qu'elles ne s'échangent pas entre microtubules mais que 4/ elles peuvent glisser le long d'un polymère donné. Nous proposons que ce phénomène de glissement pourrait s'expliquer par des propriétés purement physico-chimiques des polymères en solution.

C'est pourquoi nous pensons que de nombreux enzymes ou organelles pourraient se comporter de la même manière et glisser le long du cytosquelette. Cette structure serait alors une véritable surface activatrice intracellulaire, augmentant considérablement l'efficacité des processus de diffusion et jouant également un rôle important dans la compartimentalisation des cellules.

Hormones and Cell Regulation, Volume 9
INSERM European Symposium
J.E. Dumont, B. Hamprecht and J. Nunez editors
© 1985 Elsevier Science Publishers B.V. (Biomedical Division)

THE INTERMEDIATE FILAMENT-DESMOSOME CYTOSKELETON COMPLEX

WERNER W. FRANKE AND ROY A. QUINLAN

Institute of Cell and Tumor Biology, German Cancer Research Center, Im
Neuenheimer Feld 280, D-6900 Heidelberg, Federal Republic of Germany

CLASSES OF INTERMEDIATE-SIZED FILAMENTS

The cytoplasm of most cells of vertebrates contains a system of characteristic unbranched filaments of diameter 7-12 nm which in many cell types represent some of the most abundant cytoplasmic proteins. The protein constituents of these "intermediate-sized filaments" (IF) are different in different
kinds of cells, and at least five different classes can be distinguished (Table
1): Cytokeratins are proteins related to epidermal keratins and are specific
for epithelial cells (1-4); vimentin IF occur in various types of non-epithelial cells, notably those of mesenchymal origin (5-7); desmin is a major
specific constitutent of most, but not all cells of myogenic differentiation
(6,8,9), glial filaments occur in astrocytes and some related cells, and
neurofilaments are characteristic of neuronal differentiation (10,11). In
addition, certain cell types can simultaneously express two classes of IF proteins. Here one has to distinguish between two different situations. The first
ist that two IF classes can be co-expressed in the same cell but occur as
separate IF. An example of this are cells co-expressing cytokeratins and
vimentin such as in numerous cultured epithelial and carcinoma cells (5,12-15),
in some suspended carcinoma cells growing in ascites form (16), in the isolated,
myxoid cells of pleiomorphic adenomas (17,18) and in the non-coherent cells
of the parietal (i.e. distal) endoderm of mouse embryos (19-21). In the
other situation, cells co-expressing vimentin and desmin (22-25) can form true
heteropolymer filaments which contain heterodimeric subunits as we have shown
by oxidative cross-linking at the single cysteine residue present in both
molecules (26). Similarly, co-expression of glial filament protein and vimentin
has been shown in astrocytes and cultured glioma cells (27-29), and here
again true hybrid polymer IF have been demonstrated by chemical cross-linking
(30). The value of the use of antibodies to these different IF proteins in
tumor diagnosis has been established in reports from a number of laboratories,
both for experimentally induced animal tumors (31) and for human neoplasms
(32-35).

The different IF proteins have been identified as distinct translational
products of specific mRNAs (36-40) which are synthesized in relation to
specific routes of cell differentiation. In the last three years cDNA clones of
vimentin (41,42), desmin (43,44), glial filament protein (45) and diverse

Table 1

Classes of Intermediate Filaments in Different Cells

Constituent Proteins	Polypeptides	Molecular weight ($M_p \times 10^{-3}$)*	Isoelectric pH (in 9.5 M urea)	Occurrence
I. Cytokeratins (α-keratins)	2-10	40-68	5-8	True epithelia and carcinoma cells; oocytes of certain amphibia and early embryonic stages of various species
II. Vimentin	1	57	5.3	Non-epithelial cells, spec. mesenchymal cells, also Sertoli cells, lens cells, some myogenic cells, sarcomas; many cultured cell lines
III. Desmin	1	53	5.4	Muscle cells, with the exception of one type of vascular smooth muscle cells
IV. Glia-Filament Protein	1	51	5.6	Astrocytes and gliomas
V. Neurofilament Protein	3	210,160,68	5.65, 5.28,5.25	Neurones, a certain type of neuroblastomas and phaeo-chromocytomas
VI. Vimentin-Desmin Hybrid	2			Certain vascular smooth muscle cells; BHK-21 cell line
VII. Vimentin-GFAP Hybrid	2			Astrocytes and some cultured glioma cells

*Values are from estimations of gel electrophoretic mobility after denaturation with SDS; actual molecular masses based on amino acid sequences may be somewhat different.

cytokeratin polypeptides (46-54) have been described. In addition, the genomic structure of vimentin (55) and several cytokeratins (56) have been presented.

The various classes of IF are members of a multigene family amd display a common principle of organization of the secondary structure as deduced from amino acid sequence data (47,48,50,55,57-59). There is a central α-helical core region of approximately 310 amino acids ("rod") which is flanked on both sides by non-α-helical regions (aminoterminal "head" and carboxyterminal "tail"). These non-α-helical regions are variable in size and, by and large, account for the different molecular weights of the various constituent proteins of the different IF classes. The α-helical rod region is interrupted in two portions by small non-α-helical regions, the positions of which are conserved in all the different IF classes. This conservation in the arrangement of the secondary structure of IF proteins also suggests a common principle for the assembly and polypeptide arrangement within the IF. In addition to this there is evidence that the number and positioning of introns in the gene is also conserved for the different IF proteins (55,56).

THE CYTOKERATIN FAMILY

The cytokeratin IF subclass displays the largest complexity for the IF family. It represents a multigene family of approximately 20 different poly-peptides (according to Moll et al., ref. 60, 19 cytokeratin polypeptides have been identified in human tissues; similar complexities have been reported for man, rodents and cow: 60-65) which are expressed in cell type-specific combinations of 2-10 polypeptides in a given cell (60). We have established, in collaboration with Drs. H. Denk and R. Krepler (Universities of Graz and Vienna, Austria) and B. Geiger (Weizmann Institute of Science, Rehovot, Israel), the tissue-specific patterns of expression of cytokeratin polypeptides for various animal and human epithelial tissues and tumors (60,66-72), with special emphasis on organs with a high cell type-complexity such as the bronchopulmonary tract (67,68), the urogenital tract (69,70) and the epidermis (71,72). Using monoclonal antibodies specific for cytokeratin poly-peptides, specific epithelial cell types as well as subclasses of carcinomas can be immunologically distinguished (73-76).

Analysis of products of translation in vitro has shown that the majority, if not all, of the various cytokeratin polypeptides found in different tissues are products of discrete mRNAs (36,40,49,51,64,77). Using poly(A)$^+$RNA from bovine snout epidermis and bovine bladder, we have produced cDNA libraries for six different epidermal and three non-epidermal cytokeratin polypeptides. Hybridization under stringent conditions has shown that these different keratin mRNAs are again expressed in cell type-specific patterns, corresponding to the patterns of the polypeptides identified in these cells, and that different mRNAs, even those of the same subfamily (cytokeratins can be divided into two large subfamilies,the basic "type II" and the acidic "type I" polypeptides; cf. 46,64,75; for relationship to type I and II wool keratin see 78), do not share regions of extended sequence identity (51-54; see, however, also 46,49). Sequence analysis of these cDNA clones allows the comparison of different cytokeratin polypeptides from both subfamilies I and II of the same species and also of cytokeratins from different species (47,48,50, 52,54,78). These sequence data comparisons show: (1) The two cytokeratin subfamilies share only ca. 30 % homology which is essentially restricted to the α-helical core domain; (2) Within one subfamily different types of keratin poly-peptides can be distinguished (e.g., "serine-rich" or "glycine-rich" carboxyterminal sequences can be distinguished in the non-helical part of type I cytokeratin); (3) equivalent keratins in different species share a remarkably high degree of sequence homology, including the organization into special domains (at least three different domains can be distinguished in the non-helical region of the carboxyterminal "tails" of the basic, i.e. type II cytokeratins from epidermis; 79) and certain peptide "signals" which are maintained in evolution (for sequence data of Xenopus cytokeratins see 54). In addition, we have isolated six different genomic clones of bovine keratins, most probably containing complete genes. Restriction analysis, S1-nuclease mapping, heteroduplex mapping and partial sequence data have revealed that the exon-intron positions in the different genes are very similar, probably identical for certain introns, and that the positions of some of the introns are identical, for example in all bovine type II cytokeratins examined and in hamster vimentin can be related to certain important points of secondary and tertiary structure of the proteins (55,56).

SUBUNIT ORGANIZATION AND ASSEMBLY

Subunit complexes of IF, including those of the cytokeratin type, can be obtained by exposure of intact IF to 4 M urea or 2 M guanidinium hydrochloride, For example, this treatment transforms IF of hepatocyte cytokeratin into a monodisperse population of molecules of ca. 5S which represent tetramers of M_r ca. 210,000 (specifically, heterotypic tetramers in the case of the cytokeratins). Such methods, combined with chemical cross-linking techniques, have allowed the identification of the cytokeratin IF subunits as obligatory tetrads of two representatives of each the basic and the acidic subfamily of cytokeratin polypeptides (80), but we can presently not decide whether

these tetramers are pairs of identical heterodimers, heterotypic pairs of
homodimers, or pairs of different heterodimers. Similar subunits have been
described for chymotrypsin-or trypsin-resistant "cores" of wool keratin
(81) and desmin (58). All data available therefore support models of an IF
subunit complex containing an even number of polypeptide chains (58,82) and
exclude triple chain arrangements (83). Different combinations of pairs of
cytokeratins can be characterized by their dissociation and re-association
behaviour in denaturing agents ("melting profile") thus showing that
different cytokeratin IF have different internal complex stability (84).

IF assembly has so far been only studied in vitro, usually starting from
denatured proteins under very unphysiological conditions (e.g., 85,86).
Because of the low solubility of IF proteins, studies of IF assembly in
living cells are very difficult. However, we have found a way to study the
assembly of cytokeratin IF in living cells by microinjection of cytokeratin
mRNAs into living cultured cells which do not express cytokeratins,
followed by visualization of the translation products in vivo using
specific antibodies (87,88). With this approach we have shown that
cytokeratin IF can be formed from products of injected mRNA at multiple
disperse sites in non-epithelial cells, without a special "IF organizing
center" (87). In addition, these experiments have shown that the newly
synthesized cytokeratin molecules assemble with each other and are not
integrated into the pre-existing vimentin IF abundant in these cells, thus
confirming that there exist mechanisms that sort out cytokeratins and
vimentin (see also ref. 15). Using the mRNA microinjection technique we
have also demonstrated that different kinds of cytokeratins can co-assemble
into common filaments, even cytokeratins which are never co-expressed in
vivo (88). These experiments indicate that (i) the cell type-specific
expression of IF proteins is regulated at a pre-translational, probably the
transcriptional level, and (ii) the reason for the cell type-specific
expression of different cytokeratins is not the incompability in co-
assembly but must be due to other regulatory mechanisms, probably those
governing cytokeratin gene expression.

We have also found dynamic changes of IF protein organization. The most
conspicuous example is the transition of IF material from typical IF into
large spheroidal aggregates of densely packed, finely filamentous material
during mitosis of many cultured epithelial and carcinoma cells (89,90) as

well as mitotic intestinal cells (91) and certain solid carcinomas (92).
Perimitotic structural changes of cytokeratin IF organization, though of a
more subtle character, have also been described in other cells such as PtK$_2$

cells, using a monoclonal antibody which recognizes a determinant accessible only during mitosis and in early G1 phase (93). Using a specific monoclonal vimentin antibody, a similar change has recently been found for vimentin IF in various mitotic cells, including human fibroblasts (94). This shows that IF, albeit so insoluble and stable, can be dynamically re-arranged during the cell cycle as well as perhaps also during differentiation processes (for an example in myogenesis see 95). Changes of IF organization can also occur in pathological processes. For example, we have found, in collaboration with H. Denk (University of Graz, Austria), that "Mallory bodies", i.e. structures characteristic of alcoholic hepatitis and other diseases of man as well as of griseofulvin intoxication of mouse hepatocytes, consist of large intracellular heaps of relatively short filament rods with a fuzzy coat and contain structurally altered forms of hepatocyte cytokeratins (96-98).

In most epithelial cells cytokeratin IF are attached to certain plasma membrane domains, i.e. the plaques of desmosomes and hemidesmosomes. These membrane-plaque complexes have been isolated with various procedures (99-102) and the major desmosomal proteins have been characterized by biochemical analyses (e.g. amino acids, carbohydrates, peptide maps) and immunological methods (103-108). Immunocytochemistry using antibodies against major desmosomal plaque polypeptides, of M_r 250,000 and 215,000 (desmoplakins I and II, both are closely related as shown by peptide maps and common antigenic determinants) facilitates the localization of desmosome-specific proteins and hence provides a non-morphological definition of the desmosome (101,103-108). Some of the desmosomal polypeptides such as components 5 (M_r 83,000) and 6 (M_r 75,000) have been identified as products of translation in vitro (109), showing that distinct mRNAs code for these proteins. Antibodies to desmosomal proteins have also been found to provide differentiation markers that are useful in tumor diagnosis (101,104,106). For example, carcinomas are generally positive for both cytokeratins and desmoplakins (104).

In addition, we have found that desmosomes and typical desmosomal proteins such as desmoplakins are also produced by certain non-epithelial cells such as myocardiac and Purkinje fiber cells of heart (101,103,106). Cardiac cells do not contain IF of the cytokeratin type but in this case desmin IF are attached to the desmoplakin-containing plaques (106,110), a finding also reported by Tokuyasu et al. (111). A third type of desmosome-IF relationship has been found in the arachnoid cells of the brain and in the tumors derived therefrom, i.e. meningiomas (112) and here vimentin IF are the only IF present which are attached to true desmosomal plaques.

The desmosome is a notoriously stable structure. However, upon dissociation

of cells and symmetrical splitting of desmosomal junctions by Ca^{2+} depletion in
the external medium, the "hemi-desmosome-like" domain of the split desmosome is
taken up by endocytosis, maintaining the organization of the specific membrane-
plaque-IF attachment (113). This illustrates the stability of the IF plaque
complex even after disruption of cell-to-cell interaction.

As the expression of IF and desmosomal proteins is so conspicuously related
to certain routes of cell differentiation during embryogenesis. The earliest
class of IF expressed are the cytokeratins A and D (equivalent to human
components Nos. 8 and 18) which are found to be abundant in trophectodermal
cells of pre-implantations embryos (114-116). In certain species such as the
toad, Xenopus laevis, cytokeratin polypeptides are already detected in the
oocyte (117). Cytokeratins are also the only IF proteins found in embryonic
ectoderm and visceral (proximal) endoderm of post-implantation mouse embryos
at, e.g., days 7 and 8 after fertilization, and cytokeratins have also been
detected in parietal endoderm (19-21,118,119). Ectodermal and visceral endo-
dermal cells also contain desmosomes and express typical desmosomal proteins
(120). The next IF protein to be expressed is vimentin which appears specifically in the primary mesenchymal cells which in turn no longer produce
cytokeratin IF and desmosomes (21,120). This illustrates that the embryonic
expression program does involve mechanisms not only of induction of one type of
IF protein but also effective clearance mechanisms of proteins of another
IF type and inhibition of their expression. Studies of subsequent tissue and
organ development should allow to determine the complexity of IF protein
expression in relation to increasing tissue and cell type complexity.

It has recently become clear that the pattern of expression of cytokeratin
polypeptides in specific epithelial cell type is not only determined by cell
differentiation in a determinative way but also can be changed by environmental
factors (121). One component known to change, both quantitatively and qualitatively, the expression of certain cytokeratins in cultured keratinocytes and
other epithelial cells is vitamin A or retinoic acid, respectively (e.g., 122-
125). However, hormonal additions to cell culture media can also influence the
expression pattern of cytokeratins as we have demonstrated for the influence of
hydrocortisone, prolactin and insulin on cultured bovine mammary gland
epithelial cells (126), for diverse programs of exposure of rat hepatoma cells
to glucocorticoids (127) and for treatments of cultured human keratinocytes
with combinations of vitamin A, hydrocortisone and EGF (128). These observations
show that the regulation of expression of cytokeratins, and probably also other
IF proteins, can be experimentally altered in vitro, thus opening a possible
experimental pathway to elucidate the mechanisms involved.

REFERENCES

1. Franke WW, Weber K, Osborn M, Schmid E, Freudenstein C (1978) Exp Cell Res 116:429-445

2. Sun T-T, Green H (1978) J Biol Chem 253:2053-2060

3. Franke WW, Appelhans B, Schmid E, Freudenstein C, Osborn M, Weber K (1979) Differentiation 15:7-25

4. Sun T-T, Shih CH, Green H (1979) Proc Natl Acad Sci USA 76:2813-2817

5. Franke WW, Schmid E, Osborn M, Weber K (1978) Proc Natl Acad Sci USA 75:5034-5038

6. Bennett GS, Fellini SA, Croop JM, Otto JJ, Holtzer H (1978) Proc Natl Acad Sci USA 75:4364-4368

7. Hynes RO, Destree AT (1978) Cell 13:151-163

8. Lazarides E, Hubbard BD (1976) Proc Natl Acad Sci USA 73:4344-4348

9. Small JV, Sobieszek A (1977) J Cell Sci 23:243-268

10. Liem RHK, Yen SH, Salmon GD, Shelanski ML (1978) J Cell Biol. 79:637-645

11. Dahl D, Bignami A (1984) In: Dowben RM, Shay JW (eds) Cell and Muscle Motility. Vol VI. Plenum Press, New York, in press

12. Franke WW, Schmid E, Breitkreutz D, Lüder M, Boukamp P, Fusenig NE, Osborn M, Weber K (1979) Differentiation 14:35-50

13. Franke WW, Schmid E, Weber K, Osborn M (1979) Exp Cell Res 118:95-109

14. Franke WW, Schmid E, Winter S, Osborn M, Weber K (1979) Exp Cell Res 123:25-76

15. Osborn M, Franke WW, Weber K (1980) Exp Cell Res 125:27-46

16. Ramaekers FCS, Haag D, Kant A, Moesker O (1983) Proc Natl Acad Sci USA 80:2618-2622

17. Caselitz J, Osborn M, Seifert G, Weber K (1981) Virchows Arch (Path Anat) 393:273-286

18. Krepler R, Denk H, Artlieb U, Moll R (1982) Differentiation 21:191-199

19. Lane EB, Hogan BLM, Kurkinen M, Garrels JI (1983) Nature 303:701-704

20. Lehtonen E, Lehton V-P, Paasivuo R, Virtanen I (1983) EMBO J 2:1023-1028

21. Franke WW, Grund C, Jackson BW, Illmensee K (1983) Differentiation 25: 121-141

22. Gard DL, Bell PB, Lazarides E (1979) Proc Natl Acad Sci USA 76:3894

23. Tuszynski GP, Frank ED, Damsky CH, Buck CA, Warren L (1979) J Biol Chem 254:6138-6143

24. Steinert PM, Idler WW, Cabral F, Gottesman MM, Goldman RD (1981) Proc Natl Acad Sci USA 78:3692-3696

25. Schmid E, Osborn M, Rungger-Brändle E, Gabbiani G, Weber K, Franke WW (1982) Exp Cell Res 137:329-340

26. Quinlan RA, Franke WW (1982) Proc Natl Acad Sci USA 79:3453-3456

27. Schnitzer J, Franke WW, Schachner M (1981) J Cell Biol 90:435-447

28. Yen SH, Fields KL (1981) J Cell Biol 88:115-126

29. Sharp G, Osborn M, Weber K (1982) Exp Cell Res 141:385-395

30. Quinlan RA, Franke WW (1983) Eur J Biochem 132:477-484

31. Bannasch P, Zerban H, Schmid E, Franke WW (1980) Proc Natl Acad Sci USA 77:4948-4952

32. Schlegel R, Banks-Schlegel S, McLeod JA, Pinkus GS (1980) Am J Pathol 101:41-49

33. Gabbiani G, Kapanci Y, Barrazone P, Franke WW (198) Am J Pathol 104: 206-216

34. Altmannsberger M, Osborn M, Hölscher A, Schauer A, Weber K (1981) Virchows Arch (Cell Pathol) 37:277-284

35. Osborn M, Weber K (1983) Lab Invest 48:372-394

36. Fuchs E, Green H (1979) Cell 17:573-582

37. Schmid E, Ghosal D, Franke WW (1980) Eur J Cell Biol 22:374

38. Czosnek H, Soifer D, Wisniewski HH (1980) J Cell Biol 85:726-734

39. O'Connor CM, Asai DJ, Flytzanis CN, Lazarides E (1981) Mol Cell Biol 1:303-309

40. Magin TM, Jorcano JL, Franke WW (1983) EMBO J 2:1387-1392

41. Quax-Jeuken YEFM, Quax WJ, Bloemendal H (1983) Proc Natl Acad Sci USA 80:3548-3552

42. Capetanaki Y.G., Ngai J, Flytzanis LN, Lazarides E (1983) Cell 35:411-420

43. Quax, WJ, Egbert WV, Hendriks W, Quax-Jeuken Y, Bloemendal H (1983) 35:215-223

44. Capetanaki YG, Ngai J, Lazarides E (1984) Proc Natl Acad Sci USA, in press

45. Lewis SA, Balcarek, JM, Krek V, Shelanski M, Cowan NJ (1984) Proc Natl Acad Sci USA 81:2743-2746

46. Fuchs EV, Coppock SM, Green H, Cleveland DW (1981) Cell 27:75-84

47. Hanukoglu I, Fuchs E (1982) Cell 31:243-252

48. Hanukoglu I, Fuchs E (1983) Cell 33:915-924

49. Roop DR, Hawley-Nelson P, Cheng CK, Yuspa SH (1983) Proc Natl Acad Sci USA 80:716-720

50. Steinert PM, Rice RH, Roop DR, Trus BL, Steven AC (1983) Nature 302: 794-800

51. Jorcano JL, Magin TM, Franke WW (1984) J Mol Biol 176:21-37

52. Jorcano JL, Rieger M, Franz JK, Schiller DL, Moll R, Franke WW (1984) J Mol Biol 179:257-281

53. Franke WW, Schiller DL, Hatzfeld M, Magin TM, Jorcano JL, Mittnacht S, Schmid E, Cohlberg JA, Quinlan RA (1984) Cold Spring Harbor Laboratory, Cancer Cells 1, The Transformed Phenotype. 177-190

54. Hoffmann W, Franz JK (1984) EMBO J 3:1301-1306

55. Quax W, Vree Egbers W, Hendriks W, Quax-Jeuken Y, Bloemendal H (1983) Cell 35:215-223

56. Lehnert ME, Jorcano JL, Zentgraf H, Blessing M, Franz JK, Franke WW (1984) EMBO J, in press

57. Geisler N, Weber K (1981) Proc Natl Acad Sci USA 78:4120-4123

58. Geisler N, Weber K (1982) EMBO J 1:1649-1656

59. Geisler N, Kaufmann E, Fischer S, Plessmann U, Weber K (1983) EMBO J 2:1295-1302

60. Moll R, Franke WW, Schiller DL, Geiger B, Krepler R (1982) Cell 31:11-24

61. Franke WW, Schiller DL, Moll R, Winter S, Schmid E, Engelbrecht I, Denk H, Krepler R, Platzer B (1981) J Mol Biol 153:933-959

132

62. Franke WW, Winter S, Grund C, Schmid E, Schiller DL, Jarasch E-D (1981) J Cell Biol 90:116-127

63. Wu Y-J, Parker LM, Binder NE, Beckett MA, Sinard JH, Griffiths CT, Rheinwald JG (1982) Cell 31:693-703

54. Schiller DL, Franke WW, Geiger B (1982) EMBO J 1:761-769

65. Sun T-T, Eichner R, Schermer A, Cooper D, Nelson WG, Weiss RA (1984) Cold Spring Harbor Laboratory, Cancer Cells 1, The Transformed Phenotype. 169-176

66. Denk H, Krepler R, Lackinger E, Artlieb U, Franke WW (1982) Lab Invest 46:584-596

67. Blobel GA, Moll R, Franke WW, Vogt-Moykopf I (1984) Virchows Arch (Cell Pathol) 45:407-429

68. Blobel G, Gould VE, Moll R, Lee I, Huszar M, Geiger B, Franke WW (1984) Lab Invest, in press

69. Moll R, Levy R, Czernobilsky B, Hohlweg-Majert P, Dallenbach-Hellweg G, Franke WW (1983) Lab Invest 49:599-610

70. Achtstätter T, Moll R, Moore B, Franke WW (1984) J Histochem Cytochem, in press

71. Moll R, Franke WW, Volc-Platzer B, Krepler R (1982) J Cell Biol 95:285-295

72. Moll R, Moll I, Franke WW (1984) Arch Dermatol Res 276:349-363

73. Debus E, Moll R, Franke WW, Weber K, Osborn M (1984) Am J Pathol 114:121-130

74. Gigi O, Geiger B, Eshhar Z, Moll R, Schmid E, Winter S, Schiller DL, Franke WW (1982) EMBO J 1:1429-1437

75. Tseng SCG, Jarvinen MJ, Nelson WG, Huang J-W, Woodcock-Mitchell J, Sun T-T (1982) Cell 30:361-372

76. Ramaekers AF, Huysmans A, Moesker O, Kant A, Jap P, Herman C, Vooijs P (1983) Lab Invest 49:353-361

77. Kim KH, Rheinwald J, Fuchs EV (1983) Mol Cell Biol 3:495-502

78. Crewther WG, Dowling LM, Steinert PM, Parry DAD (1983) Int J Biol Macromol 5:267-274

79. Jorcano JL, Franz JK, Franke WW (1984) Differentiation, in press

80. Quinlan RA, Cohlberg JA, Schiller DL, Hatzfeld M, Franke WW (1984) J Mol Biol 178:365-388

81. Gruen LC, Woods EF (1983) Biochem J 209:587-593

82. Weber, Geisler N Weber K (1984) Cold Spring Harbor Laboratory Cancer Cells 1, The Transformed Phenotype. 153-159

83. Steinert PM, Idler WW, Goldman RD (1980) Proc Natl Acad Sci USA 77:4534-5438

84. Franke WW, Schiller DL, Hatzfeld M, Winter S (1983) Proc Natl Acad Sci USA 80:7113-7117

85. Steinert PM, Idler WW, Aynardi-Whitman M, Zackroff R, Goldman RD (1982) Cold Spring Harbor Symp Quant Biol 46:465-474

86. Renner W, Franke WW, Schmid E, Geisler N, Weber K, Mandelkow E (1981) J Mol Biol 149:285-306

87. Kreis TE, Geiger B, Schmid E, Jorcano JL, Franke WW (1983) Cell 32:1125-1137

88. Franke WW, Schmid E, Mittnacht S, Grund C, Jorcano JL (1984) Cell 36:813-825

89. Franke WW, Schmid E, Grund C, Geiger B (1982) Cell 30:103-113

90. Lane EB, Goodman SL, Trejdosiewicsz LK (1982) EMBO J 1:1365-1372

91. Brown DT, Anderton BH, Wylie CC (1983) Cell Tissue Res 233:619-628

92. Geiger B, Kreis TE, Gigi O, Schmid E, Mittnacht S, Jorcano JL, von Bassewitz DB, Franke WW (1984) Cold Spring Harbor Laboratory, Cancer Cells 1, The Transformed Phenotype. 201-215

93. Franke WW, Schmid E, Wellsteed J, Grund C, Gigi O, Geiger B (1983) J Cell Biol 97:1255-1260

94. Franke WW, Grund C, Kuhn C, Lehto V-P, Virtanen I (1984) Exp Cell Res 154:567-580

95. Danto SI, Fischman DA (1984) J Cell Biol 98:2179-2191

96. Franke WW, Denk H, Schmid E, Osborn M, Weber K (1979) Lab Invest 40:207-220

97. Denk H, Franke WW, Eckerstorfer R, Schmid E, Kerjaschki D (1979) Proc Natl Acad Sci USA 76:4112-4116

98. Denk H, Franke WW, Kerjaschki D, Eckerstorfer R (1979) Int Rev Exp Pathol 20:77-121

99. Skerrow CJ, Matoltsy AG (1974) J Cell Biol 63:524-530

100. Drochmans P, Freudenstein C, Wanson JC, Laurent L, Keenan TW, Stadler J, Leloup R, Franke WW (1978) J Cell Biol 79:427-443

101. Franke WW, Schmid E, Grund C, Müller H, Engelbrecht I, Moll R, Stadler, J, Jarasch E-D (1981) Differentiation 20:217-241

102. Franke WW, Kapprell H-P, Müller H (1983) Eur J Cell Biol 32:117-130

103. Müller H, Franke WW (1983) J Mol Biol 163:647-671

104. Franke WW, Moll R, Müller H, Schmid E, Kuhn C, Krepler R, Artlieb U, Denk H (1983) Proc Natl Acad Sci USA 80:543-547

105. Geiger B, Schmid E, Franke WW (1983) Differentiation 23:189-205

106. Franke WW, Moll R, Schiller DL, Schmid E, Kartenbeck J, Müller H (1982) Differentiation 23:115-127

107. Cowin P, Garrod DR (1983) Nature 302:148-150

108. Cohen SM, Gorbsky G, Steinberg S, (1983) J Biol Chem 258:2621-2627

109. Franke WW, Müller H, Mittnacht S, Kapprell H-P, Jorcano JL (1983) EMBO J 2:2211-2215

110. Kartenbeck J, Franke WW, Moser JG, Stoffels U (1983) EMBO J 2:735-742

111. Tokuyasu KT, Dutton AH, Singer SJ (1983) J Cell Biol 96:1736-1742

112. Kartenbeck J, Schwechheimer K, Moll R, Franke WW (1984) J Cell Biol 98:1072-1081

113. Kartenbeck J, Schmid E, Franke WW, Geiger B (1982) EMBO J 1:725-732

114. Jackson BW, Grund C, Schmid E, Bürki K, Franke WW, Illmensee K (1980) Differentiation 17:161-179

115. Paulin D, Babinet C, Weber K, Osborn M (1980) Exp Cell Res 130:297-304

116. Oshima RG, Howe WE, Klier FG, Adamson ED, Shevinsky LH (1983) Dev Biol 99:447-455

117. Franz JK, Gall L, Williams MA, Picheral B, Franke WW (1983) Proc Natl Acad Sci USA 80:6254-6258

118. Jackson BW, Grund C, Winter S, Franke WW, Illmensee K (1981) Differentiation 20:203-216

119. Franke WW, Schmid E, Schiller DL, Winter S, Jarasch ED, Moll R, Denk H, Jackson BW, Illmensee K (1982) Cold Spring Harbor Symp Quant Biol 46:431-453

120. Franke WW, Grund C, Kuhn C, Jackson BW, Illmensee K (1982) Differentiation 23:43-59

121. Doran TI, Vidrich A, Sun T-T (1980) Cell 22:17-25

122. Fuchs E, Green H (1981) Cell 25:617-625

123. Eckert RL, Green H (1984) Proc Natl Acad Sci USA 81:4321-4325

124. Kim KH, Schwartz F, Fuchs E (1984) Proc Natl Acad Sci USA 81:4280-4284

125. Ramaekers F, Schaap H, Mulder M, Huysmans A, Vooijs P (1984) Cell Biol Int Rep 8:721-730

126. Schmid E, Schiller DL, Grund C, Stadler J, Franke WW (1983) J Cell Biol 96:37-50

127. Venetianer A, Schiller DL, Magin T, Franke WW (1983) Nature 305:730-733

128. Eichner R, Bonitz P, Sun T-T (1984) J Cell Biol 98:1388-1396

RESUME

Les filaments intermédiaires (FI) sont différents dans différents types cellu-laires. Il en existe au moins 5 classes qui diffèrent par la nature de la protéine monomère : c'est ainsi que dans les cellules épithéliales, les FI sont formés de cytokératines; dans les cellules provenant du mésenchyme, de vimentine; dans les cellules du muscle, de desmine; dans les astrocytes et les gliomes, de "glia-filament protein" et dans les neurones ainsi que dans certains types de neuroblastomes, de "neurofilament protein". Nous avons montré que certains types cellulaires peuvent exprimer les protéines constitutrices de 2 classes de FI, ces protéines pouvant former des FI séparés ou encore des hétéropolymères conte-nant les 2 types de protéines. Des anticorps dirigés contre ces différentes pro-téines sont utilisés dans le diagnostic des tumeurs. C'est la famille des cyto-kératines qui est la plus complexe. On connaît 20 cytokératines différentes qui sont exprimées dans des combinaisons spécifiques de 2 à 10 cytokératines selon le type cellulaire. Nous avons établi les profils spécifiques d'expression des cytokératines dans différents tissus épithéliaux et dans des tumeurs chez l'animal et chez l'homme. L'utilisation d'anticorps monoclonaux spécifiques des cytokératines nous permet de distinguer, du point de vue immunologique, différents types de cellules épithéliales ainsi que des sous-classes de carcinomes. Dans la plupart des cellules épithéliales, les FI constitués de cytokératines sont attachés à des domaines de la membrane plasmique appelés plaques de desmosomes et d'hémidesmosomes. Des anticorps contre les protéines qui consti-tuent ces plaques sont des marqueurs de différenciation également utiles dans le diagnostic des tumeurs.

Hormones and Cell Regulation, Volume 9
INSERM European Symposium
J.E. Dumont, B. Hamprecht and J. Nunez editors
© 1985 Elsevier Science Publishers B.V. (Biomedical Division)

DYNAMICS OF MICROFILAMENT ORGANIZATION AND MEMBRANE ANCHORAGE

BRIGITTE M. JOCKUSCH, ANNETTE FÜCHTBAUER, CHRISTIANE WIEGAND and BERND HÖNER
Developmental Biology Unit, University of Bielefeld, 48 Bielefeld, FRG

INTRODUCTION

The multiple functions of microfilaments in animal cells are generally ascribed to a variety of complex organization patterns, which comprise actin filaments interacting with "actin-binding" proteins (1,2). In contrast to the muscle system, microfilaments in nonmuscle cells have been found to be dynamic structures which change "according to need": locomoting, spreading or dividing cells express various microfilament patterns different from well-spread, stationary cells adhered to a substratum (cf. 3,4,5). Little is known on the control mechanism used by the cell to accomplish the dynamics of microfilament organization: there are, however, data suggesting that Ca^{2+} (6,7,8), posttranslational modification of acting-binding proteins (7,8,9) and growth factors binding to the cell surface and exerting a hormone-like effect (10,11) may be involved.

Our own work described in this article concentrates on the most complex microfilament pattern in nonmuscle cells, the stress fiber system. Stress fibers are parallel microfilament bundles expressed in monolayers of fibroblastic, endothelial or epithelial cells in tissue culture (12) as well as in vivo (13,14). Their construction is reminiscent of myofibrils: they consist of filamentous actin, tropomyosin, myosin and α-actinin arranged in repetitive units (15). However, in contrast to the latter, their function is probably not so much correlated with cellular movement but with adhesion and contact formation between cells and between cells and their matrix. Thus, while myofibrils in general undergo isotonic contractions, stress fibers probably exert isometric contractions: they generate tension (5). To do so, they form reversibly point-like connections with the inside of the plasma membrane. On the underside of the cell, the same regions have been identified as the areas where the cell actually meets with or approaches the substratum (focal contacts, 16). While very little is known about the organization of the cellular surface within the focal contact, several components have been identified at the inner face of the plasma membrane: convergent actin filaments, and the structural proteins vinculin (17), talin (18) and α-actinin (19).

To gain some insight in the correlation between the expression of stress fibers, focal contacts, cellular morphology and adherence, we have microinjected cells with substances specifically interfering with structural components of stress fibers. Our results indicate that injection of purified actin-binding proteins or specific antibodies against them is a valuable method to learn more about microfilament organization and function. To complement the microinjection studies, we have tested the hypothesis that tyrosine-specific phosphorylation of vinculin is incompatible with stress fiber expression and cellular adherence (20), by biochemical and structural analysis of cells protected by interferon against the attack of tumorvirus associated tyrosine specific kinase.

MATERIAL AND METHODS

Cells. For microinjection, mouse fibroblasts as well as epithelial cells from mouse and from rat kangaroo (PtK$_2$) were used. For biochemical and morphological analysis of transformed cells, Rous Sarcoma Virus (RSV) infected chicken embryo fibroblasts, with or without previous treatment with homologous interferon, were used.

Structural proteins, antibodies. The following substances were introduced into cells by microinjection: (1) Actin capping proteins: Brain Capping Protein (BCP) (21,22,23), CAP-42 (7,8). (2) Actin severing proteins: Actin Depolymerizing Factor (ADF, 24), gelsolin (25). (3) Affinity purified (specific) polyclonal antibodies against smooth muscle actin, vinculin, myosin and pig brain tubulin.

Microinjection. Microinjection was performed with glass capillaries (26). Injection of the buffer actin-binding proteins and antibodies were kept in had no effect on stress fibers or focal contacts.

Biochemical analysis. RSV-associated tyrosine phosphorylation as a function of interferon pretreatment was determined after ^{32}P-labeling of cells. Total cellular proteins and vinculin immunoprecipitated with specific antibody were subjected to acid hydrolysis and subsequent two dimensional chromatography of amino acids.

Microscopy. Stress fiber expression and integrity was monitored in indirect immunofluorescence. For double labeling fluorochrome-coupled phallacidin (specific for filamentous actin) was combined with antibodies. Reflection contrast microscopy and electron microscopy of cellular sections nearest to the substratum were employed to analyze focal contact regions outside and inside of the plasma membrane.

RESULTS AND DISCUSSION

Actin capping proteins are defined as proteins which bind in vitro to the ends of actin filaments. Since the actin filament is polar, with one end being preferred in energy-dependent addition of subunits, capping proteins with high affinity for this ("plus") end have a profound effect on filament growth and length (see A. Weeds et al., Actin Capping Proteins, this volume). Proteins of this class are Brain Capping Protein(BCP) isolated from mammalian brain (21,22) and CAP 42, isolated from the cellular slime mold Physarum polycephalum (7,8). Injection of these proteins into living cells has dramatic consequences for focal contacts and stress fibers: focal contacts disappear within minutes, followed by disintergration of stress fibers (27,28). In fibroblasts, this is paralleled by a change in cellular morphology: cells, having apparently lost firm adhesion to the substratum, contract and assume an overall shape reminiscent of locomoting cells like granulocytes or tumor cells (27). In epithelial cells, the intercellular junctions as well as the circumferential microfilament bundles are not affected by the capping proteins, thus, cellular morphology remains unaltered. However, the radial stress fibers and focal contacts disintegrate exactly like in fibroblasts (Fig. 1). These effects are fully reversible: the injected cells completely recover within a few hours (27,28,Fig.1). The time point of maximal effect and the time period needed for recovery depend on the number of capping protein molecules introduced into the cell. For example, with an intracellular concentration of BCP after microinjection, corresponding roughly to a molar ratio of BCP: actin monomers $1:10^4$, virtually all cells lost their stress fibers within 30 minutes, and 50% of cells recovered within 2 hours. The effect of capping proteins in such low concentrations as related to actin is probably due to two reasons: they need to act only on the "plus end" of actin filaments and, since they probably also bind to "F-actin monomers", they interfere with the cellular equilibrium between actin filaments and their subunits. This view is supported by the finding that capping proteins disrupt only the microfilament system of living cells: in detergent-extracted cell models, stress fibers proved resistent to even high concentrations of BCP.

Another class of actin-binding proteins has been characterized by their ability not only to "cap" actin filaments at their ends but also to cut (sever) existing filaments intermittently. They are therefore called actin severing proteins (cf. 2,3, Weeds et al., this volume). We injected two of such proteins, Actin Depolymerizing Factor (ADF) from human and gelsolin, a closely related protein, from pig plasma (24,25). Again, stress fibers and focal contacts were rapidly destroyed, the effect being concentration dependent and reversible. To obtain similar time courses, the intracellular concentrations of the severing proteins had to be similar to the data obtained for the actin capping proteins. This seemed initially surprising, since one might assume such cutting factors to act in much lower concentrations. However, since little is known about the affinities of severing proteins for actin subunits within the filament as compared to the filament ends (the latter being amplified in number by the severing activity), the interpretation of our microinjection data in a quantitative way is very difficult. There are, however, two distinct differences in the effect of severing proteins as compared to capping proteins in living cells: (1) Stress fibers are disrupted starting at the site of injection in the case of severing proteins, whereas with capping proteins, they seem to disintegrate from their distal ends. (2) Actin severing proteins act on detergent-extracted cells as well as on living cells. The actin filament "backbones" on stress fibers in living cells as well as in cell models are attacked and cut down by severing proteins. However, the fate of other structural stress fiber components differs in the two systems: while for example myosin remains within detergent-treated stress fibers when actin filaments have been destroyed by severing proteins, the disintegration process of stress fibers in the living cell is not confined solely to actin: myosin leaves the stress fibers after severing protein-induced actin filament breakdown.

When we analyzed the effect of microinjected specific antibodies against cytoskeletal proteins on stress fibers and cell-substratum adherence, we found that anti-actin, even in high concentrations, did not disturb the cell. This is probably explained by the high concentration of actin outside the stress fiber system. An antibody against tubulin, which very effectively disrupts cytoplasmic microtubules after microinjection, had also no effect on either stress fibers or focal contacts. More recently, we have obtained an antibody against smooth muscle myosin which crossreacts with nonmuscle myosin, selectively removing myosin from stress fibers. Interestingly, in this case the actin filament "backbones" stay intact, as seen in double staining experiments. The action of this antibody can be overcome by the cell: after several hours, myosin is restored in the proper positions within the stress fibers. This antibody may become a valuable tool to answer questions about the role of myosin in the motility of ruffling membranes, microspikes etc. (29).

Vinculin, one of the structural components of microfilament termini within the area of focal contacts, has been found to bundle actin filaments in vitro (30,31). When we injected specific antibodies against vinculin into fibroblasts or epithelial cells, they interfered inside the cell with the integrity of focal contacts (28,32). Although this is superficially reminiscent of the results described above for actin capping and severing proteins, there are important differences:

Fig. 1. The effect of microinjected brain capping protein (BCP, 23) on the microfilament system of PtK$_2$ cells. (a) Control, injected with imidazole buffer, (b-h): cells injected with BCP at 0.1 mg/ml in imidazole buffer, approximately 10^{-7}ul per cell. Injected cells (arrows) were fixed and stained for actin in indirect immunofluorescence, at 10 min (a,b), 1 h (c), 1.5 h (d), 2.5 h (e), 5 h (f) and 7 h (g,h) after injection. Already 10 min after injection, radial stress fibers in the injected cells start to disintegrate. At 2.5 h, they have competely disappeared, while the circumferential microfilament bundles seem unaffected. After 5 h, restauration of stress fibers begins and is completed at 7 h. Bar: 50 um.

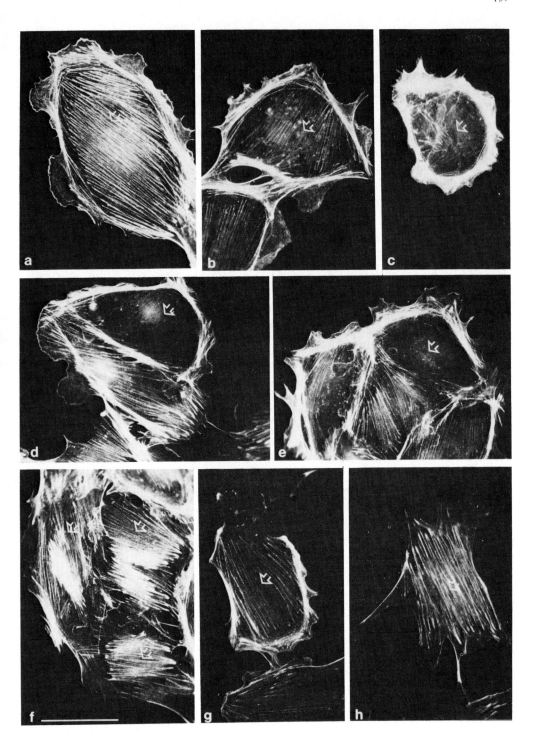

(1) The cells loose their adhesion to the substratum, but fibroblasts do not contract, they keep their flat shape, (2) the effect is irreversible, cells do not recover, (3) within the former area of focal contacts, the termini of microfilaments change from a point-like to a "flattened" configuration, with vinculin and actin remaining in that area. Stress fibers may disintegrate slowly, but not completely. These findings suggest that this antibody interferes with vinculin: actin binding and that this interaction is necessary for maintenance and function of focal contacts.

From studies with RSV-transformed chicken embryo fibroblasts, it was speculated that vinculin looses its function in organizing the distal ends of microfilaments into the configuration required for interaction with the plasma membrane, upon tyrosine phosphorylation by an RSV-specific kinase (20). This hypothesis seems, for technical reasons, difficult to prove directly. However, we could at least establish that indeed there is an inverse correlation between the phosphorylation of tyrosine residues in vinculin and the expression of stress fibers and focal contacts: when cells are treated with homologous interferon prior to infection with RSV, tyrosine phosphorylation of vinculin (as analyzed in immuno precipitates) is suppressed, and cells display a "normal", flattened morphology with well developed stress fibers and focal contacts (33,34). The studies described here suggest that there is a tight correlation between cellular morphology, cellular adhesion and microfilament organization. A complete unravelling of the fine structure and regulation mechanisms of the focal contact area seems necessary to understand this correlation, and to understand the changes observed during the cell cycle or during transformation towards malignancy. We hope that the combination of in vitro studies, biochemical analysis and structural studies on single cells after microinjection which we pursue, will faciliate progress in this field.

ACKNOWLEDGEMENTS

Supported by the Deutsche Forschungsgemeinschaft and the Stiftung Volkswagenwerk. We are indebted to Drs. G. Gabbiani, A. Weeds and A. Wegner for generous supplies of ADF, gelsolin and BCP, respectively.

REFERENCES

1. Weeds, A (1982) Nature 296: 811-816
2. Jockusch, B M (1983) Mol Cell Endocrinol 29: 1-19
3. Sanger, J W (1975) Proc Natl Acad Sci USA 72: 1913-1916
4. Lazanides, E (1976) J Supramol Struct 5: 531-563
5. Burridge, K (1981) Nature 294:691-692
6. Glenney, JR, Bretscher, A, Weber, K (1980) Proc Natl Acad Sci USA 77:6458-6462
7. Maruta, H, Isenberg, G, Schreckenbach T, Hallmann, R, Risse, G, Shibayama, T, Hesse, J (1983) J Biol Chem 258:10144-10150

8. Maruta, H, Isenberg, G (1983) J Biol Chem 258:10151-10158

9. Sefton, BM, Hunter, T, Ball, EH, Singer, SJ (1981) Cell 24: 165-174

10. Schlessinger, J, Geiger, B (1981) Expl Cell Res 134:273-279

11. Westermark, B, Lindberg, U (1983) J Muscle Res Cell Motil 4:589-609

12. Goldman, RD (1971) J Cell Biol 51:752-762

13. Byers, HR, Fujiwara, K (1982) J Cell Biol 93:804-811

14. Gabbiani, G, Gabbiani, F, Lombardi, D, Schwartz, SM (1983) Proc Natl Acad Sci USA 80:2361-2364

15. Sanger, JW, Sanger, JM, Jockusch, BM (1983) J Cell Biol 96:961-969

16. Abercrombie, M, Dunn, GA (1975) Expl Cell Res 92:57-62

17. Geiger, B (1979) Cell 18: 193-205

18. Burridge, K, Connell, L (1983) J Cell Biol 97:359-367

19. Lazarides, E, Burridge K (1975) Cell 6:289-298

20. Hunter, T (1980) Cell 22:647-648

21. Kilimann, MW, Isenberg, G (1982) EMBO J 1: 889-894

22. Isenberg, G, Jockusch, BM (1982) In: Sakai, H, Mori, H, Borisy, GG (eds) Biological functions of microtubules and related structures. Academic Press, Japan, pp 275-283

23. Wanger, M, Wegner, A (1984) submitted

24. Chaponnier, C, Borgia, R, Rungger-Brändle, E, Weil, R, Gabbiani, G (1979) Experientia 35: 1039-1041

25. Harris, HE, Bamburg, JR, Bernstein, BW, Weeds, AG (1982) Anal Biochem 119: 102-114

26. Graessmann, M, Graessmann, A (1976) Proc Natl Acad Sci USA 73:366-370

27. Füchtbauer, A, Jockusch, BM, Maruta, M, Kilimann, MW, Isenberg, G (1983) Nature 304: 361-362

28. Jockusch, BM, Füchtbauer, A (1983) Cell Motility 3:391-397

29. Höner, B, Jockusch BM, Eur. Muscle Club 1984, (abstr)

30. Jockusch, BM, Isenberg, G (1981) Proc Natl Acad Sci USA 78:3005-3009

31. Isenberg, G, Leonard, K, Jockusch, BM (1982) J Mol Biol 158:231-249

32. Füchtbauer, A, Jockusch, BM (1984) J submicrosc Cytol 16:109-110

33. Strube, W, Jungwirth, Ch, Ziemicki, A, Jockusch, BM (1984) Eur J Cell Biol 33:36 (abstr)

34. Strube, W, Jungwirth, Ch, Jockusch, BM (1984) submitted

142

RESUME

Le travail décrit dans cet article se rapporte à la corrélation entre l'expression de faisceaux ordonnés de microfilaments ("stress fibers"), les contacts focaux, la morphologie cellulaire et l'adhésion des cellules en culture. L'approche que nous avons utilisée pour étudier ces phénomènes est la microinjection dans des cellules vivantes de différentes protéines qui se lient à l'actine filamenteuse ou à des protéines associées à l'actine. Nous avons trouvé que l'injection de protéines qui interfèrent avec l'élongation du filament d'actine ou avec son intégrité conduit à la désintégration des faisceaux de microfilaments avec désorganisation des contacts focaux. Ces effets sont réversibles et donc compatibles avec des fonctions cellulaires normales. Par contre, une interférence avec la vinculine (protéine qui stabilise les faisceaux d'actine filamenteuse) par microinjection d'anti-vinculine est caractérisée par une disparition irréversible des contacts focaux, ceci étant accompagné d'une perte rapide de l'adhésion cellulaire. Il semble donc que cet anticorps perturbe les interactions vinculine:actine qui sont nécessaires pour le maintient et la fonction des contacts focaux. Cette conclusion est corroborée par des études sur la phosphorylation de la vinculine induite par des virus dans des fibroblastes transformés par ces virus.

Hormones and Cell Regulation, Volume 9
INSERM European Symposium
J.E. Dumont, B. Hamprecht and J. Nunez editors
© 1985 Elsevier Science Publishers B.V. (Biomedical Division)

ACTIN CAPPING PROTEINS

A.G. WEEDS, H.E. HARRIS, J. GOOCH and B. POPE.
MRC Laboratory of Molecular Biology, Hills Road, Cambridge CB2 2QH,
(England)

INTRODUCTION

Actin is a major component of eukaryotic cells, where it is
involved in many types of motile activity including cytoplasmic
streaming, phagocytosis, pseudopod formation and cytokinesis. It
exists in a variety of supramolecular structures in addition to
free monomeric and filamentous forms. Some of these are highly
ordered as for example in the acrosomal process of Limulus sperm
(1) or the microvilli of intestinal epithelium (2). Elsewhere the
organisation is less evident, for example in the cortical cytoplasm
of cultured cells where dense actin meshworks are cross-linked into
an apparently amorphous gel(3). Motile events within cells such
as those described above require both temporal and spatial
re-arrangement of actin-containing structures. There must be
mechanisms for the rapid disassembly of actin bundles or
cross-linked gels and for fragmentation of filaments into more
readily diffusible forms. In addition, the rapid reassembly of
filaments at specific locations and with specific orientation in
the cell may require nucleating sites often associated with
membranes or organelles within the cell. Capping proteins,
(proteins that bind selectively to one or other end of an actin
filament), may play a part in both disassembly and reassembly
processes. However, their precise functions remain unclear.

Actin assembly and polarity

Before describing capping proteins and their properties, it is
important to understand some of the properties of actin itself.
These have been discussed in detail elsewhere (4,5), but the
important features relevant to this discussion are summarised
below:-

F-actin is a polar structure, which is readily shown when
filaments are "decorated" with myosin subfragment-1. Using
fragments decorated in this way as nuclei for polymerization,
monomeric actin adds preferentially to their "barbed" ends (6).

The rate of filament elongation dCp/dt, where Cp is the concentration of subunits in the polymer, may be expressed as follows:-

$$dCp/dt = (k_+^b + k_+^p) [N][A] - (k_-^b + k_-^p)[N] \qquad (1)$$

where [A] is the actin monomer concentration, [N] the number concentration of filaments, and k_+ and k_- the association and dissociation rate constants at the two ends (barbed = b and pointed = p). Rate constants have been measured for assembly at both ends as a function of actin concentration using actin bundles from either intestinal microvilli (7) or acrosomal processes from Limulus sperm (8) under a variety of ionic conditions. Results showed that the association constant at the barbed end (about 10 $\mu M^{-1} s^{-1}$) is about 10 times that at the pointed end (1-2 $\mu M^{-1} s^{-1}$). Extrapolation of these data to zero actin concentration gives the corresponding dissociation rate constants, which are about 1.5-2 s^{-1} at either end.

When polymerization stops (where dCp/dT = 0), Equation 1 gives:-

$$[A_\alpha] = (k_-^b + k_-^p)/(k_+^b + k_+^p)$$

where $[A_\alpha]$ is the steady-state concentration of actin monomer (the critical concentration). The critical concentration of G-actin in 1 mM $MgCl_2$ and 100 mM KCl is about 0.2 μM. Polymerisation of actin is accompanied by the hydrolysis of ATP bound to the subunits in the filament, i.e. it involves an irreversible chemical reaction. For this reason the dissociation constants at the two ends need not be identical. Calculation of the dissociation constants at the two ends from the rate constants gives a value of about 0.1 μM at the barbed end and 0.5 μM at the pointed end (8,9). A consequence of the non-identity of the dissociation constants at the two ends is that monomers may add to the preferred assembly end while at the same time dissociating from non-preferred end (i.e. may be transferred from the pointed to barbed ends of the filaments). This has been termed treadmilling (10).

While it is now relatively simple to measure elongation rates using actin filament fragments or bundles of actin as nuclei, the study of bulk polymerization from G-actin is much more complex. Spontaneous polymerization of G-actin by addition of $MgCl_2$ and KCl shows a sigmoidal time course with an initial lag phase. This lag time reflects the slow formation of a stable nucleus. Kasai et al., (11) suggested that the smallest stable nucleus is a trimer and more recent and detailed experiments have supported this conclusion (12,13). A further factor contributing to the assembly kinetics is fragmentation, which markedly increases the rate of polymerization. This occurs spontaneously but to a much greater extent when there is mechanical agitation.

Predicted properties of capping proteins

Based on this information about actin, certain properties may be predicted for capping proteins:-

1) Blocking an end of an actin filament will inhibit elongation of that filament. The inhibition of elongation rate will be more marked if the barbed end is blocked because this is the preferred assembly end.

2) Blocking the barbed end of a filament will increase the critical concentration of actin monomers (8,9) and this will affect the stability of the capped filaments in a situation where capped and uncapped filaments exist together, at least under conditions where the critical concentrations at the two ends are different. As was first discussed by Kirschner (14), because of the difference in critical concentration at the two ends of a filament, capping the barbed ends of filaments will promote dissociation of monomers from the pointed ends of these filaments. Under conditions where not all filaments are blocked at their barbed ends, the monomers released will bind to the barbed ends of unblocked filaments. Similarly in a situation where a proportion of the filaments are capped at their pointed ends, there will be a transfer of monomers onto these filaments from uncapped ones. Thus filaments capped at their pointed ends will grow at the expense of uncapped filaments or those capped at their barbed ends. This is shown schematically in Fig. 1.

3) If the capping protein binds to two or more actin monomers to form a nucleus, polymerization of G-actin will be accelerated. This will be true even when the capping protein blocks the preferred assembly end because spontaneous nucleation is so very slow.

4) Capping proteins will prevent the re-annealing of filaments.

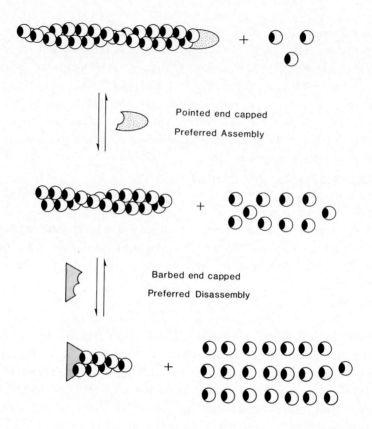

Fig. 1. Schematic representation of the effects of capping the ends of actin filaments on the critical concentration of monomers. The polarity is marked by the dark zones at the barbed ends.

General classification of capping proteins

All of these properties have been reported for a variety of different capping proteins, which appear to fall into two distinct classes:- (i) those that block an end but do not sever; (ii) those that sever filaments and block an end in the process. Among the former class are β-actinin (15), Acanthamoeba capping protein (16), acumentin (17), brain capping protein (18) and two proteins from Physarum (19). The latter class includes macrophage gelsolin (20,21) and related proteins from platelets (22,23) and plasma (24-28), villin from the intestinal microvillus (29,30), fragmin from Physarum (31,32) and severin from Dictyostelium (33). A number of other proteins have also been described but these are less well characterized.

The general classification of actin capping proteins has been reviewed elsewhere (4,34,35). More information is available since these reviews were written, some of which will be described here, but the fundamental objective of understanding their roles in cell function and motility has yet to be achieved. Much effort has been directed towards understanding the properties of capping proteins in vitro, in particular how the interactions with monomeric and filamentous actins are controlled. Immunological studies have been carried out to demonstrate the presence of these proteins in different cell types and to explore their cellular locations. Relationships between the different proteins have also been examined by these techniques. More recently, experiments have been carried out using micro-injection techniques to analyse the effects of some of these proteins on actin-containing structures within cells.

Although the first capping protein to be discovered was β-actinin (15), macrophage gelsolin is generally accepted as the paradigm of this class of proteins. It was first reported in 1979 by Yin and Stossel (20) and has been characterized in a series of subsequent papers (21 and refs therein). It severs actin filaments and inhibits elongation of F-actin at the barbed end. In addition, it accelerates the polymerization of G-actin in the presence of calcium by forming stable complexes which act as nuclei. All its activities are dependent on the presence of micromolar calcium ions concentrations.

Villin is a protein that is closely related to gelsolin in its capping and severing activities, but has the additional property that it bundles actin filaments in the absence of calcium ions. When calcium is added to these bundles, they are disrupted and the actin filaments severed into small pieces. The severing and bundling activities of villin can be separated using limited proteolysis by V8 protease. The core fragment of 90,000 Mr behaves like gelsolin in retaining calcium-dependent severing, capping and nucleating activities (36). This simple "ON-OFF" mechanism in response to calcium ions is not seen however with a number of other capping proteins as will be shown below.

PIG PLASMA GELSOLIN

Discovery and Purification

We first identified plasma gelsolin (formerly called actin

depolymerizing factor) in the course of studies on platelet actin in the presence of plasma (26,27). Purification was achieved by ion exchange chromatography, using both cation and anion exchangers to isolate the protein from the major plasma components, albumin and γ-globulin (37,38). Yields were not good and degradation by plasma proteases a major problem. Unlike villin, the protein cannot be purified on Sepharose-DNase columns to which actin has been non-covalently bound. Although plasma gelsolin binds to these columns in the presence of calcium ions, elution with buffers containing EGTA does not release significant amounts of gelsolin. Addition of denaturing agents like urea or guanidine-HCl removes both gelsolin and actin (38). However, improved purification has been achieved using actin directly coupled to Sepharose. Our current purification scheme uses precipitation from plasma with polyethylene glycol (5-15%), carboxymethyl-cellulose chromatography and affinity chromatography on Sepharose-actin. The gelsolin is eluted with 4M $MgCl_2$ and immediately desalted on G-10 Sephadex.

Methods for studying interactions with actin.

Until recently the main methods used to analyse the effects of capping proteins on filamentous actin were viscometry and flow birefringence. These are limited in their applications since they can be used only at low molar ratios of capping protein to actin subunits and quantitative analysis is difficult. Furthermore, a molecular interpretation of the measurements is not always possible. Sedimentation of F-actin in a Beckman Airfuge is a useful method to measure binding of capping proteins to F-actin and sensitivity can be increased using radiolabelled capping proteins or actin. Electron microscopy provides a means of visualising shortened filaments but is limited in the size of fragment that can be detected. None of these methods can be used at high molar ratios of capping protein:actin subunits (1:1 to 1:10).

DNAase inhibition assay. A more quantitative assay for the interaction of plasma gelsolin and actin was developed based on the fact that G-actin and complexes of actin with gelsolin inhibit pancreatic DNAase I (39). This assay is very useful for analysing the interaction of gelsolin with F-actin at much lower molar ratios (1:1 to 1:20, gelsolin:actin subunits), but is not suitable for much higher ratios for the following reasons:- (i) monomeric actin in the F-actin preparations gives a significant background inhibition of DNAase and inhibition by added gelsolin must be

measured against this background inhibition; (ii) the amounts of
DNAase that can be used are limited because hydrolysis of DNA is
very fast and the Km for substrate high (39); (iii) the
hyperchromicity assay measures a small change in absorbance on top
of a large background level so that substrate concentration itself
is limited by the sensitivity of the spectrophotometer. Although
the assay can be carried out at higher wavelengths where larger
concentrations of DNA can be used, there is often interference at
these wavelengths from protein absorbance. Nevertheless, the
assay was shown to correlate well with other methods including
sedimentation (39) and it also provides a useful means for
measuring the concentration of active gelsolin in both pure and
partially purified samples (37).

Actin overlay method. New methods have been developed in recent
years to study actin binding proteins. These include ^{125}I-actin
overlay of polyacrylamide gels (40). We have used this very
sensitive method under the conditions described by Snabes and
Bryan, but found that it is sometimes misleading because proteins
renatured from sodium dodecyl sulphate can be "sticky" and bind
actin adventitiously. (In the course of our experiments on actin
binding to fragments of gelsolin, a band of about 10,000 Mr was
labelled strongly. This turned out to be one of the polypeptides
chains of chymotrypsin).

^{125}I-gelsolin. In order to measure binding of plasma gelsolin
to F-actin directly, gelsolin has been iodinated with ^{125}-I using
the method of Bolton and Hunter (41). Reaction is carried out at
0.1 to 1 mg/ml in 0.1 M sodium borate pH 8.4 and excess reagent
removed by gel filtration on G-10 Sephadex in 10 mM Tris-HCl pH
8.0, 0.1 M NaCl, 5 mM $MgCl_2$, 3 mM NaN_3 and with gelatin added as a
carrier protein at 1 mg/ml. Gelsolin bound to F-actin can be
measured following centrifugation in a Beckman Airfuge at 10^5 g for
20 min.

Fluorescence Assays. Two fluorescent probes have been used to
monitor actin polymerization and depolymerization and the
interaction of actin with capping proteins:-

N-pyrenyl iodoacetamide labelled actin (PI-actin). N-pyrenyl
iodoacetamide labels Cys 374 and shows a 20-30 fold enhancement of
fluorescence when the actin is polymerized (42). PI-actin has
been used in a number of laboratories to analyse the kinetics of
polymerization (43,44) and we have used rabbit muscle actin

labelled with this reagent to monitor the effects of plasma
gelsolin on actin assembly and disassembly (38). (It is important
to note that chemical modification of actin may alter the
properties of actin (45), so that wherever possible the fluorescent
material should be added in tracer quantities or other experiments
carried out with unlabelled actin.)

Experiments have been carried out at either 10^{-4}M calcium or <
10^{-8} M calcium (addition of 0.25 mM EGTA to 0.1 mM calcium reduces
the free concentration to $<10^{-8}$ M). In this paper these two
conditions are defined as $+Ca^{2+}$ and $-Ca^{2+}$ respectively. When
G-actin is induced to polymerize by added salt, there is an
increase in fluorescence with time (excitation wavelength = 366 nm
emission wavelength = 384 nm). Following an initial lag, this
increase follows an approximately exponential time course. The
rate constant, k_{obs} is the negative slope of plots of log (F_{α} -
F_t), where F_{α} is the fluorescence of polymerized actin at the end
of the reaction and F_t is the fluorescence at time t (38). The
value of k_{obs} is considerably greater when this polymerization is
initiated with 0.1 M KCl or 0.1 M KCl + 1 mM $MgCl_2$ ($-Ca^{2+}$) compared
to the same ionic conditions $+Ca^{2+}$. This is because calcium ions
stabilize the monomeric form of actin and inhibit nucleation
(12,13), but removal of all divalent cations causes denaturation
(4). In order to minimise the differences in assembly rates due to
calcium removal, the $MgCl_2$ was increased to 5 mM. Under these
conditions, the rate of assembly was the same $\pm Ca^{2+}$. These
conditions result in rapid spontaneous nucleation of actin and fast
assembly rates.

Experiments were also carried out using PI actin in an attempt
to measure binding of gelsolin to monomeric actin at equimolar
ratios. Little enhancement of fluorescence (<20%) was detected
with PI-actin but NBD-actin gives a much greater signal in these
experiments (see below).

7-chloro-4-nitrobenzeno-2-oxa-1,3 diazole labelled actin
(NBD-actin). Actin can be labelled with this reagent on Lys 373
after first blocking Cys 374 with N-ethyl maleimide (46). Bryan
and Kurth have shown that there is a 2.5 fold enhancement of
fluorescence of NBD-actin when mixed with platelet gelsolin (47).
We have carried out binding experiments with plasma gelsolin and
observed similar fluorescence enhancement.

Gel filtration. Small complexes of gelsolin and actin have been

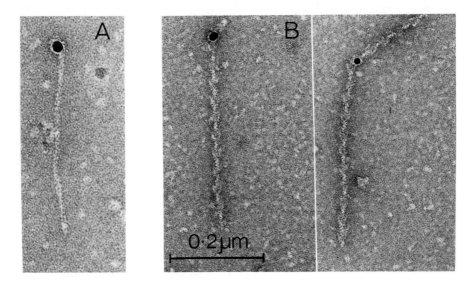

Fig. 2. A) Gelsolin labelled with colloidal gold attached to an actin filament. B) The same, decorated with myosin subfragment-1 to show polarity.

examined by gel filtration on Sephacryl S-200 columns in 20 mM Tris-HCl pH 8.0, 0.1 M KCl, 5 mM $MgCl_2$, 0.2 mM ATP, 0.2 mM dithiothreitol, 1 mM NaN_3, 0.1 mM $CaCl_2$ (± 0.25 mM EGTA). The column was calibrated with suitable proteins to estimate the apparent Mr of the complexes. Polyacrylamide gel electrophoresis followed by densitometry of Coomassie blue stained actin and gelsolin provided a means of determining the molar ratios of the two proteins.

Electron microscopy. Electron microscopy was carried out as described previously (38). Binding of gelsolin to filament ends has been visualised using colloidal gold labelled gelsolin Gelsolin (80 nM) bound to colloidal gold was used to nucleate the polymerization of 2.5 µM actin. Samples were applied to electron microscope grids and washed with 1.7 µM myosin subfragment-1 (48).

Results.

Gelsolin binds to the barbed ends of actin filaments. Direct visualization of gelsolin on the barbed ends of actin filaments is shown in Fig. 2. The dense gold particles are exclusively at the barbed ends of the filaments. Inhibition by gelsolin of growth of actin filaments at their barbed ends was also demonstrated using fragments of Limulus sperm (48). These results confirm that

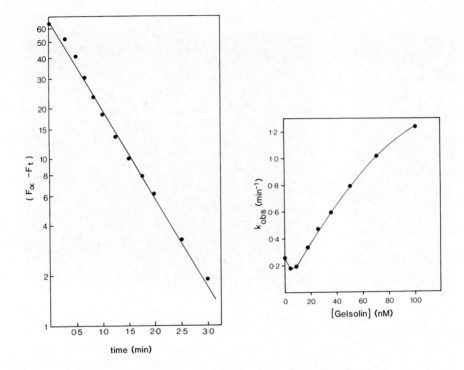

Fig. 3. A) First order plot of the change in fluorescence intensity during approach to equilibrium (F_α - F_t) for 12 µM PI-actin polymerized with 100 nM gelsolin. B) Dependence of polymerization rate, k_{obs}, on gelsolin concentration, using 12 µM PI-actin.

plasma gelsolin, like macrophage and platelet gelsolins, binds to the barbed ends of actin filaments.

Gelsolin promotes actin polymerization. As described earlier (38) gelsolin increases the rate of actin polymerization as measured by enhancement of fluorescence with PI-actin. This effect is much more marked when calcium ions are present. For example, when 5 µM actin was polymerized in the presence of 140 nM gelsolin, k_{obs} was 0.76 min^{-1} in calcium and 0.04 min^{-1} in the absence of calcium, compared to control values without gelsolin of 0.04 min^{-1}. Fig. 3A shows a first order plot for the fluorescence approach to steady-state when 12 µM actin is polymerized in the presence of 100 nM gelsolin +Ca^{2+}. The first order rate constant (k_{obs}) = 1.24 min^{-1}. The dependence of k_{obs} on gelsolin concentration is shown in Fig. 3B. Very low concentrations of gelsolin (4 nM and 9 nM) appear to reduce k_{obs} below the value estimated for control samples, which would be expected if nuclei formed spontaneously were able to be capped at their barbed ends and to elongate only at their non-preferred end. At higher gelsolin concentrations there is an approximately linear increase in k_{obs}, though the values appear to reach a plateau at high gelsolin/actin ratios (38). Even in the absence of calcium, polymerization occurs at an enhanced rate at high gelsolin concentrations presumably because gelsolin:actin complexes form nuclei more readily than monomeric actin (38).

Gelsolin binds to monomeric actin $\pm Ca^{2+}$. There is a marked increase in the fluorescence of NBD-actin when it is mixed with gelsolin (Fig. 4). These experiments were carried out at actin concentrations well below the critical concentration and show that the fluorescence rises almost linearly to a plateau at molar ratios of 1 actin per 1-1.3 gelsolin in the presence of calcium ions . The binding is extremely tight (Kd about 1-5 nM). Similar experiments have been carried out at 20 and 80 nM actin showing the same 1:1 stoichiometry for maximal fluorescence enhancement. In the absence of calcium the extent of the fluorescence change is similar, but the binding much weaker (Fig 4B). The data points approximate a Kd of 50 nM.

Gel filtration of complexes prepared by mixing gelsolin and actin at concentrations in the range 1-4 µM and at 1:1 molar ratios shows that the complex chromatographed in the presence of calcium elutes at a higher apparent Mr than the complex chromatographed in the absence of calcium. The apparent molecular weight values were:-
complex $(+Ca^{2+})$ = 230,000; complex $(-Ca^{2+})$ = 160,000; gelsolin $(\pm Ca^{2+})$ 110,000; actin = 48,000.
These results suggest that the complex in calcium contains approximately two actin monomers while that prepared in the absence of calcium contains only one. These conclusions were supported by densitometry of actin and gelsolin bands on polyacrylamide gels.

When the complex purified in calcium was re-chromatographed on the same column in the absence of calcium, its position changed to that of the $-Ca^{2+}$ complex. However, rechromatography of $-Ca^{2+}$ complex in calcium-containing buffer showed no change in its elution position. These observations indicate that the difference in mobility $\pm Ca^{2+}$ is unlikely to be due to shape changes in the complexes, a conclusion supported by experiments using the

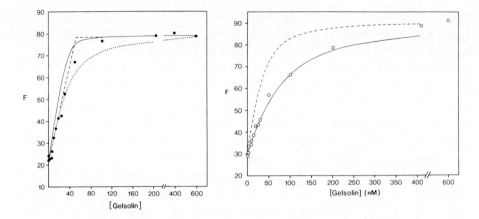

Fig. 4.   A) Fluorescence intensity of 40 nM NBD-actin mixed with gelsolin in 100 μM calcium.   Solid line shows theoretical binding based on Kd = 1 nM, dotted line for Kd = 10 nM.   Dashed line shows the equivalence point used for calculation of stoichiometry (52 nM gelsolin).   B) Similar experiments in < 0.01 μM calcium.   Solid line based on Kd = 50 nM, dotted line Kd = 10 nM.

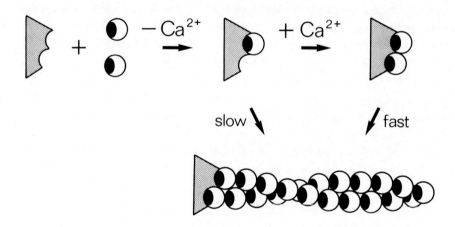

Fig. 5.   Schematic representation of nucleating effects of gelsolin in the presence and absence of calcium, based on complexes formed under these two conditions.   Polarity indicated as in Fig. 1.

analytical ultracentrifuge.

Taken together these results suggest that plasma gelsolin rapidly forms a nucleus for polymerization when calcium is present, but that nucleation is much less efficient in the absence of calcium. A plausable model which explains these results is that complex of 2 actins/gelsolin behaves as a stable nucleus for polymerization but the binary complex requires interaction with at least one more actin monomer before elongation can occur (Fig. 5). Both gel filtration experiments and chemical cross-linking experiments (H.E. Harris, unpublished work) support this conclusion that the complex formed in the presence of calcium contains two actin subunits.

In the light of these observations it is difficult to account for the stoichiometry obtained from fluorescence experiments with NBD-actin +Ca^{2+}. Binding is very tight since the fluorescence measurements appear to fit a curve with Kd = 1-5 nM. It should therefore be possible to obtain an accurate estimate of the stoichiometry and in all experiments the value appears to be 1:1. These apparently contradictory conclusions might be reconciled if the second actin subunit has a much weaker binding constant than the first and does not produce any further fluorescence enhancement when it binds. Bryan and Kurth (45) have also suggested "silent" binding sites in the binding of platelet gelsolin to NBD-actin but in their experiments the binding of the first actin was "silent". Complexes of gelsolin and actin isolated from platelets produced fluorescence enhancement when mixed with NBD actin +Ca^{2+}, but gelsolin purified from these complexes gave no fluorescence increase in the absence of calcium nor was there any evidence for a complex using gel filtration under these conditions. However, complexes formed by reconstitution in calcium could not be dissociated when the calcium was removed (49). Further experiments are in progress to try to resolve these paradoxes, but on the evidence available to date it appears that there are substantial differences in the calcium sensitivity of the interaction of actin with gelsolins from macrophages, platelets and plasma.

Effects of gelsolin on the critical concentration of actin. Fig. 6A shows the effects of gelsolin on the critical concentration of PI-actin. (Because of the marked difference in the quantum yield of G- and F-actins, the critical concentration can readily be obtained by measuring fluorescence at different actin concentrations and extrapolating the regression lines calculated for F-actin to their intersection with that for G-actin). As can be seen from Fig. 6B, there is an increase in critical concentration at the lowest gelsolin concentration used (2 nM). At

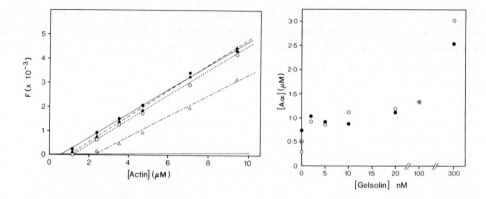

Fig. 6. A) Measurement of critical concentration using PI-action. (●) no gelsolin; (▲) 2 nM gelsolin; (○) 20 nM gelsolin; (△) 300 nM gelsolin. The line just above the abscissa indicates the fluorescence of G-actin. B) Values for the critical concentration obtained from these results in 100 µM calcium (●) and <0.01 µM calcium (○).

higher concentrations of gelsolin the critical concentration appeared to increase further, and this effect was particularly marked at very high concentrations (see 300 nM gelsolin, Fig. 6B). In contrast to similar measurements using villin (30) and Acanthamoeba capping protein (50), where the critical concentration reaches a plateau at <7 nM of the relevant capping protein, the results here and elsewhere (38) consistently show an increase in the critical concentration with increasing gelsolin by this method. However, our observations are consistent with the recent observations using sonicated F-actin that the critical concentration is dependent on the number concentration of actin filaments (51).

Calcium sensitivity of gelsolin binding to F-actin. In our earlier experiments based on fluorescence, viscometry and electron microscopy, we showed that F-actin was fragmented by plasma gelsolin irrespective of the calcium ion concentration. This was in marked contrast to reports for macrophage gelsolin and other capping proteins (21,29,31,33,36). It is now clear that there are differences in the extent of binding and severing when plasma gelsolin is mixed with F-actin ±Ca^{2+}, but these differences are detected only at gelsolin:actin subunit ratios of <1:100, where the fluorescence changes are very small indeed. Viscometry provides a sensitive assay for severing and measurements have been carried out using 12 µM actin and gelsolin:actin subunit ratios between 1:1000 and 1:100. In all cases the viscosity was significantly lower when calcium was present.

Gelsolin binding to F-actin can be measured quantitatively using radiolabelled gelsolin in sedimentation assays. Experiments carried out with 8 µM actin and ^{125}I-gelsolin at concentrations between 1 nM and 2 µM showed that at very low gelsolin concentrations (1-10 nM), virtually all the gelsolin was retained in the pellet when calcium was present but only about 50% when calcium was absent. The proportion of radiolabel in the pellet decreased with increasing gelsolin concentration both $\pm Ca^{2+}$, but the decrease was greater $+Ca^{2+}$. Thus calcium sensitivity was lost at about 80 nM gelsolin (1:100 gelsolin:actin subunits) and under these conditions there is no decrease in the proportion of actin that is pelleted. At higher gelsolin concentrations there is an increasing amount of actin in the supernatant and the proportion of counts bound to the pelleted actin drops to about 20%. Radioautography of samples of supernatant and pelleted material confirm that these differences in radioactivity correspond to the 90,000 Mr gelsolin band.

Experiments to measure gelsolin binding to F-actin at different calcium concentrations showed that the half-maximal change occurs at 0.5 µM free calcium. This is the same calcium concentration range that was previously shown for actin polymerization by gelsolin. Measurements of calcium binding to gelsolin by equilibrium dialysis using ^{45}-calcium also show a dissociation constant of about 0.5 µM.

These experiments show that plasma gelsolin binds to F-actin $\pm Ca^{2+}$. However, there is a difference in the proportion of gelsolin bound $\pm Ca^{2+}$ at low ratios of gelsolin:actin subunits and a difference in the extent of severing as evidenced by viscometric methods. At ratios of gelsolin:actin subunits > 1:100 these differences disappear. Most of the label is found in the supernatant after centrifugation in an Airfuge and this is accompanied by an increase in the proportion of non-sedimentable actin.

If the difference in the binding of gelsolin to F-actin $\pm Ca^{2+}$ were due simply to differences in the binding constants, then at very high actin concentrations these differences should be eliminated. Pelleting experiments have been carried out over a wide range of actin concentrations upto 140 μM using 4.4 nM ^{125}I-gelsolin. The proportion of label bound to the pelleted actin -Ca^{2+} remained 50-60% of the value +Ca^{2+} even at the highest actin concentrations. The results suggest that there may be two forms of gelsolin present within these preparations, one of which only binds actin when calcium is present, while the other binds ±Ca^{2+}.

In an attempt to distinguish two possible forms of plasma gelsolin, purified protein was subjected to 2-dimensional polyacrylamide gel electrophoresis and the gels overlayed with ^{125}I-actin ±Ca^{2+}. There were no differences in either the Coomassie blue stained gels themselves or the radioautographs. Thus if two forms exist, they cannot be separated by isoelectric focussing or SDS gel electrophoresis. It is possible that one form can been generated from the other by alterations in molecular conformation and we have yet to explore the effects of controlled denaturation and renaturation on our preparations.

DISCUSSION
Distribution and homology with other capping proteins.

There is strong immunological cross-reactivity between plasma and intracellular forms of gelsolin. Antisera to rabbit macrophage gelsolin cross-react with a 90,000 Mr protein from a variety of cultured cells and tissue samples, including heart, uterus, bladder, brain and thyroid, and also a protein in human serum (52). Using antisera to pig plasma gelsolin, we have used similar methods to show cross-reactivity with a 90,000 Mr protein from platelets, liver, lung, heart, bladder, thymus and bone marrow. Thus gelsolins appears to have a very wide tissue distribution.

Recently it has been reported that there is extensive amino acid sequence homology between human plasma gelsolin and rabbit macrophage gelsolin at the N-terminus (53). The human plasma protein contains an addition 25 residues at its N-terminus, which accounts for the difference in mobility between these two proteins on SDS gel electrophoresis. Comparative studies on pig plasma gelsolin in our own Laboratory have shown nearly 100% homology with the first 15 residues of the common sequence of macrophage and human plasma gelsolins, but the pig plasma protein contains only 9 additional residues at its N-terminus, compared to 25 for the human protein (unpublished work).

There is no evidence from our own work of immunological cross-reactivity between villin and gelsolin using antibodies to both plasma gelsolin and villin, but recent sequence analysis of fragments of villin have revealed homologies between the two proteins (Matsudaira and Jakes, unpublished work). These comparisons are restricted to the region of published sequence at the N-terminus of gelsolin (52), but they suggest that there exists a family of homologous proteins in a variety of tissues with similar functions.

Effects of gelsolins and other capping proteins within cells.

Microinjection methods have been used to explore the effects of capping proteins in cellular function. Injection of fragmin-actin complexes from Physarum into Amoeba proteus caused changes in shape and behaviour (54). There was suppression of pseudopod formation and constriction of the cell surface, which were interpreted as a relaxation of the advancing cell region due to a reduction in the polymerizable actin pool. Disruption of microfilament organization was observed when brain capping protein was injected into cultured epithelial or fibroblastic cells (55). The observations showed that microfilament bundles were disintegrated progressively from their distal ends but the process was reversible as the microfilament bundles were restored fully within 4-5 hr. Macrophage gelsolin blocks the axonal transport of membraneous organelles as seen in perfusion experiments with isolated axoplasm, but only in the presence of micromolar calcium (56). All sizes of organelles including 50 nm vesicles and mitochondria were affected in both orthograde and retrograde directions.

Recently Drs. Jockusch and Fuchtbauer have explored the effects of microinjection of plasma gelsolin into rat kangaroo ptK$_2$ cells. As shown in Fig. 7, anti-actin staining reveals that radial actin cables are fragmented and disintegrate. As shown previously for brain capping protein, the effect is fully reversible with time. Taken together all these results confirm that capping proteins affect cytoskeletal organisation and function within cells. It is of interest that both severing and non-severing forms of capping proteins have disrupting effects but while the brain capping protein affected marginal lamellae and focal contacts at the periphery of the cell (55), gelsolin showed its effects radially from the site of micro-injection (Fig. 7). These observations are

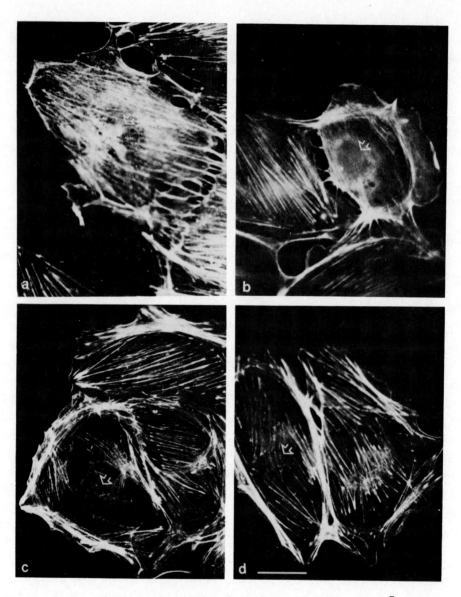

Fig. 7. PtK cells injected with 2.5 µg/ml (approx. 10^{-7} µl/cell) pig plasma gelsolin. The injected cells are marked by arrows. Cells were fixed and stained with anti-actin at 0.5 (a), 1 (b), 2.5 (c) and 3.5 hr (d) after microinjection. Radial cables are fragmented within the first 30 min after injection and disintegrate subsequently. This effect is reversible: after 2.5 hr, the injected cells show partially restored, short cables. At 3.5 hrs, the injected cells show fully expressed radial actin cables indistinguishable from controls. Bar: 25 µm (We thank Drs. A. Fuchtbauer and B.M. Jockusch for providing these results).

consistent with the properties described for these proteins.

Possible mechanisms of action and their amplification.

Gelsolins show two distinct but related activities: (i) they act as nucleators for polymerization of monomeric actin; (ii) they block filament ends and sever F-actin. Both activities will lead to the production of shorter filaments which cannot anneal and to an increase in the critical concentration of monomers.

One possible function for these proteins could be to provide nucleating sites for filament formation within cells. Although gelsolin appears to be localised in punctate regions at the cell surface and in spots in the leading edge of forming pseudopods of transformed rat fibroblasts, as evidenced by antibody staining (57), its role there has yet to be established. The protein is water soluble and therefore unlikely to be associated directly with membranes. Studies on microfilament bundles in microvilli in vitro (58) and Limulus acrosomes in vivo (59) have shown that elongation occurs at their barbed ends. Tilney has boldly claimed that "in no system is there any evidence for end-on attachment of actin filaments to membranes" (59). On the basis of these arguments and those of Kirschner concerning the stability of filaments capped at their barbed ends (14), it seems unlikely that gelsolins play a major part in site specific nucleation. It is more probable that the primary function of gelsolins is to reduce the rigidity of the cytoplasm by shortening and destabilizing actin filaments.

One mechanism for stabilizing monomeric actin within cells is the formation of a complex with profilin (60). Profilins have been found in abundance in a variety of cells, e.g. in platelets there is sufficient profilin to complex over one-third of the actin present (61). The combined effects of profilin and capping proteins will profoundly affect actin polymerization as shown with villin (62). A model which depends on the differences in critical concentration at the two ends of F-actin has been proposed to account for the action of profilin in enhancing the effects of barbed end capping on actin depolymerization (63,64). It provides a potential amplifying mechanism and can be reversed when the capping protein is removed.

Regulatory mechanisms

It is obviously essential that the interaction of capping protein and actin can be regulated in cells. Macrophage gelsolin and villin provide examples of fully reversible calcium dependent interactions. Complete reversibility of binding is not universally observed with other barbed end capping proteins. Dimeric complexes formed between fragmin and actin (32), severin and actin (33), platelet gelsolin and actin (49) are not dissociated when the calcium concentration is reduced below 10^{-8} M and this has also been shown in the work described here for plasma gelsolin. Gel filtration experiments have shown that platelet gelsolin migrates as a 90,000 Mr monomer when isolated from non-stimulated platelets in EGTA, but once the platelets have been stimulated, the gelsolin forms a stable dimeric complex with actin (49). Although platelet activation in haemostasis is a one-way process and might not require reversible regulation, it is possible in vitro to demonstrate reversible activation in response to ADP and other agonists. It will therefore be of interest to find out whether under these conditions complexes have been formed during the activation phase and whether on subsequent reversal these complexes can be dissociated within the cell.

Further insights into regulatory mechanisms have come from recent studies on actin capping proteins of Physarum. In addition to fragmin, these cells contain a calcium-dependent capping complex, which is a dimer of two polypeptides of 42,000 Mr termed Cap 42 (a + b) (19). This protein has no severing activity but induces rapid depolymerization of actin monomers from the pointed ends of filaments. The complex shows calcium sensitivity only when phosphorylated by a specific kinase; calcium sensitivity is lost following dephosphorylation by acid phosphatase. The two components have been separated (65). Cap 42 (a) shows calcium dependent actin binding but is not phosphorylated, while Cap 42 (b) has properties similar to the native complex in showing calcium sensitivity only after phosphorylation. Thus calcium sensitivity can itself be regulated by phosphorylation. However, there is no evidence in the case of gelsolins for regulation of activity by phosphorylation (57).

SUMMARY

There is now considerable evidence for the existence of actin capping proteins in many kinds of cells. Most of those described to date bind to the barbed ends of actin filaments and many have severing activity. Their properties have been analysed in vitro and their localization and distribution studied in cells. Micro-injection provides a means of exploring their effects on cellular structures and morphology. As yet we can only speculate about their function. In the case of plasma gelsolin, which is clearly a secreted variant of the intracellular form (53), present in plasma at relatively high concentrations (27), the function may be to scavenge actin filaments released accidentally into the bloodstream.

REFERENCES

1. DeRosier D, Mandelkow E, Silliman A, Tilney L, Kane R (1977) J Mol Biol 113:679-695

2. Matsudaira P, Mandelkow E, Renner W, Hesterberg LK, Weber K (1983) Nature 301:209-214

3. Hoglund AS, Karlsson R, Arro E, Fredriksson BA, Lindberg U (1980) Journal of Muscle Research and Cell Motility 1:127-146

4. Korn E (1982) Physiol Rev 62:672-737

5. Pollard TD, Craig SW (1982) Trends in Biochemical Sciences 7:55-58

6. Woodrum DT, Rich SA, Pollard TD (1975) J Cell Biol 67:231-237

7. Pollard TD, Mooseker MS (1981) J Cell Biol 88:654-659

8. Bonder EM, Fishkind DJ, Mooseker MS (1983) Cell 34:491-501

9. Wegner A (1982) J Mol Biol 161:607-615

10. Wegner A (1976) J Mol Biol 108:139-150

11. Kasai M, Kawashima H, Oosawa F (1964) J Polymer Sci 44:51-69

12. Tobacman LS, Korn ED (1983) J Biol Chem 258:3207-3214

13. Cooper JA, Buhle EL, Walker SB, Tsong TY, Pollard TD (1983) Biochemistry 22:2193-2202.

14. Kirschner MW (1980) J Cell Biol 86:330-334

15. Maruyama K, Kimura S, Ishii T, Kuroda M, Ohashi K, Maramatsu S (1977) J Biochem (Japan) 81:215-232

16. Isenberg G, Aebi U, Pollard TD (1980) Nature 288:455-459

17. Southwick FS, Hartwig JH (1982) Nature 297:303-307

18. Kilimann MW, Isenberg G (1982) EMBO Journal 1:889-894

19. Maruta H, Isenberg G (1983) J Biol Chem 258:10151-10158

20. Yin HL, Stossel TP (1979) Nature 281:583-586

21. Yin HL, Hartwig JH, Maruyama K, Stossel TP (1981) J Biol Chem 256:9693-9697

22. Wang L-L, Bryan J (1981) Cell 25:637-649

23. Markey F, Persson T, Lindberg U (1982) Biochim Biophys Acta 709:122-133

24. Chaponnier C, Borgia R, Rungger-Brandle E, Weil R, Gabbiani G (1979) Experientia 35:1039-1041

25. Norberg R, Thorstensson R, Utter G & Fagraeus A (1979) Eur J Biochem 100:575-583

26. Harris HE, Bamburg JR, Weeds AG (1979) in The Cytoskeleton: membranes and movement, p. 28, Cold Spring Harbor NY.

27. Harris HE, Bamburg JR, Weeds AG (1980) FEBS Lett 121:175-177.

28. Harris DA, Schwartz JH (1981) Proc Natl Acad Sci 78:6798-6802

29. Bretscher A, Weber K (1980) Cell 20:839-847

30. Walsh TP, Weber A, Higgins J, Bonder EM, Mooseker MS, Biochemistry 23:2613-2621

31. Hasegawa T, Takahashi S, Hayashi H, Hatano S (1980) Biochemistry 19:2677-2683

32. Hinssen H (1981) Eur J Cell Biol 23:225-240

33. Giffard RG, Weeds AG, Spudich JA (1984) J Cell Biol 98:1796-1803

34. Craig SW, Pollard TD (1982) Trends in Biochemical Sciences 7:88-92

35. Weeds AG (1982) Nature 296:811-816

36. Glenney Jr. JR, Kaulfus P, Weber K (1981) Cell 24:471-480

37. Harris HE, Gooch J (1981) FEBS Lett 123:49-53

38. Harris HE, Weeds AG (1983) Biochemistry 22:2728-2741

39. Harris HE, Bamburg JR, Bernstein BW, Weeds AG (1982) Anal Biochem 119:102-114

40. Snabes MC, Boyd AE, Bryan J (1981) J Cell Biol 90:809-812

41. Bolton AE, Hunter WM (1973) Biochem J 133:529-539

42. Kouyama J, Mihashi K (1981) Eur J Biochem 114:33-38

43. Tellam R, Frieden C (1982) Biochemistry 21:3207-3214

44. Pollard TD (1983) Anal Biochem 134:406-412

45. Malm B, Larrson H, Lindberg U (1983) J Muscle Res and Cell Motility 4:569-588

46. Detmers P, Weber A, Elzinga M, Stephens RE (1981) J Biol Chem 256:99-105

47. Bryan J, Kurth MC (1984) J Biol Chem 259:7480-7487

48. Harris HE, Weeds AG FEBS Lett (submitted)

49. Kurth MC, Bryan J (1984) J Biol Chem 259:7473-7479

50. Cooper JA, Blum JD, Pollard TD (1984) J Cell Biol 99:217-225

51. Pantaloni D, Carlier M-F, Coue M, Lal AA, Brenner SL, Korn ED (1984) J Biol Chem 259:6274-6283

52. Yin HL, Albrecht JH, Fattoum A (1981) J Cell Biol 91:901-906

53. Yin HL, Kwiatkowski DJ, Mole JE, Sessions Cole F (1984) J Biol Chem 259:5271-5276

54. Gawlitta W, Hinssen H, Stockem W (1980) Eur J Cell Biol 23:43-52

55. Fuchtbauer A, Jockusch B, Maruta H, Kilimann MW, Isenberg G (1983) Nature 304:361-364.

56.   Brady S, Lasek RJ, Allen RD, Yin HL, Stossel TP (1984) Nature
       310:56-58

57.   Wang E, Yin HL, Krueger JG, Caliguiri LA, Tamm I (1984) J
       Cell Biol 98:761-771

58.   Mooseker MS, Pollard TD, Wharton KA (1982) J Cell Biol
       95:223-233

59.   Tilney LG, Bonder EM, DeRosier DJ (1981) J Cell Biol
       90:485-494

60.   Carlsson L, Mystrom L-E, Sundkvist I, Markey F, Lindberg U
       (1977) J Mol Biol 115:465-483

61.   Harris HE, Weeds AG (1978) FEBS Lett 90:84-88

62.   Markey F, Larsson H, Weber K, Lindberg U (1982) Biochim
       Biophys Acta 704:43-51

63.   Tobacman LS, Korn ED (1982) J Biol Chem 257:4166-4170

64.   Tseng PC-H, Pollard TD (1982) J Cell Biol 94:213-218

65.   Maruta H, Isenberg G (1984) J Biol Chem 259:5208-5213

RESUME

   Il existe actuellement d'innombrables évidences témoignant de l'existence de
protéines induisant le "capping" de l'actine dans de nombreux types cellulaires.
La majorité de celles décrites à ce jour se lient à l'extrémité "barbed" des
filaments d'actine et peuvent avoir une activité séparatrice. Leurs propriétés
ont été analysées in vitro et leur localisation et distribution étudiées dans les
cellules. La technique de microinjection nous fournit un moyen d'explorer leurs
effets sur les structures et la morphologie cellulaires. Cependant, on ne peut
que spéculer au sujet de leur fonction. Dans le cas de la gelsoline plasmatique,
qui est clairement une variante sécrétée, de la forme intracellulaire, et est
présente dans le plasma à concentration relativement élevée, la fonction pourrait
être d'éliminer les filaments d'actine libérés accidentellement dans la circulation sanguine.

SECOND MESSENGER SYSTEMS
SYSTEMES DE SECOND MESSAGERS

Hormones and Cell Regulation, Volume 9
INSERM European Symposium
J.E. Dumont, B. Hamprecht and J. Nunez editors
© 1985 Elsevier Science Publishers B.V. (Biomedical Division)   169

THE CONTROL MECHANISMS OF CYCLIC NUCLEOTIDE PHOSPHODIESTERASE ACTIVITIES :
REGULATION POTENTIAL OF cAMP CATABOLISM.

CHRISTOPHE ERNEUX, FRANCOISE MIOT & JACQUES E. DUMONT
Institute of Interdisciplinary Research (IRIBHN), School of Medicine, Free
University of Brussels, Campus Erasme, Route de Lennik 808, B-1070 Brussels,
(Belgium).

1. INTRODUCTION

The phosphodiesterase reaction was initially described by Sutherland and
Rall in 1958 for cAMP catabolism (1). The crude enzyme, characterized in dog
heart extracts, catalyzed the hydrolysis of cAMP into 5'-AMP, was activated by
magnesium ions and inhibited by the methylxanthines, caffeine or theophylline.
In the seventies, it was shown that cyclic nucleotide phosphodiesterases
consisted in a system of multiple forms of "high" and "low Km" for cAMP and
cGMP and of different molecular weight (2). Distinction was also made between
soluble and particulate activities and an insulin-stimulation of a "low Km" cAMP
particulate phosphodiesterase in adipose tissue was reported by Loten and
Sneyd (3). Recently, three prominent phosphodiesterases have been purified to
apparent homogeneity (4). Those are now recognized as different non-intercon-
vertible molecular entities (5, 6). On the other hand, on purified enzymes,
more and more evidence for specific control mechanisms has been demonstrated
(7, 8).

All tissues hydrolyse cyclic nucleotides by a phosphodiesterase reaction.
The only other known pathway for cyclic nucleotide catabolism is the efflux of
cAMP (or cGMP). This pathway is much less characterized but in the few cases
where the intracellular turnover is known with precision its role is quantitatively small compared to phosphodiesterase activity. For example, it was
estimated that in fibroblasts (WI-38 cells), the escape of cAMP contributes to
no more than 15 % of the total turnover of cAMP (i.e., hydrolysis plus escape,
ref. 9). Therefore in the intact cell, cAMP accumulation is mainly controlled
by the activities of adenylate cyclase and the phosphodiesterases. Phosphodiesterase mutiple forms are rather complex enzymes regulated through the participation
of many hormones as well as target signal molecules as Ca^{2+}, calmodulin or
cGMP (7, 8). The purpose of this article is to review briefly the mechanisms
involved in the control of phosphodiesterase activity. Emphasis will be
placed on the description of multiple forms of cyclic nucleotide phosphodiesterases and current evidence for separated controls on these purified enzymes.

Abbreviations : TSH, thyroid-stimulating hormone; SDS, sodium dodecyl sulfate;
MIX, 1-methyl-3-isobutylxanthine; 7-benzyl MIX, 1-methyl-3-isobutyl-7-benzyl-
xanthine; IIX, 1-isoamyl-3-isobutylxanthine; Ro 20-1724, 4-(3-butoxy-4-methoxy-
benzyl)-2-imidazolidinone.

TABLE 1.

CHARACTERISTICS OF THE THREE PROMINENT PHOSPHODIESTERASE FORMS

Enzyme property	[a] Calmodulin-sensitive phosphodiesterase soluble (11-16)	[b] cGMP-stimulated phosphodiesterase soluble (17-21)	[c] cAMP phosphodiesterase	
			soluble (22)	particulate (23)
Cyclic nucleotide specificity	cAMP and cGMP	cAMP and cGMP	cAMP	cAMP
Preferential substrate	cGMP	cGMP	cAMP	cAMP
Activators	calmodulin fatty acids, phospholipids proteases	cGMP methylxanthines	proteases	insulin
Inhibitors	K+ polyamines calmodulin-binding proteins	fatty acids	—	—
Subunit composition	dimer	dimer	monomer	monomer
Subunit M_r (SDS gel electrophoresis)	59,000 (bovine brain) 61,000 (bovine heart)	105,000	60,000	52,000
Substrate-velocity relationship	Michaelis-Menten (in the presence of calmodulin)	positive cooperativity (in the absence of 3 µM cGMP)	Michaelis-Menten	Negative cooperativity

a. Bovine brain calmodulin-sensitive phosphodiesterase has been shown to exist in isozymes (24).

b. Stimulation of activity by cGMP (and positive cooperativity) is detected with micromolar cAMP as substrate (20-21).

c. High affinity cAMP phosphodiesterases (22, 23).

Evidence suggesting that individual controls are operating in the intact cell
will also be presented.

2. THE CATABOLISM OF CYCLIC NUCLEOTIDES INVOLVES A RATHER COMPLEX ENZYMATIC SYSTEM OF MULTIPLE ENZYMES.

2.1. Three major phosphodiesterase activities.

Cyclic nucleotide phosphodiesterases comprise distinct enzymes and isozymes
that can be distinguished from each other on the basis of substrate specificity,
physical properties, subcellular localization or control mechanisms (2, 4, 8).
Three prominent soluble activities have been characterized in a large number
of tissues and cell types. For example, in dog thyroid crude soluble fraction, the three enzymes can be separated by DEAE-cellulose chromatography at
pH 7.4 with an exponential gradient of $(NH_4)_2SO_4$ from 0 to 0.5 M. In the
typical profile shown in Fig. 1, the first peak of activity is stimulated by
excess calmodulin in the presence of 2 mM Ca^{2+} (Ca^{2+}/calmodulin), it is
therefore related to as the calmodulin-sensitive phosphodiesterase. cAMP
hydrolysis of the second peak is stimulated by micromolar cGMP (3 μM) and
related to as the cGMP-stimulated phosphodiesterase. These two forms also
degrade cGMP as well as cAMP but always with a lower Km for cGMP. The third
peak of activity is neither stimulated by cGMP nor by calmodulin but specifically hydrolyses cAMP (5, 10). The three phosphodiesterases found in thyroid
tissue have been purified to homogeneity from other sources and some of their
properties outlined in Table 1.

2.2. Evidence showing that the three major phosphodiesterase activities are different proteins.

Three pieces of evidence definitely indicate that the three major phosphodiesterase activities are different proteins : i) the inhibition sequences of
cyclic nucleotide analogs competitive at the catalytic site for the calmodulin-
sensitive form and the cGMP-stimulated form are rather distinct. This suggests
that their catalytic sites must be different (25); ii) monoclonal antibodies
to individual phosphodiesterases have been produced and used for specific
identification: none of the antibodies to either the calmodulin-sensitive form
or the cGMP-stimulated form appear to recognize antigenic determinants on one
of the other enzyme (26); this defines antigenic sites on two isolated phosphodiesterases; iii) finally, the peptide maps of purified phosphodiesterases
(the calmodulin-sensitive form and the insulin-sensitive peripheral plasma
membrane form) are not comparable which further establishes that the enzymes
are distinct molecular entities (27).

2.3. <u>The distribution of major phosphodiesterases is variable from one cell</u>
<u>type to another</u>.

The profile of cyclic nucleotide phosphodiesterase activities always
differs depending on the nature of the tissue and species. In contrast to the
dog thyroid profile, bovine adrenal phosphodiesterase could be resolved by the
same type of DEAE-cellulose chromatography as in Fig. 1, in a single major
peak of activity identified to as the cGMP-stimulated phosphodiesterase
(Fig. 2). Other tissues e.g., bovine testis or cerebral cortex contain only
little amounts of cGMP-stimulated phosphodiesterase as determined by immunoti-
tration analysis (26).

2.4. <u>Criteria for identification of phosphodiesterases</u>.

Because monoclonal antibodies usually recognize single antigenic determinants, they are very powerful probes to identify a given phosphodiesterase in
crude preparations (26). Another powerful criterium for such an identification
is to take advantage of the specificity of phosphodiesterase(s) inhibition
pattern(s) by pharmacological compounds or substrate analogs (25,28). For the
sake of simplicity, this is not the way generally used to characterize phosphodiesterase multiple forms; they are often classified according to particular
biochemical properties, kinetic behavior or responsiveness to endogenous
effectors (e.g. calmodulin or cGMP). Individual criteria are often inadequate
to identify a given phosphodiesterase. For example, enzymes with both "high"
and "low Km" for cAMP as substrate are stimulated by calmodulin showing that,
at least, two types of phosphodiesterase are calmodulin-sensitive (4).   In
another report, Weber and Appleman characterized an insulin-stimulated cAMP
phosphodiesterase in rat adipose tissue; one of their criteria was potent
inhibition of cGMP upon cAMP hydrolysis (29). However, other phosphodiesterases,
not being insulin-sensitive also present this sensitivity to cGMP (e.g. in S49
mouse lymphoma cells illustrated in ref. 30). Thus multiple criteria are
often required in order to compare phosphodiesterases and definitely conclude
to the presence of identical proteins in different cell types.

2.5. <u>Tissue specific-cyclic nucleotide phosphodiesterases : the example of a</u>
<u>genetically determined alteration of phosphodiesterase activity in S49</u>
<u>mouse lymphoma cells</u>.

Multiple other phosphodiesterases e.g. in rod outer segments (31) or in
platelets (32) appear to be different tissue-selective enzymes compared to the
prominent phosphodiesterase forms (Table 1). They differ by multiple criteria : substrate specificity, sensitivity to Ca^{2+}/calmodulin or to cGMP,

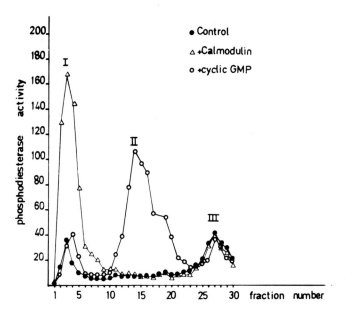

Fig. 1. Separation of dog thyroid phosphodiesterase activities by DEAE-cellulose chromatography at pH 7.4, following the procedure detailed in ref. 5. Hydrolysis of 3 μM cAMP was measured in the absence (control, ●) and in the presence of excess Ca^{2+}/calmodulin (△) or 3 μM cGMP (O). Activity is expressed as pmol cAMP hydrolyzed. 20 $\mu \ell^{-1}.30$ min^{-1}.

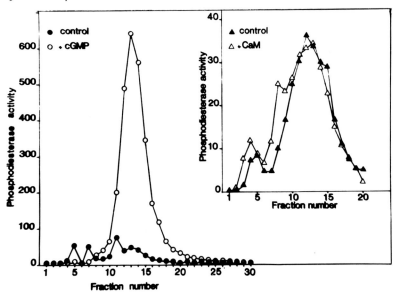

Fig. 2. Separation of bovine adrenal phosphodiesterase activities following the same procedure and assay conditions as in Fig. 1. Activity is expressed as pmol cAMP hydrolyzed. 20 $\mu M^{-1}.10$ min^{-1}.

kinetic behavior or physical properties. For example, a mutant of S49 mouse lymphoma cells, termed K30a, is characterized by an increased capacity to degrade cAMP and cGMP as compared to wild type (33). K30a phosphodiesterase (predominant form of activity in K30a extracts) is distinct from the calmodulin-sensitive enzyme or the cGMP-stimulated form and is also not cAMP specific. On the contrary, it hydrolyses both cAMP and cGMP but preferentially cGMP (Km 0.04 μM), is not sensitive to the synthetic imidazolidinone Ro 20-1724 but inhibited by cGMP (with cAMP 0.2 μM as substrate). It is not an important form of activity in wild type extracts where an enzyme cAMP specific and highly sensitive to inhibition by Ro 20-1724 is predominant. Among multiple interpretations, the K30a phenotype could have resulted from an increased transcription of a gene (or genes) coding for a distinctive K30a phosphodiesterase (33). The authors did not rule out the possibility that such a gene might be expressed at a low level in wild type S49 cells or several other normal tissues.

2.6. cGMP-binding cyclic nucleotide phosphodiesterases.

cGMP-binding cyclic nucleotide phosphodiesterases in rat platelet or lung, rod outer segment and the bovine adrenal cGMP-stimulated phosphodiesterase are characterized by a cGMP-binding activity that is copurified with a phosphodiesterase activity (17,34-37). None of them are cAMP-specific phosphodiesterases or sensitive to Ca^{2+}/calmodulin. Until now, the common characteristic between those enzymes is to share the existence of noncatalytic cGMP-binding sites unrelated to protein kinase activity. This conclusion is drawn from the observation that competitive inhibitors of the phosphodiesterase reaction (e.g. MIX) do stimulate cGMP-binding activity associated to all those enzymes (ref. 32 and Miot, unpublished). Only for the cGMP-stimulated enzyme, cGMP-binding sites are indentified to as the "allosteric" binding sites* controlling the hydrolysis of cAMP or cGMP (Erneux, unpublished). As a matter of fact, enzymes in platelet or lung and in the rod have different molecular weight or kinetic behavior compared to the cGMP-stimulated form. Enzymes in platelet or lung have not been purified to homogeneity and it is still possible that their phosphodiesterase activity is specifically controlled (e.g. by phosphorylation) by a mechanism independent of cGMP-binding. Further studies will be required to determine eventual structural or functional relationships between those enzymes. For example, the question of how similar the specificities of their noncatalytic cGMP-binding sites are has not been adequately addressed.

*Based on substrate-velocity relationships, we have an allosteric cGMP-stimulated phosphodiesterase and a positive allosteric effector, cGMP (21).

3. THE CONTROL OF PHOSPHODIESTERASE ACTIVITY INVOLVES AT LEAST THREE FUNDAMEN-
   TALLY DISTINCT MECHANISMS DEMONSTRATED ON PURIFIED ENZYME (TABLE 2).

TABLE 2.
THE CONTROL OF PHOSPHODIESTERASE ACTIVITY : MULTIPLE MECHANISMS.[a]

---

1. Activation by Ca^{2+}/calmodulin (PDE_I)

   Ca^{2+}-calmodulin (0.1 µM) + PDE_I $\longrightarrow$ (Ca^{2+}-calmodulin-PDE_I) [b]

                   basal   fully activated

                   activity

2. Allosteric activation by cGMP (PDE_{II})

           cGMP (1µM) + PDE_{II} $\longrightarrow$ (cGMP . PDE_{II})[c]

                   basal   fully activated

                   activity

3. Activation through covalent modification : phosphorylation and proteolysis

   The insulin-sensitive rat liver peripheral plasma membrane "low Km" cAMP
   PDE stimulated by phosphorylation (38).

(a) *References in the text.*

(b) *PDE_I, PDE_{II} and PDE_{III} are distinct phosphodiesterase enzymes.*

(c) *cGMP is bound to allosteric binding sites.*

3.1. Stimulation by Ca^{2+}/calmodulin.

   The widely distributed calmodulin-sensitive cyclic nucleotide phosphodiesterase, has now been purified to homogeneity from bovine heart and brain tissues
(11-16, 24). In its native state, it exists as a dimeric globular protein of
Mr $\sim$ 120,000. SDS/polyacrylamide gel electrophoresis shows a single protein-
staining band with an apparent molecular weight of 58,000 to 63,000 depending
on the tissue and species sources (4). The stoichiometry of the interaction
of calmodulin with phosphodiesterase is two mol. of calmodulin per mol. of
enzyme i.e., one for each of the two subunits of that phosphodiesterase (12).
In the absence of calmodulin, the enzyme has a "high Km" for cAMP ($\sim$200 µM)
and a "low Km" for cGMP ($\sim$10 µM). In the absence of Ca^{2+}, calmodulin does
not bind to the enzyme nor does it activate it; stimulation in the presence of
Ca^{2+} and calmodulin appears to be mainly the result of an increase in maximal
velocity (11). In dog thyroid, when intracellular Ca^{2+} is elevated by carbachol, the calmodulin-sensitive enzyme becomes fully active and thereby cAMP
accumulation stimulated by TSH is depressed (see below).

3.2. Stimulation by cGMP.

The cGMP-sensitive (or-stimulated) cyclic nucleotide phosphodiesterase is now purified to homogeneity from bovine adrenal or heart tissues and also from calf liver (17,18). SDS/polyacrylamide gel electrophoresis shows a single molecular weight polypeptide of Mr 105,000. Gradient gel electrophoresis under non denaturating conditions indicates a native enzyme of Mr $\sim$ 240,000. The purified enzyme demonstrates positive homotropic cooperativity with respect to cAMP or cGMP as substrate. cGMP is acting as a positive allosteric effector as proven by the use of cyclic nucleotide analogs as probes of catalytic and effector binding sites (20, 21). Until now, the involvement of this enzyme in a negative control of cAMP levels in intact cells has never been proven. However, in cultured hepatoma cells, incubated with dexamethasone for 48 to 96 hours, the cGMP-sensitive phosphodiesterase activity, is specifically decreased whereas activity of the "low Km" cAMP phosphodiesterase is unchanged (39).

3.3. Activation through covalent modification : phosphorylation and proteolysis

Many reports suggest that phosphorylation mechanisms are involved in the net increase of "low Km" cAMP phosphodiesterase activities by agents which increase cAMP (for reviews see refs. 7,8). This is shown, for example, in S49 mouse lymphoma cells when cAMP is elevated by isoproterenol. A mutant isolated from those cells and deficient in cAMP dependent protein kinase activity does not show enhanced phosphodiesterase activity (40). It has been reported that activity of the insulin-sensitive peripheral plasma membrane phosphodiesterase is increased in cell free system when liver membranes are incubated in the presence of insulin, cAMP and ATP (38). A phosphorylated-stimulated enzyme could be purified to homogeneity from this material (41).

"Low Km" cAMP phosphodiesterases, in partially purified fractions, as until now, are very susceptible to proteolysis. In rat kidney, or in liver, the cAMP high affinity soluble enzyme is activated by a cathepsin L protease (42). Proteolysis also solubilized and stimulated the insulin-sensitive particulate "low Km" cAMP phosphodiesterase in liver and adipose tissue (43). Whether such an irreversible mechanism of activation occurs in the intact cell is doubtful.

4. THE CONTROLS OF CYCLIC NUCLEOTIDE PHOSPHODIESTERASE ACTIVITY IN INTACT CELLS

4.1. Regulatory potential of the catabolism of cAMP.

cAMP is hydrolyzed by a system of multiple enzymes of "high" and "low Km" and of different maximal velocities for cAMP. Such systems, by definition, present a negative cooperativity (in the purely phenomenological meaning of the word, ref. 44). It follows that the rate of cAMP catabolism will not be first order (at least in those systems in which a high affinity cAMP phosphodiesterase is predominant, e.g. in mouse lymphoma S49 cells, discussed in ref. 45). A theoretical analysis of a negatively cooperative phosphodiesterase system shows that this property may exert significant quantitative effects on the accumulation of cAMP stimulated by adenylate cyclase-activating ligands (46): i) it increases the cooperativity of the relation between the cAMP concentration and the concentration of activating ligand, thus increasing the sensitivity to variations of ligand concentration, ii) it enhances the sensitivity of cAMP accumulation to slight variations of phosphodiesterase maximal velocity, which makes this negatively cooperative enzyme an efficient device for regulating cAMP level within the cell.

Evidence for the first prediction is illustrated in different systems such as in the thyroid (44) or in mouse lymphoma S49 cells (45). The K30A mutant of the same cells (30) associated to increased high affinity cAMP phosphodiesterase activity is more definitely upwards concave as expected from the simulations. Evidence for the second prediction is suggested in the case of insulin-sensitive high affinity cAMP phosphodiesterase activities in liver or adipose tissue (3,47,48). The effects of insulin on phosphodiesterase maximal velocity are small with regard to the effects on cAMP. This apparent discrepancy can be explained by the negative cooperativity of phosphodiesterase activity-insulin sensitive; since the cAMP hydrolysis is not first order, small effects on phosphodiesterase maximal velocity (23,41,47,48) provoke important changes in cAMP levels.

4.2. Effect of hormones and other agents on phosphodiesterase activity : short term (non transcriptional) and long term (transcriptional) controls.

Many hormones (insulin, glucagon, catecholamines, growth hormone, etc.) activate particulate phosphodiesterase(s) with high affinity for cAMP by mechanisms modifying the activity or the level of the enzyme(s) (for reviews see refs. 7,8). Insulin can decrease hepatic cAMP concentrations by activating two "low Km" cAMP phosphodiesterase activities : a peripheral plasma membrane enzyme and a "dense vesicle" or "dense microsomal" enzyme. Based on molecular weight, affinity for cGMP and sensitivity to glucagon, the two enzymes are distinct. The mechanisms involved appear non transcriptional. The peripheral plasma membrane enzyme purified by Marchmont et al. (23,41) is specifically insulin-sensitive and not activated by glucagon (48). The purified enzyme is apparently monomeric (Mr~52,000), diplays a "low Km" (0.7 µM) for cAMP and apparent negative cooperativity. cGMP acts as a poor substrate (Km =120 µM) or inhibitor of cAMP hydrolysis. As stated before, activation by insulin is attributed to an insulin-triggered phosphorylation of the enzyme which shifts its kinetic parameters to a higher maximal velocity and a decreased Hill coefficient (23,41). The "dense microsomal" enzyme has been partially purified by Loten et al. (43,47). It can be activated by insulin but also by glucagon or dibutyryl cAMP (48). Insulin and glucagon together, each at maximal effective concentration, have a greater effect on phosphodiesterase activity that does either hormone alone. The enzyme has a molecular weight of about 120,000, "low Km" values for cAMP and cGMP (0.3 µM) with apparent negative cooperativity and a higher maximal velocity for cAMP than for cGMP (47). In contrast to the action of insulin, short term incubation of human diploid fibroblast (WI-38 cells) with prostaglandin E_1 causes total cAMP phosphodiesterase activity in homogenates to decrease. The prostaglandin-provoked decline in enzyme activity is immediate and coincides with the rapid increase of intra-cellular cAMP levels (49).

In C6 glioma cells, stimulation of β-adrenergic receptor lead to the induction of a specific form of cAMP phosphodiesterase that is not sensitive to calmodulin. As this increase is blocked by specific inhibitors of RNA polymerase II (α-amanitin), the authors have suggested that new mRNA synthesis is required for this effect (50). This is an example of a transcriptional control mechanism.

5. DIRECT STIMULATION BY CALCIUM OF THE CALMODULIN-SENSITIVE PHOSPHODIESTE-RASE IN THE INTACT CELL : THE EXAMPLE OF DOG THYROID SLICES EXPOSED TO ACETYLCHOLINE OR CARBACHOL.

5.1. The control of cAMP accumulation by acetylcholine or carbachol in dog thyroid is Ca^{2+}-mediated and relieved by MIX.

Activation of muscarinic cholinergic receptors leads to a decrease in cAMP levels in many tissues. It has been reported than in rabbit myocardium (51) and rat parotid (52) for instance, receptor-mediated inhibition of adenylate cyclase is involved. As activation of muscarinic receptors also increases phosphatidylinositol turnover and calcium influx in the cytoplasm (from the exterior or from intracellular stores), two additional mechanisms could be the

direct inhibition of adenylate cyclase by Ca^{2+} (53) or the stimulation of cAMP
catabolism (cAMP efflux and/or hydrolysis).

The characteristics of the negative control of carbachol on cAMP demonstrated in dog thyroid are the following (54, 55): i) the effect of carbachol
is abolished in calcium depleted slices as well as in presence of calcium
competitors; ii) it is mimicked by the divalent cation ionophore A23187 in the
presence of Ca^{2+} as well as by high concentrations of calcium itself and iii)
it is relieved by MIX. The carbachol effect, in dog thyroid, is thus clearly
Ca^{2+}-dependent, in contrast to the negative control of norepinephrine on cAMP
demonstrated in the same system which does not require extracellular calcium
and is not relieved by MIX (56). As in other cells (51), norepinephrine, by
an α_2-receptor, directly inhibits dog thyroid-prepared adenylate cyclase.

5.2. The Ca^{2+}/calmodulin-sensitive phosphodiesterase is involved in the negative control of carbachol on cAMP levels in dog thyroid slices.

The Ca^{2+}-mediated inhibition of cAMP accumulation in dog thyroid could
result from various causes : i) a direct inhibition of adenylate cyclase by
Ca^{2+}, ii) a stimulation of the efflux of cAMP, iii) an activation of phosphodiesterase activity.

Dog thyroid-adenylate cyclase is inhibited by high (10-100 µM) concentrations of calcium. This effect however was not relieved by MIX (57). Moreover
it had been shown, in intact cell, that the carbachol Ca^{2+}-mediated decrease
in cAMP bears on the catabolism rather than on the synthesis of cAMP (54).
The efflux of cAMP has not been measured in dog thyroid but if operating it
should be quantitatively important relative to phosphodiesterase activity to
account for the mechanism of carbachol action. If similar to the efflux in
WI-38 cells, it should be inhibited by 7-benzyl MIX but not by MIX or IIX
(58). The sequence of inhibitors relieving the carbachol effect on cAMP in
dog thyroid is clearly different (Fig. 3).

Three phosphodiesterase activities are separated from a dog thyroid crude
soluble fraction : a Ca^{2+}/calmodulin form, a cGMP-stimulated enzyme and a cAMP-
specific enzyme (Fig. 1). Thus phosphodiesterase activity could be enhanced
by at least three distinct mechanisms (Table 2). As carbachol increases cGMP
accumulation, the cGMP-stimulated phosphodiesterase could be evoked. However,
sodium nitroprusside and $MnCl_2$, which independently of Ca^{2+}, increases cGMP do
not depress cAMP accumulation. Moreover sodium nitroprusside does not relieve
the inhibitory effect of carbachol (55, 57). Such data argue against a cGMP
direct stimulation on cAMP hydrolysis in dog thyroid slices exposed to carbachol.

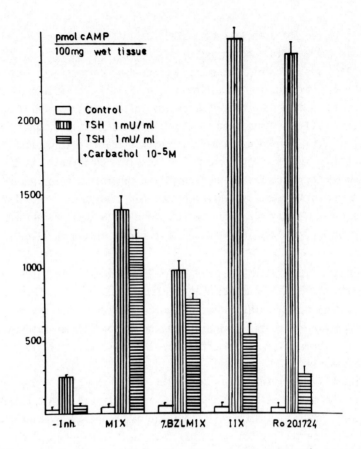

Fig. 3. cAMP accumulation in dog thyroid slices incubated in the presence of TSH (1 mU/ml) ± 10 μM carbachol. Details of experimental procedure are given in ref. 59. Inhibitors are at 50 μM. cAMP expressed as pmol of cAMP per 100mg of wet tissue ± SEM.

TABLE 3.

INHIBITION OF PURIFIED DOG THYROID PHOSPHODIESTERASES

	calmodulin-sensitive enzyme	$I_{50}^{(a)}$ cAMP-specific enzyme
	μM	μM
MIX	3.2 + 0.8	12.8 + 0.3
7-benzyl MIX	6.3 + 1.2	62.0 + 6.0
IIX	> 100	7.1 + 1.9
Ro 20-1724	> 100	5.2 + 0.6

(a) I_{50} = concentration of inhibitor required to inhibit 50 % the hydrolysis of cAMP at 1 μM.

With the alkylated xanthines and the synthetic imidazolidinone Ro 20-1724, we could show that IIX and Ro 20-1724 are potent inhibitors of the cAMP-specific phosphodiesterase and do not inhibit the calmodulin-sensitive form but that in contrast, MIX and 7-benzyl MIX are relatively specific for the calmodulin-sensitive enzyme (Table 3). The rank order of inhibition of the cAMP-specific phosphodiesterase (IIX and Ro 20-1724) corresponds to the rank order of potentiation of TSH-enhanced cAMP accumulation in intact cells. These data suggest that in the absence of carbachol (thus at basal Ca^{2+} level), the cAMP accumulated in response to TSH is mainly hydrolyzed by the "low Km" cAMP phosphodiesterase and that the calmodulin-sensitive form does not contribute significantly. On the contrary, those inhibitors which potently inhibit the calmodulin-sensitive enzyme (MIX and 7-benzyl MIX) also relieve the carbachol negative control on cAMP, while the two relatively potent inhibitors of the cAMP specific form (IIX and Ro 20-1724) do not (Fig. 3). The use of a larger series of inhibitors of phosphodiesterase has further confirmed this conclusion (10,59). The data suggest that in dog thyroid slices exposed to carbachol, the intracellular Ca^{2+} rise activates the calmodulin and thereby the calmodulin-sensitive phosphodiesterase; this enzyme now becomes the dominant form in controlling cAMP levels.

6. CONCLUSIONS.

Cyclic nucleotide phosphodiesterases consist in a system of multiple enzymes of "low Km" and "high Km" values for cAMP and cGMP. Both non transcriptional and transcriptional control mechanisms of phoshodiesterase activity have been characterized. Three fundamentally distinct controls are now characterized on purified enzymes. Those directly involve Ca^{2+}/calmodulin stimulation, a cGMP control and a mechanism of enzyme phosphorylation on three different enzymes. In liver, insulin-stimulation of two "low Km" cAMP phosphodiesterases has been proposed to explain how insulin depresses glucagon-induced increases in cAMP (60). In dog thyroid, carbachol through a muscarinic receptor, decreases cAMP in TSH-stimulated slices (54, 55,57). This effect results from the Ca^{2+}-stimulation of a calmodulin-sensitive phosphodiesterase in this tissue (10,59). This mechanism is the first evidence that a "high Km" cAMP enzyme is involved in a negative and immediate cAMP regulation. In two other cell types*, human astrocytoma cells (61) and rat prostatic tissue (62), the carbachol provoked decrease in cAMP could be perhaps attributed to a similar mechanism. The latter at the level of cAMP catabolism, is obviously distinct from the well known direct inhibition of adenylate cyclase by

*in these cells, the carbachol provoked decrease in cAMP shows some common characteristics with the effect observed in dog thyroid.

muscarinic and α-adrenergic agents (51). These two mechanisms do not exclude each other and might on the contrary complement each other in some cells. This further exemplifies the rule that when particular controls are important for the cell, they are generally multiple.

ACKNOWLEDGMENTS

We are grateful to Dr J. Wells for having provided the alkylated xanthines. We wish to thank Mrs Colette Moreau for her expert technical assistance and Mrs Danièle Leemans for the preparation of the manuscript. This work was realized under contract of the Ministère de la Politique Scientifique (Actions Concertées). C. Erneux is Chargé de Recherches au Fonds National de la Recherche Scientifique.

REFERENCES

1. Sutherland EW, Rall TW (1958) J Biol Chem 232:1077-1091
2. Thompson WJ, Appleman MM (1971) J Biol Chem 246:3145-3150
3. Loten EG, Sneyd JGT (1970) Biochem J 120:187-193
4. Beavo JA, Hansen RS, Harrison SA, Hurwitz RL, Martins TJ, Mumby MC (1982) Mol Cell Endocrinol 28:387-410
5. Erneux C, Couchie D, Dumont JE (1980) Eur J Biochem 104:297-304
6. Keravis TM, Wells JN, Hardman JG (1980) Biochim Biophys Acta 613:116-129
7. Wells JN, Hardman JG (1977) Adv Cyclic Nuc Res 8:119-143
8. Manganiello VC, Yamamoto T, Elks M, Lin MC, Vaughan M (1984) Adv Cyclic Nucleotide & Phosphorylation Res 16:291-301
9. Barber R, Butcher RW (1983) Adv Cyclic Nucleotide Res 15:119-138
10. Miot F, Dumont JE, Erneux C (1983) FEBS Lett 151:273-276
11. Klee CB, Crouch TH, Krinks MH (1979) Biochemistry 18:722-729
12. La Porte DC, Toscano WA, Storm DR (1979) Biochemistry 18:2820-2825
13. Tucker MM, Robinson JB, Stellwagen E (1981) J Biol Chem 256:9081-9058
14. Sharma RK, Wang TH, Wirch E, Wang JH (1980) J Biol Chem 255:5916-5923
15. Davis CW and Daly JW (1978) J Biol Chem 253:8683-8686
16. Kincaid RL, Manganiello VC, Vaughan M (1979) J Biol Chem 245:4970-4973
17. Martins TJ, Mumby MC, Beavo JA (1982) J Biol Chem 257:1973-1982
18. Yamamoto T, Manganiello VC, Vaughan M (1983) J Biol Chem 258:12526-12533
19. Yamamoto T, Yamamoto S, Osborne JC, Manganiello VC, Vaughan M, Hidaka H (1983) J Biol Chem 258:14173-14177
20. Erneux C, Couchie D, Dumont JE, Baraniak J, Stec WJ, Garcia Abbad E, Petridis G, Jastorff B (1981) Eur J Biochem 115:503-510

21. Erneux C, Couchie D, Dumont JE, Jastorff B (1984) Adv Cyclic Nucleotide & Prot Phosphorylation Res 16:107-118

22. Thompson WJ, Epstein PM, Strada SJ (1979) Biochemistry 18:5228-5237

23. Marchmont RJ, Ayad SR, Houslay MD (1981) Biochem J 195:645-652

24. Sharma RK, Adachi AM, Adachi K, Wang JH (1984) J Biol Chem 259:9248-9254

25. Couchie D, Petridis G, Jastorff B, Erneux C (1983) Eur J Biochem 136: 571-575

26. Hurwitz RL, Hansen RS, Harrison SA, Martins TJ, Mumby MC, Beavo JA (1984) Adv Cyclic Nucl & Prot Phosphorylation Res 16:89-106

27. Takemoto DJ, Hansen J, Takemoto LJ, Houslay MD (1982) J Biol Chem 257: 14597-14599

28. Kramer GL, Garst JE, Mitchel SS, Wells JN (1977) Biochemistry 16:3316-3321

29. Weber HW, Appleman MM (1982) J Biol Chem 257:5339-5341

30. Brothers VM, Walker N, Bourne HR (1982) J Biol Chem 257:9349-9355

31. Baehr W, Devlin MJ, Applebury ML (1979) J Biol Chem 254:11669-11677

32. Hamet P, Coquil JF, Bousseau-Lafortune S, Franks DJ, Tremblay J (1984) Adv Cyclic Nucleotide & Prot Phosphorylation Res 16:119-136

33. Bourne HR, Brothers VM, Kaslow HR, Groppi V, Walker N, Steinberg F (1984) Adv. Cyclic Nucleotide & Prot Phosphorylation Res 16:185-194

34. Hamet P, Coquil JF (1978) J Cyclic Nucleotide Res 4:281-290

35. Coquil JF, Franks DJ, Wells JN, Dupuis M, Hamet P (1980) Biochim Biophys Acta 631:148-165

36. Francis SH, Lincoln TM, Corbin JD (1980) J Biol Chem 255:620-626

37. Yamasaki A, Sen I, Bitensky MW, Casnellie JE, Greengard P (1980) J Biol Chem 255:11619-11624

38. Marchmont RJ, Houslay MD (1980) Nature 286:904-906

39. Ross PS, Manganiello VC, Vaughan M (1977) J Biol Chem 252:1448-1452

40. Bourne HR, Coffino P, Melmon KL, Tomkins GM, Weinstein Y (1975) Adv Cyclic Nucl Res 5:771-786

41. Marchmont RJ, Houslay MD (1981) Biochem J 195:653-660

42. Strewler GJ, Manganiello VC (1979) J Biol Chem 254:11891-11898

43. Loten EG, Francis SH, Corbin JD (1980) J Biol Chem 255:7838-7844

44. Boeynaems JM, Van Sande J, Pochet R, Dumont JE (1974) Mol Cell Endocrinol 1:139-155

45. Butcher RW (1984) Adv Cyclic Nucleotide & Prot Phosphorylation Res 16: 1-12

46. Erneux C, Boeynaems JM, Dumont JE (1980) Biochem J 192:241-246

47. Loten EG, Assimacopoulos-Jeannet FD, Exton JH, Park CR (1978) J Biol Chem 253:746-757

48. Houslay MD, Wallace AV, Wilson SR, Marchmont RJ, Heyworth CM (1983) Horm Cell Regul 7:105-120

49. Nemecek GN, Ray KP, Butcher RW (1979) J Biol Chem 254:598-601

50. Schwartz JP, Onali P (1984) Adv Cyclic Nucleotide & Prot Phosphorylation Res 16:195-203

51. Jakobs KH, Aktories K, Lasch P, Wilhem S, Schultz G (1980) Horm Cell Regul 4:89-106

52. Oron Y, Kellogg J, Larner J (1978) FEBS Lett 94:331-334

53. Piascik MT, Lewis Wisler P, Johnson CL, Potter JD (1980) J Biol Chem 255:4176-4181

54. Van Sande J, Erneux C, Dumont JE (1977) J Cyclic Nucleotide Res 3:335-345

55. Decoster C, Mockel J, Van sande J, Unger J, Dumont JE (1980) Eur J Biochem 104:199-208

56. Cochaux P, Van Sande J, Dumont JE (1982) Biochim Biophys Acta 721:39-46

57. Dumont JE, Miot F, Erneux C, Couchie D, Cochaux P, Gervy-Decoster C, Van Sande J, Wells JN (1984) Adv Cyclic Nucleotide & Phosphorylation Res 16:325-336

58. Nemecek GM, Wells JN, Butcher RW (1980) Mol Pharmacol 18:57-64

59. Miot F, Erneux C, Wells JN, Dumont JE (1984) Mol Pharmacol 25:261-266

60. Jefferson LS, Exton JH, Butcher RW, Sutherland EW, Park CR (1968) J Biol Chem 243:1031-1038

61. Meeker RB, Harden TK (1982) Mol Pharmacol 22:310-319

62. Shima S, Komoriyama K, Hirai M, Kouyama H (1983) Biochem Pharmacol 32:529-533.

RESUME

Les phosphodiestérases des nucléotides cycliques se composent de plusieurs enzymes à "haut" et "bas Km" pour l'AMP cyclique et le GMP cyclique. Tous les tissus et les cellules de mammifères hydrolysent les nucléotides cycliques en 5'-nucléotides par une (ou plusieurs) phosphodiestérase(s). Cette voie métabolique est contrôlée par l'intermédiaire de plusieurs hormones (l'insuline et le glucagon, par exemple) ainsi que par des molécules-signaux intracellulaires comme le Ca^{2+}, la calmoduline ou le GMP cyclique. Les contrôles se font dans certains cas au niveau de la transcription. Trois mécanismes de contrôle direct ont été décrits en système acellulaire sur trois enzymes distincts et font intervenir une stimulation par le Ca^{2+} (et la calmoduline), un contrôle par le GMP-c et un mécanisme de phosphorylation. Un contrôle par le GMP-c n'a pu être démontré en système intact mais des cellules d'hépatome de rat en culture, incubées en présence de dexaméthasone, ont une activité réduite pour l'enzyme stimulé par le GMP-c. Dans le foie, plusieurs auteurs ont proposé que l'insuline stimule l'activité de deux phosphodiestérases pour rendre compte de l'effet anti-glucagon sur l'AMP-c. Enfin, dans la thyroïde de chien, le carbachol inhibe le taux d'AMP cyclique dans des tranches stimulées à la thyrotropine. Cet effet résulte d'une stimulation du catabolisme de l'AMP cyclique, par l'effet activateur du Ca^{2+} sur une phosphodiestérase dépendante de la calmoduline.

Hormones and Cell Regulation, Volume 9
INSERM European Symposium
J.E. Dumont, B. Hamprecht and J. Nunez editors
© 1985 Elsevier Science Publishers B.V. (Biomedical Division)

PHOSPHOLIPID/Ca²⁺-DEPENDENT PROTEIN KINASE, CELL DIFFERENTIATION AND TUMOR PROMOTION.

M. CASTAGNA[x], C. PAVOINE[x] , S. BAZGAR[x], A. COUTURIER[x], M. CHEVALIER[x], and M. FISZMAN[xx].

[x] Institut de Recherches Scientifiques sur le Cancer, BP 8, 94802 Villejuif Cedex
[xx] Institut Pasteur, 25, rue du Docteur Roux 75724 Paris Cedex 15

INTRODUCTION

During the past few decades clinical as well as experimental evidence has accumulated indicating that cancer results from a multistep and multifactorial process. From studies on animal models and more recently on cell cultures it has been concluded that transformation of cells requires at least two steps (1-3). The initial step is a frequent event which gives rise to potentially transformed or "initiated" cells. The second step is a rare event which may occur each time an initiated cell divides.

Tumor promoters are defined as chemical or physical agents that are not carcinogenic by themselves but most likely increase the probability of the second event required for transformation to occur.

The structures of tumor promoters are very diverse. Those from the series of phorbol esters which have been initially isolated from plant and identified by Hecker (as reviewed in 4), have been implicated in some human neoplasia (5).These agents, which are the most currently used in experimental carcinogenesis, are amphiphatic compounds which may incorporate into phospholipids (6).

Phorbol esters active in promoting transformation elicit in culture cells a pleiotropic response which largely depends on the target cell. 12-0-tetradecanoyl phorbol 13-acetate (TPA) reversibly evokes in some cell lines several properties associated with the tumor phenotype such as changes in cell morphology, high saturation density and low calcium requirement and/or inhibition of terminal differentiation (7-9).For instance when TPA was applied either to normal quail myoblasts or RSV ts

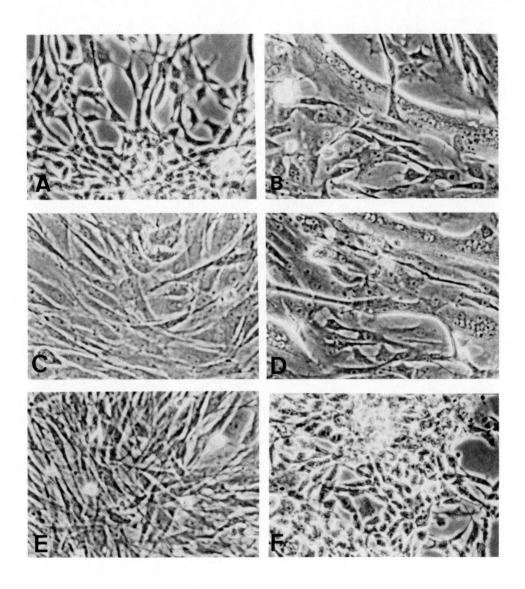

Fig. 1 : Photomicrographs of quail embryo myoblasts. A) RSV ts68 transformed quail myoblasts maintained at 35°C ; B) TSV ts68 transformed quail myoblasts shifted to 41° for 48 hrs ; C) Normal quail myoblasts 15 hrs after plating ; D) Normal quail myoblasts 48 hrs after plating ; E) Normal quail myoblasts treated for 48 hrs with 20ng/ml TPA ; F) RSV ts68 transformed quail myoblasts incubated for 48 hrs at 41°C in the presence of 20ng/ml TPA.

NY68 mutant transformed cells at the non-permissive temperature (41°), morphological changes occurred and myogenic differentiation was prevented (fig.1). Alternatively, reciprocal effects were observed in some other cell lines, mostly malignant, where TPA induced cells to differentiate and inhibited cell growth : a human promyelocytic cell line HL60 which undergoes monocyte/macrophage differentiation has been used as a model system. The opposite effects of TPA in a variety of differentiating cell systems are reviewed in ref. 7 and 10.

In 1980 Driedger and Blumberg discovered saturable, specific, high affinity binding sites for phorbol esters which were mostly located in membranes (11). The receptor sites were present in species ranging from sponges to humans and in all tested cell types except red blood cells. The structure-activity relationship was in good agreement with promoting potencies within the series of phorbol esters. Consequently, it was inferred that phorbol esters exert their cellular effects through interaction with their specific receptor.

Therefore, the postulated molecular target for phorbol esters should be an ubiquitous membrane component, able upon interaction, to elicit a pleiotropic response and increase either the intracellular calcium concentration or the calcium sensitivity of some process presumably involved in mitogenesis to account for the low calcium requirement associated with phorbol ester treatment.

I met Dr. Nishizuka at a meeting held in Brussels in July 1980. He has introduced me to the recent discovery from his group : a new species of protein kinase, the phospholipid/Ca^{2+} -dependent protein kinase which he refers to as protein kinase C. We agreed that the enzyme was a good candidate for the molecular target I was looking for. The purpose of this paper is to show that this hypothesis proved to be right, and provide some evidence supporting the possibility that protein kinase C may be implicated in cell differentiation and transformation.

PHOSPHOLIPID/Ca^{2+}- DEPENDENT PROTEIN KINASE

1/ Physical and catalytic properties

This ubiquitous enzyme transfers phosphate from ATP to serine and threonine residues in various exogenous and endogenous

protein substrates. Physical, kinetic and catalytic properties of
this enzyme have been recently reviewed (12) and some of them
will be only briefly outlined in this report. The activity is
strictly dependent on the presence of Ca^{2+} and phospholipid. The
order of potency for phospholipid in supporting enzyme activity
is as follows : phosphatidylserine> phosphatidic acid >
phosphatidylinositol> phosphatidylethanolamine> sphingomyelin >
phosphatidylcholine. Thus anionic phospholipids were the most
potent   which indicated a role for surface potential in the
interaction between enzyme and phospholipid. Protein kinase C,
which has been isolated to homogeneity from different tissues,
appears well conserved over evolution indicating that the enzyme
plays a key role in cell survival (13-15). The molecular weight
of rat brain enzyme is 77,000, Stokes radius 42Å, S value 5.1 and
isoelectric point 5.6. The enzyme consists of a single
polypeptide chain composed of a hydrophobic domain which probably
interacts with the membrane and a hydrophilic domain containing
the catalytic site. A Ca^{2+}-dependent neutral protease active at
micromolar calcium concentrations can split the two domains as
long as the enzyme is tightly bound to the membrane. A
proteolytic fragment of molecular weight 51,000 is formed which
displays a full catalytic activity without requiring either
calcium or phospholipid. The physiological significance of such a
fragment has to be demonstrated.

Nishizuka and coworkers have shown that diacylglycerol, a
neutral lipid pratically absent in resting cells, dramatically
increased reaction velocity and decreased the activation constant of
phospholipid,as well as calcium,in such a way that the enzyme
could be activated at physiological concentrations of this cation
(0.1uM). The activation was highly specific since unsaturated
diacylglycerol esterified in positions 1 and 2 was by far the
best effector when compared with 1,3 unsaturated or 1,2 and 1,3
saturated diacylglycerol. Results presented in fig. 2 are
obtained with the two isomeric forms of diolein and distearin.
Other neutral lipids such as monoacylglycerol, triacylglycerol
and free fatty acid were ineffective.

2/ Involvement in transmembrane signalling

The great breakthrough from Nishizuka's group was to show
evidence for the coupling of protein kinase C activation to

receptor-linked inositol phospholipid breakdown. Substantial
evidence has been obtained, mostly from platelets, that
diglyceride which transiently accumulated as a result of signal-
triggered inositol phospholipid turnover serves as a natural
effector of the enzyme. A large class of extracellular
signals,many of which are listed in Table I, enhance
phosphoinositide breakdown upon interaction with specific
receptors and protein kinase C appears as a keystone in this
transducing pathway. The primary substrates from the inositol
phospholipid cycle seems to be phosphatidylinositol 4-phosphate
and phosphatidylinositol 4,5-bisphosphate which are rapidly
hydrolysed following signal interaction (16). Phospholipase C
catalyzes the rapid formation of inositol 1,4-diphosphate and
inositol 1,4,5 trisphosphate (17,18). Evidence has been
presented which suggests that inositol trisphosphate (IP_3)
mobilizes Ca^{2+} from its internal stores eventually giving rise to
the signal-triggered increase of ionized Ca^{2+} (19). A possible
model for signal activation is drawn in fig. 10.

TABLE I
EXTRACELLULAR SIGNALS ENHANCING PHOSPHOINOSITIDE METABOLISM.

Stimulus	Tissue
- Cholinergic neurotransmitters (muscarinic type)	Various
- α-adrenergic agonists	Various
- Dopamine D2	Cerebral cortex
- Histamine H1	Gastric mucosa
- Thyrotropin	Thyroid gland
- Corticotrophin-releasing hormone	Anterior pituitary
- Pancreazymin/Cholecystokinin	Pancreas
- Caerulein	Pancreas
- Thrombin	Platelets
- Platelet-activating factor	Platelets
- Collagen	Platelets
- Thromboxan A_2	Platelets
- Vasopressin	Liver
- Angiotensin II	Liver
- Glucose	Islets of Langerhans
- Mitogenic Lectins	Lymphocytes
- Growth factors (EGF, PDGF)	Fibroblastes
- F-Met-Leu-Phe	Leucocytes

The diacylglycerol/IP$_3$-mediated activation has to be under the control of regulatory mechanisms which appears to vary from one cell type to another. For instance, cyclic nucleotides inhibit the signal activation in lymphocytes, platelets and leucocytes although they are ineffective in adipocytes or liver (12) and cyclic AMP levels and inositol phospholipid turnover rates were concurrently increased in response to various agents as reviewed in ref. 20.

Protein kinase C present in a soluble form or loosely bound to membranes under resting conditions, presumably tightly binds to membranes and becomes activated as a result of the interaction of external signals with specific surface receptors.

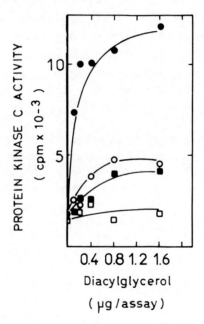

Fig.2. Protein kinase C activation by various diacylglycerols. Activity was assyed as described in ref 25 in the presence of 5uM CaCl$_2$. (●) 1,2 diolein, (○) 1,2 distearin, (■)1,3 diolein, (□)1,3 distearin.

PHORBOL ESTER-MEDIATED PROTEIN KINASE C ACTIVATION

1/ Evidence for direct activation of protein kinase C.

Tumor-promoting phorbol esters rapidly evoke in platelets a release-associated aggregation (21) similarly to receptor-mediated physiological ligands such as thrombin, collagen and

platelet aggregating factor. At low concentrations, TPA enhanced
the phosphorylation of a number of platelet proteins, especially
the 40Kd protein, the phosphorylation of which has been
associated with serotonin release (22). In contrast with thrombin
induced-platelet aggregation, TPA did not cause either an
increased rate of inositol phospholipid turnover or a
diacylglycerol accumulation (fig. 3).

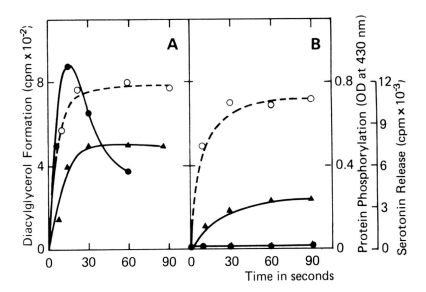

Fig. 3. Time-courses of diacylglycerol formation (—•—),
40Kd protein phosphorylation (--o--) and serotonine release
(—▲—) in platelets treated with thrombin (A) or TPA (B).
Experimental conditions are described elsewhere (21).

This finding suggested that tumor promoters may substitute, to
some extent for diacylglycerol. Kinetic analysis conducted on the
brain enzyme have indicated that TPA at nanomolar concentrations
behaved indeed like diacylglycerol at micromolar concentrations
in directly activating protein kinase C. A dose-response curve is
shown in fig. 4. At submaximal concentrations the effects of both
activators were additive.

2/ Properties of protein kinase C activation
Ca^{2+}-dependency of diglyceride-and TPA-mediated protein
kinase C activation has been studied in the presence of

increasing doses of EGTA. Free calcium concentrations in the reaction
mixture were calculated from the total amount of added EGTA and
either contaminating or added calcium. Total calcium
concentrations were measured by atomic absorption
spectrophotometry. Results from a typical experiment which are
presented in table II show that both TPA and 1,2 diolein
activated the enzyme in the presence of EGTA up to 5mM that is at
trace concentrations of ionized calcium.

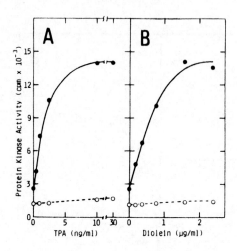

Fig. 4. Dose-dependent activation of protein kinase C by TPA (A)
and diolein (B). The enzyme was assayed as described in ref. 21
in the presence of 10uM $CaCl_2$ plus phospholipid at 20ug/ml(—•—)
or without (--o--) phospholipid.

The order of potency for various phospholipid species in
supporting TPA-mediated protein kinase C activation was examined in
the presence of 5uM $CaCl_2$ or 0.5mM EGTA. The most efficient
phospholipid species to activate enzyme activity were the most
potent in supporting enzyme activation regardless of the calcium
concentration (fig. 5). It should be pointed out that the active
species are mainly located within the inner half of the plasma
membrane. Therefore membrane assymetry favors enzyme activation.
Protein kinase C activity was inhibited by local
anesthetics, antipsychotic and other phospholipid-interacting
drugs including dibucaine, tetracaine, trifluoperazine,

chlorpromazine, triflupromazine, mellitin and
alkyllysophosphatidylcholine (ALP). In contrast with cyclic
nucleotide-dependent protein kinases, myosin light chain-
dependent protein kinase was also susceptible to these drugs,
which all compete with phospholipid (23,24). As shown in table
III the inhibitory action was partially overcome by calcium to
varying extent. Similarly TPA partially counteracted action of
the drugs and increased the IC_{50} as previously reported (25).

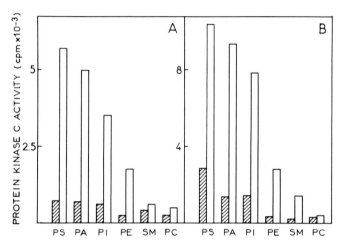

Fig. 5.   Relative potencies of phospholipid to support basal ▨
and TPA-stimulated ▭ protein kinase C activity ; plus 0.5mM
EGTA (A), plus 5uM $CaCl_2$. Reprinted from ref. 25.

     Following the initial evidence showing that TPA was able to
trigger protein kinase C activation, several studies have led to
the assumption that the phospholipid-associated enzyme was the
high affinity receptor for phorbol esters (26-29). Thus it has
been shown that the level of protein kinase C activity is
parallel to that of the phorbol ester binding activity in all
tested mammalian tissues and that both activities copurified in
different chromatographic systems. In addition, it has been shown
that diacylglycerol competes with phorbol dibutyrate (PDBu), a
phorbol ester less hydrophobic than TPA, on its binding site.
Scatchard analysis of PDBu binding to a homogeneous preparation
of the enzyme has revealed that the apparent dissociation binding

constant of PDBu was 8nM, this value being equal to its activation constant for the enzyme. The stoichiometry of the binding reaction was found to be 1:1. Kikkawa _et al._ (29) have evaluated the amount of protein kinase C in crude extracts of rat brain homogenates at 16 pmol/mg protein, an amount equivalent to the total amount of PDBu which binds to the same tissue. These results suggest that the phospholipid/protein kinase complex is probably the only high affinity receptor site for phorbol esters.

Based on the foregoing evidence it is reasonable to propose that TPA, through its interaction with its receptor site, usurps the role of the physiological effector of protein kinase C, as illustrated in fig. 10, and may thus mediate at least some of these pleiotropic actions.

TABLE II
CA^{2+} DEPENDENCY OF PROTEIN KINASE C ACTIVATION

Free calcium concentrations were calculated knowing total concentration of EGTA and calcium present or added in the assay mixture with $K_{CA}=4.10^6M^{-1}$. Total amounts of calcium which were present in added salts and various components were measured using an atomic absorption Spectrophotometer Perkin-Elmer 373. In our usual conditions total calcium concentration present in the assay mixture in the presence of 0.5mM EGTA was found approx. 6uM. The assay for enzyme activity was performed according a technique previously described (25) except that free magnesium concentration was brought up to 4mM in EGTA-containing samples.

Addition	Ca^{2+}*	TPA**	1,2-DG**
10uM CaCl$_2$	-	8185	8450
0.5mM EGTA	2.7 nM	8520	8050
2mM EGTA	0.6 nM	8290	8010
5mM EGTA	0.2 nM	7925	7895

* Calcium present in EGTA was neglected.
** TPA was added 50ng/ml and 1,2 diolein (1,2-DG) at 0.4ug/assay

PROTEIN KINASE C AND DIFFERENTIATION OF HL60, A PROMYELOCYTIC CELL LINE.

Tumor promoting phorbol esters elicit in HL60 cells a macrophage-like phenotype at nanomolar concentrations associated with the occurrence of various enzymatic and antigenic markers, morphological maturation and a high phagocytic index. As briefly outlined above, a number of phospholipid-interacting drugs inhibit basal and TPA-stimulated protein kinase C activity. The potency of these various drugs to inhibit TPA-induced differentiation as well as that of Mn^{2+}, a potent inhibitor of protein kinase C (unpublished data), has been examined. Fig. 6 shows that all the drugs inhibit TPA-induced cell adhesion to plates and, it is of interest to note, that the IC_{50} was generally close to that measured in the in vitro assay for protein kinase C activity with the exception of ALP which appears surprisingly more sensitive to inhibit cell adhesion than the kinase activity in vitro. Under our experimental conditions the drugs did not affect growth rate in TPA-treated cells.

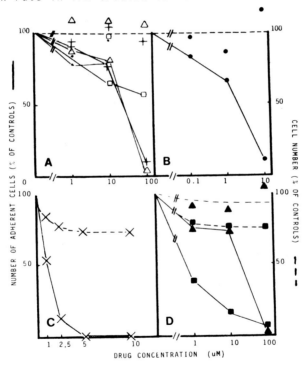

Fig. 6. Dose-response curve of drug inhibition of TPA-induced cell adhesion to Petri dishes. Drugs were added with TPA at 3nM to cultures of exponentially growing HL60 cells. The number of cell adherent or in suspension were scored 48h later. Panel A : chlorpromazine (•), trifluoperazine (Δ), triflupromazine (+), tetracaine (□). Panel B : mellitin (●). Panel C :alkyllyso-phosphatidylcholine (✕). Panel D : manganese (■) palmitoyl-carnitin (▲). Cell growth was expressed by the total number of cells (adherent + free), cell adhesion by the number of adherent cells.

196

2) Protein kinase C activation and cell differentiation

The effect on protein kinase C activity of various inducers of differentiation in HL60 cells has been studied. As shown in tableIV, dimethylsulfoxide and retinoic acid, which both induce granulocyte differentiation, did not affect enzyme activity. Neither did 1∝,25-dihydroxycholecalciferol, the active metabolite of vitamin D3, which induces some markers of macrophage differentiation (30). Diglycerides have also been tested for their ability as inducers of differentiation. Although synthetic diglyceride 1-oleoyl-2-acetyl-glycerol, dilaurin and dicaprylin revealed as potent activators of the kinase, we failed to detect a significant cell adhesion to substratum. The possibility that the metabolism of these compounds is too fast to induce differentiation can not be excluded.

TABLE III

INHIBITORY ACTION OF SOME PHOSPHOLIPID-INTERACTING DRUGS.

TPA was added to the reaction mixture at 10ng/ml, EGTA at 0.5mM and CaCl$_2$ at 0.1mM. Results taken from ref. 25.

| Drugs | IC_{50} (uM) | | | |
| | Basal activity | | Stimulated activity | |
	EGTA	Ca	EGTA	Ca
Trifluoperazine	25	38	62	64
Chlorpromazine	12	71	57	90
Triflupromazine	60	91	67	80
Dibucaine	20	200	400	480
Tetracaine	170	310	320	250
Mellitin	2	5	3	6
ALP	30	80	120	250

From the preliminary data presented herein it is not possible to definitely implicate protein kinase C activation in the induction of the macrophage phenotype in HL60 cells. The precise role of protein kinase C in cell differentiation, if any, requires further investigation.

PROTEIN KINASE C AND TUMOR PROMOTION.

1) Aberrant and physiological transmembrane signalling

As emphasized above and in (31) TPA substitutes for

diacylglycerol which acts as a second messenger of inositol
phospholipid turnover linked-extracellular signals. However the
resulting protein kinase C activation was different in some
respects from that occurring in response to physiological
stimuli. It should be stressed that the half-life of
diacylglycerol was less than one minute,whereas phorbol esters
like TPA are slowly metabolized in tissues (32) indicating that
these agents may trigger a sustained activation of the pathway.
In addition, as briefly mentioned above, the diacylglycerol/IP3-
mediated pathway seems to be under the control of regulatory
mechanisms. The cyclic nucleotide-mediated negative control of
this pathway has been particularly studied in platelets (12).
Activation resulting from physiological stimuli was inhibited
whereas TPA-induced activation was independent of such control.
As illustrated in fig. 7 the phosphorylation of 40Kd protein, the
preferential substrate for protein kinase C in platelets, was not
affected by increasing intracellular levels of cyclic nucleotides
in TPA-treated cells whereas thrombin-mediated phosphorylation
was inhibited.

TABLE IV

EFFECT OF VARIOUS KNOWN OR POSTULATED INDUCERS OF DIFFERENTIATION
ON PROTEIN KINASE C ACTIVATION

Various inducers were tested on brain enzyme and the assay was
performed as already described (25). In contrast to TPA, dihydro-
xyvitamin D3 (1α,25 (OH)$_2$ D3), dimethylsulfoxide (DMSO) and retinoic acid which were added as solutions, diglycerides were mixed
with phosphatidylserine prior to addition to incubation mixture as
sonicated suspensions.

Inducers	Concentration	Type of differentiation	Protein kinase C activator
TPA	5nM	macrophage/monocyte	+
1α 25 (OH)$_2$D$_3$	0.1uM	"	-
DMSO	1.3%	granulocyte	-
Retinoic acid	10uM	"	-
OAG	10→200ug/ml	not detectable	+
Dicaprylin	50→500ug/ml	" "	+
Dilaurin	50→500ug/ml	" "	+

In summary, TPA mediates a sustained activation of protein
kinase C in platelets independent of physiological regulatory
mechanisms, some of them acting in response to external signals.
Although the cell system used to clarify the properties of TPA-
induced protein kinase C activation may not be the most relevant
in studying cell transformation, it is remarkable that these
properties may account for two characteristics of cell
transformation which are the low calcium requirement for growth
and progressively acquired refractoriness to the environment.It may
be worthwhile to point out that cells should be repeatedly
exposed to TPA over weeks or months in order to promote
transformation and this suggests that the prolonged cell
activation may ultimately lead to the irreversible expression of
the tumor phenotype.

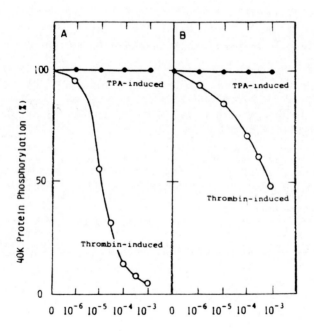

Fig. 7. Effects of cyclic nucleotides on protein kinase C activation in response to physiological stimuli of phorbol esters. ^{32}P-
prelabelled human platelets were stimulated by either thrombin or
TPA in the presence of increasing concentrations of dibutyryl
cyclic AMP (A) or 8-bromo cyclic GMP (B). Reprinted from ref. 12.

2) Phorbol ester-unrelated tumor promoters

As shown in fig. 8 the structural requirements of various phorbol esters for tumor promotion in the two-stage model of chemical carcinogenesis on mouse skin are similar to those for protein kinase activation. However mezerein , a daphnane-derived ester, which displays most of the biological properties of TPA in cultured cells, although it is a weak tumor promoter on mouse skin (8), was roughly as potent as TPA in activating protein kinase C in the reconstituted system.

In order to test mezerein for its in vivo effects the pattern of phosphoproteins was compared in ^{32}P-prelabelled mouse thymocytes treated with either TPA,1-oleoyl-2-acetyl-glycerol (OAG) or mezerein. The phosphorylation of a number of soluble proteins with molecular weights of 65Kd, 45Kd and 27Kd was enhanced by TPA and OAG. The phosphorylation of those proteins was similarly increased in the presence of mezerein at the same concentrations within a 10 min. treatment (fig. 9)

Tumor promotion occurs in various tissues in response to a number of structurally unrelated chemicals. Tumor promoters other than those of the series of phorbol esters have been tested for their potency in activating protein kinase C. These agents include teleocidin, anthralin, tetrachlorodibenzodioxin, phenobarbital, palytoxin, benzoylperoxide and 1α ,25-dihydroxyvitamin D3. Results presented in table V have shown that with the exception of teleocidin, the tested tumor promoters did not activate the enzyme at the used concentrations. Interesting to note that the ineffective compounds did not compete with the high affinity binding of phorbol esters. Studies are in progress in order to explore the possibility that these tumor promoters which have no effect on protein kinase C in vitro may indirectly activate the enzyme through, for instance, some chemical or physical modifications of its phospholipid environment. Alternatively, the possibility that activation of protein kinase C is not a general mechanism of action for tumor promoters, should be considered.

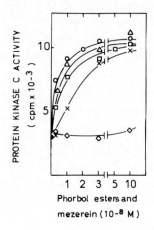

Fig. 8. Structural requirements for protein kinase C activation. Activity was assayed as described in ref. 25 in the presence of 5uM CaCl$_2$. TPA (o), mezerein (Δ), phorbol dibutyrate ($\square$), phorbol dibenzoate ($\times$), 4 α phorbol didecanoate ($\Diamond$).

TABLE V

PROTEIN KINASE ACTIVATION BY VARIOUS TUMOR PROMOTERS UNRELATED TO MACROCYCLIC DITERPENE ESTERS.

Enzyme was assayed in the presence of 5uM CaCl$_2$ as described in ref. 25. Enzyme activity was expressed in cpm transferred to histone under our experimental conditions. Benzoyl peroxide was mixed with phosphatidylserine and added to assay mixture as a soni- cated suspension in contrast with other additions of tumor promo- ters made as solution.

Added compounds	Concentration		Enzyme activity
none	-		2,600
Teleocidin	10	nM	6,320
	100	nM	9,260
Anthralin	1	mM	2,200
Tetrachlorodibenzodioxin	2	nM	2,740
	20	nM	2,490
Phenobarbital	1	mM	2,650
Palytoxin	4	ng/ml	2,580
	20	ng/ml	2,530
Benzoyl peroxide	0.6ug/assay		2,630
	6	"	2,200
	30	"	2,220

In summary, the results presented suggest that some
potent structurally dissimilar tumor promoters such as phorbol
esters and teleocidin are able to persistently activate a
transmembrane pathway triggering specific cell functions and
proliferation. Investigations are in progress to determine whether
protein kinase C triggering initiates all the phorbol ester-
mediated effects, or whether some other molecular targets are
also critical. However, the diacylglycerol/IP$_3$-mediated pathway is
also activated in response to several mitogens (see table I).
Indeed, recent evidence that this pathway may be the site of
action of several oncogene products encoded by C-sis, erb B, sarc
and ros genes, as shown in fig. 10 , emphasises its key role in
cancerogenesis (33-36).

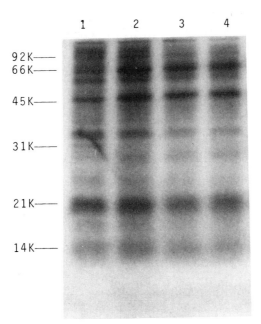

Fig. 9. Analysis of ^{32}P-labelled proteins of TPA, 1,oleoyl-2,acetyl
glycerol (OAG) or mezerein-treated mouse T lymphocytes by SDS-PAGE.
Cells were labelled at 37°C for 60 min. in MEM without phosphate
prior to the addition of either 0.2% DMSO (lane 1), 10^{-8}M TPA
(lane 2), 10^{-8}M mezerein (lane 3), or 150ug/ml OAG (lane 4). The
incubation proceeded then for 10 min. The nuclei were removed by
centrifugation and the resulting supernatant centrifugated at
43,000g for 1h. An aliquot of the supernatant fraction were taken
up in SDS sample buffer according to Laemmli, boiled for 3 min.
and loaded on 12% SDS slab gel. After electrophoresis the dried
gels were exposed for autoradiography using Kodak XAR 5 Omat film.

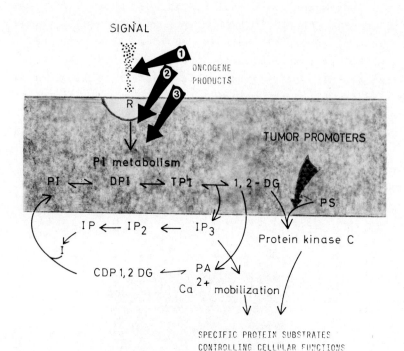

Fig 10. Postulated model for activation pathway of protein
kinase C in response to physiological signals and tumor promoters.
PI : phosphatidylinositol ; DPI : phosphatidylinositol 4-phosphate;
TPI, phosphatidylinositol 4,5-bisphosphate ; 1,2-DG : 1,2-diacyl-
glycerol ; PS : phosphatidylserine ; IP_3:inositol   1,4,5-tris-
phosphate ; IP_2 : inositol 1,4-bisphosphate ; IP : inositol phosphate ; I : inositol ; CDP 1,2-DG : CDP 1,2 diacylglycerol ; PA :
phosphatidic acid. The possible site of action of C-sis, erb B and
sarc/ros gene products have been indicated with arrows labelled
1,2 and 3 respectively.

CONCLUSION

     The purpose of this presentation was to show evidence
supporting the implication of protein kinase C, a keystone enzyme
of the diacylglycerol/IP_3-mediated signalling pathway, in phorbol
ester-induced cell differentiation and tumor promotion. The data
presented do not seem to favor the possibility that the direct
activation of the enzyme is a common mechanism of action for all

inducers of differentiation and tumor promoters. Nevertheless it would be attractive to suggest that the activation of this pathway is a requirement for tumor promotion and cell differentiation assuming that the various inducers and tumor promoters are acting at different steps of the pathway. Further investigation will be undertaken to clarify this point. It can be expected that the precise role of protein kinase C will be elucidated over the next few years. One wonders which hypothesis will account for the mechanism of action of the enzyme in tumor promotion and cell differentiation. The enzyme may act directly in the nucleus or indirectly through a cascade of phosphorylations involving tyrosine specific protein kinases to modify gene expression.

An alternative possibility would be that the enzyme acts via the modulation of tyrosine kinase activity of surface receptors since several groups reported the phosphorylation of a number of receptors for key ligands such as insulin, transferrin, and growth factors most probably catalysed by protein kinase C (37-39).

In conclusion it should be noted that phorbol esters which yield clues to cancerogenesis are likely to be valuable tools in the molecular characterization of the control of cell growth and differentiation.

ACKNOWLEDGMENTS

The authors are indebted to Dr. Jolivet for help in atomic absorption spectrophotometry, Dr. Claret for helful discussions and Drs. Moulé and Daya-Grosjean for critical reading of the manuscript. Also, the authors want to express their special gratitude to Pr. Nishizuka and his collaborators who cooperate with us to part of the presented data. The valuable assistance in preparing the manuscript of G. Pisapia, C. Marchant and J. Bannelier is greatly acknowledged. The authors thank also Drs. Poland, Thomopoulos, Fujiki and Sugimura for generously providing us with various chemicals.

The research was supported by grants from INSERM, ARC and Ministère de l'Industrie et de la Recherche.

REFERENCES

1. Berenblum I (1941) Cancer Res. 1: 44-48

2. Lasne C, Gentil A, Chouroulinkov I (1974) Nature 247:490-491

3. Kennedy AR, Cairns J, Little JB (1984) Nature 307:85-86

4. Hecker E (1971)Methods Cancer Res; 6:439-484

5. Hecker E, Lutz D, Weber J, Goettler K and Morton JF (1983) In : Liss AR 13th International Cancer Congress. Part B. Biology of Cancer, 1, New York : 219-238

6. Tran PL, Terminassian-Saraga L, Madelmont G, Castagna M (1983) Bioch. Biophys. Acta 727 : 31-38

7. Weinstein IB, Mufson RA, Lee LS, Fisher PB, Laskin J, Horowitz AD, Ivanovic V (1980) In : Pullman B Ts'o PUP and Gelboin H (eds) Carcinogenesis : Fundamental Mechanisms Environmental Effects, Amsterdam, pp 543-561

8. Blumberg PM (1980) C.R.C. Crit. Rev. Toxicol. 8 : 199-234

9. Diamond L, O'Brien TG, Baird WM (1980). Adv. Cancer Res. 32: 1-75

10. Slaga TJ, Sivak A, Boutwell RK (eds) (1978) in : Mecanisms of tumor promotion and cocarcinogenesis, 2.

11. Driedger PE, Blumberg PM (1980) Proc. Natl. Acad. Sci. 77 : 567-571

12. Nishizuka Y, Takai Y, Kishimoto A, Kikkawa U, Kaibuchi K (1984) Recent Progr. Horm. Res., 40 : 301-345

13. Schatzman RC, Raynor RL, Fritz RB, Kuo JF (1983) Biochem. J. 209 : 435

14. Kikkawa U, Takai Y, Minakuchi R, Inohara S, Nishizuka Y (1982) J. Biol. Chem. 257 : 13341-13348

15. Parker PJ, Stabel S, Waterfield MD (1984) EMBO J. 3 : 953-959

16. Putney JW, Burgers GM, Malenda SP, McKinney JS, Rubin RP (1983) Biochem. J. 212 : 483-488

17. Agranoff BW, Murthy P, Seguin EB (1983) J. Biol. Chem. 258 : 2076-2078

18. Berridge MJ (1983) J. Biochem. 212 : 849-858

19. Steb H, Irvine RF, Berridge MJ, Schultz I (1983) Nature 306: 67-69

20. Farese RV (1983). Endocrine Reviews 4 : 78-95

21. Castagna M, Takai Y, Kaibuchi K, Sano K, Kikkawa U, Nishizuka (1982) J. Chem. Biol. 257 : 7847-7851

22. Haslam RJ, Lynham JA (1977) Biochem. Biophys. Res. Commun. 77 714-722

23. Schatzman RC, Wise BC, Kuo JF (1981) Biochem. Biophys. Res. Commun. 98 : 669-676

24. Helfman DM, Barnes KC, Kinkade JM, Vogler WR, SHoji M, Kuo JF (1983) Cancer Res. 43 : 2955:2960

25. Couturier A, Bazgar S, Castagna M (1984) Biochem. Biophys. Res. Commun 121 : 443-455

26. Ashendel CL, Staller JM, Boutwell RK (1983) Biochem. Biophys. Res. Commun 111 : 340-345

27. Niedel JE, Kuhn LJ, Van der Bark GR (1983) Proc. Natl. Acad. Sci. 80 : 36-40

28. Sharkey NA, Leach KL, Blumberg PM (1984). Proc. Natl. Acad. Sci. 81 : 607-610

29. Kikkawa U, Takai Y, Tanaka Y, Miyake R, Nishizuka Y (1983) J. Biol. Chem 258 : 11442-11445

30. Castagna M (1983) Biomed. Pharmacotherapy 37 : 380-386

31. Murao SI, Gemmel MA, Callaham MF, Aderson NL, Huberman E (1983) Cancer Res. 43 : 4989-4996

32. Berry DI, Lieber MR, Fisher SM, Slaga TJ (1977) Cancer Lett. 3 : 125-132

33. Weiss R (1983) Nature 304 : 12

34. Macara IG, Marinetti GV, Balouzzi PC (1984).Proc. Natl. Acad. Sci 81 : 2728-2732

35. Sugimoto Y, Whitman M, Cantley LC, Erikson RL (1984) Proc. Natl. Acad. Sci. 81, 2117-2121

36. Downward J, Yardeny, Mayes E, Scrace G, Totty N, Stockwell P, Ullrich J, Schlessinger J, Waterfield MD (1984) Nature 307 : 521-527

37. May WS, Jacobs S, CuatreCajas P (1984) Proc. Natl. Acad. Sci. 81 : 2016-2020

38. Rovis M, Thomopoulos P, Postel-Vinay MC, Testau, Guyda HJ, Posner BI (1984) Molec. Physiol. 5 : 123-130

39. Cochet C, Gill GN, Meisenhelder J, Cooper JA, Huntert (1984) J. Biol. Chem. 259 : 2553-2558

RESUME

La transduction des signaux extracellulaires qui accélèrent la dégradation des phosphatidyl-inositides est assurée via la mobilisation du Ca^{2+} et l'activation de la protéine kinase C. Les 1,2 diglycérides insaturés sont les activateurs physiologiques de l'enzyme et les esters de phorbol à activité promotrice de tumeurs, qui se substituent à ces lipides neutres apparaissent comme leur contre-partie pathologique. A concentrations nanomolaires de Ca^{2+}ionisé, le 1,2 dioleine et l'ester de phorbol TPA activent totalement l'enzyme. De plus, le 1,2 dioleine et le TPA augmentent, comme le fait le Ca^{2+}, le IC_{50} des inhibiteurs de l'enzyme qui s'incorporent dans les phospholipides. Ces drogues inhibent aussi la différenciation induite par le TPA de la lignée promyelocytique HL60 dans des conditions qui n'affectent pas la croissance cellulaire. Certains inducteurs de différenciation tels que le dihydroxyvitamine D_3, le DMSO et l'acide rétinoique, ainsi que certains promoteurs de tumeurs tels que l'anthraline, la palytoxine, le benzoyl peroxyde n'activent pas directement l'enzyme. A l'opposé des ligands physiologiques, les esters de phorbol produident dans les plaquettes une activation soutenue de la protéine kinase C, indépendante de la régulation exercée par les nucléotides cycliques sur cette enzyme. Il est suggéré que l'activation persistente de la voie de transduction diglycéride/IP_3 puisse conduire à l'expression irréversible du phénotype tumoral.

CATIONS: TRANSPORT AND CHANNELS

CATIONS: TRANSPORT ET CANAUX

Hormones and Cell Regulation, Volume 9
INSERM European Symposium
J.E. Dumont, B. Hamprecht and J. Nunez editors
© 1985 Elsevier Science Publishers B.V. (Biomedical Division)

MECHANISM OF ACTION OF ALDOSTERONE : A PLEIOTROPIC RESPONSE.

BERNARD C. ROSSIER, MARIE PASCALE PACCOLAT, FRANCOIS VERREY, JEAN-PIERRE
KRAEHENBÜHL[*] AND KATHI GEERING.
 Institut de Pharmacologie de l'Université, rue du Bugnon 21, CH-1011 Lausanne.
*Institut de Biochimie de l'Université de Lausanne, chemin des Boveresses,
 CH-1066 Epalinges.

INTRODUCTION

   Aldosterone increases both sodium reabsorption and potassium and hydrogen secretion in a variety of high resistance (so called "tight") epithelia such as
the distal part of the colon or of the nephron (1,2). The urinary bladder of the
toad (3), and more recently epithelial cell lines in culture (4) have been useful
experimental models for studying the molecular mechanism of aldosterone action.
Until now, the mechanisms which underlie effects of aldosterone on the potassium
and the hydrogen transport are poorly understood. Due to the existence of suitable experimental models (specially the amphibian model), more is known about
the sodium transport response. For this reason, we will review data from the
literature and from our own laboratory with special emphasis on the amphibian
model. Since Edelman proposed his classic model (17) we hope to give new insights
into this question.

ALDOSTERONE AND Na$^+$ TRANSPORT : A MODEL.

   In the urinary bladder of the toad Bufo marinus (Fig. 1) the physiologic response to aldosterone (80 nM) can be divided into three distinct phases : (i) a
latent period (about 45 minutes), (ii) an early response (up to 2.5h) during
which sodium transport increases rapidly and total (transepithelial) electrical
resistance falls concomitantly, and (iii) a late response (up to 24h) during
which sodium transport further increases with no further change in resistance
(5,6). Like other steroid hormones, aldosterone exerts its action by interacting
with "cytoplasmic" (soluble) receptors. These subsequently bind to chromatin
acceptor sites, thereby controlling the expression of various genes that encode
proteins, namely the aldosterone-induced proteins (AIPs) or aldosterone-repressed
proteins (ARPs). Like glucocorticoids (7,57) aldosterone appears to control the
expression of a number of proteins localized in various cell organelles (8-13, 15,
18) thus constituting a typical pleiotropic response (7,57). Steroid hormones and
other hormones such as thyroid hormone) can induce or repress the expression of

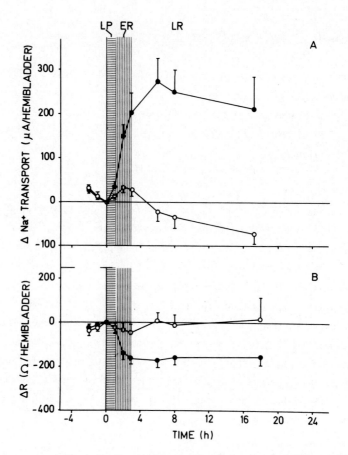

Fig. 1. Time course of the effect of aldosterone on transepithelial sodium transport (upper panel) and electrical resistance (lower panel) in toad urinary bla-
ders incubated in vitro at 25°C (as described in ref. 5). At t.$_o$, aldosterone
(80 nM) or the diluent were added (mucosal and serosal sides).
Panel A. Data are expressed as the mean ΔSCC ± SEM = SCC$_t$ - SCC$_o$) (microamperes
per hemibladder). Aldo (●). Control (O).
Panel B. Data are expressed as the mean ΔR ± SEM = (R$_t$ - R$_o$) (ohm x hemibladder).
Aldo (●). Control (O).
LP : latent period; ER : early response; LR : late response. See text for explanation.

specific proteins, thus defining an _hormonal domain_. Induction and repression are
difined in our terminology as an increase or a decrease in the rate of synthesis
of individual proteins (57). Pleiotropic responses at the translational level are
best analysed by two-dimensional (2D) gel analysis as developped by O'Farrel (14)
and recently applied to the amphibian model (15, 16). An example of such analysis
is shown in _Figure 2_. This figure shows a portion of a 2D gel which demonstrates
the existence of AIPs (circled) and ARPs (squared). The magnitude of the induc-

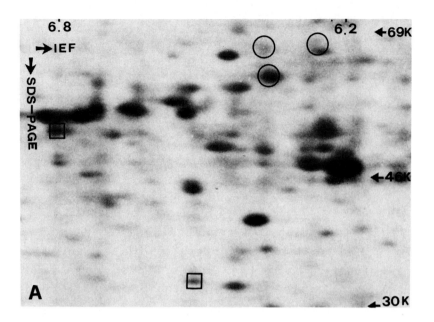

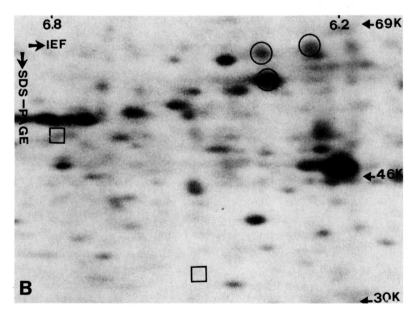

Fig. 2. Two dimensional gel autoradiogramms from toad urinary bladder cells pre-treated with or without aldosterone (80 nM) for 18 h., using equilibrium focusing for first dimension (IEF) and SDS-PAGE for second dimension.

Paired pools of epithelial cells were labeled with ^{35}S-methionine (250 μCi/mℓ) for 60 min. and lysed at 25°C according to O'Farrell (15), and samples (250'000 cpm) were run accordingly. Panel A : control. Panel B : aldosterone. Three AIPs are circled and two ARPs are squared.

tion can be very large starting from almost undetectable level in the unstimulated state. Alternatively, the induction can be small starting from conspicuous
levels of synthesis in the control situation. Likewise, the repression can be
total or partial.

The relationship between induced (or repressed) proteins and the physiologic
response is by no means easy to establish. It is, however, tempting to speculate
that various AIPs and ARPs act at different intracellular sites mediating transepithelial sodium transport. As illustrated in the model of Figure 3, one could
divide AIPs (and ARPs) into two categories : those involved in a constitutive
pathway for sodium transport and those involved in a regulatory pathway. The
constitutive pathway is directly implicated in the transport of sodium across
the epithelial cell at three critical points; namely, the apical membrane (locus
of the amiloride-sensitive sodium channel), the basolateral membrane (locus of
the ouabain-sensitive sodium pump) and the tight junctions which control the
overall leakiness of the tissue. The regulatory pathway involves a set of molecular events (e.g. phosphorylation - dephosphorylation, acylation, transmethylation) that modulate the constitutive pathway. Three sites of regulation could be
envisaged. First, the insertion (and/or the expression) of the sodium channel at
the apical membrane (site 2). Second, the insertion (and/or the expression) of
the sodium pump at the basolateral membrane (site 4) could be regulated by AIPs
or ARPs. Third, sodium transport could also be regulated by controlling either
the energy supply (ATP) to the sodium pump or the ATP/ADP ratio which in turn
could modify the apical sodium permeability. Mitochondrial enzymes such as citrate synthase which are involved in ATP production belong to the category of
regulatory proteins. Finally, one should consider the possibility that the time
course of induction (and/or repression) may well vary for each protein under
aldosterone control. Thus, the early or the late mineralocorticoid response could
be sequentially controlled by different AIPs or ARPs.

In the next sections of this chapter, we would like to review data which are
consistent with such a pleiotropic response. We will emphasize the effects of
aldosterone on the basolateral membrane since we have studied this aspect of the
question in more detail.

EFFECTS OF ALDOSTERONE AT THE APICAL MEMBRANE.

A number of recent studies using different methodological approaches clearly
indicate that aldosterone is able to modify the sodium permeability at the apical
border (19-23, 41, 56). This effect could be one of the major actions of the

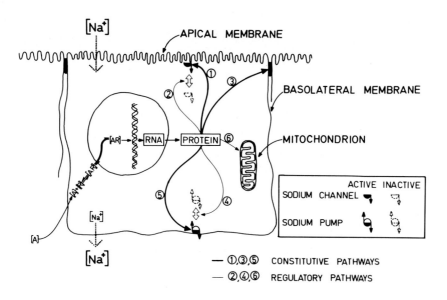

Fig. 3 *Model of the mechanisms of action of aldosterone. An epithelial cell is schematically represented. Aldosterone (A) crosses the plasma membrane and binds to its "cytosolic" (soluble) receptor (R). The complex (AR) is "activated" and binds to chromatin acceptor sites in the nucleus. Upon this interaction transcription is induced or repressed. Induced or repressed proteins (AIPs or ARPs) mediate an increased transepithelial ion transport at different sites (sites 1 to 6). See explanations in the text.*

hormone during the early response. So far it is not known how aldosterone can generate such an effect. Studies using an amiloride-sensitive trypsinization (20) or chemical modifications (41) of the apical sodium channel would argue against *de novo* synthesis of a sodium channel. The data presented are more in favour of the induction of a regulatory protein controlling the expression of the sodium channel at the apical border. Along these lines, a recent paper suggests that the action of aldosterone is mediated by a transmethylation reaction (24). Further caracterization of AIPs or ARPs controlling the apical border is therefore required before a final conclusion can be reached. Within this context, we have recently observed (44) that aldosterone is rapidly able to reveal (within 90 min.) antigenic determinants sitting in the apical membrane of the toad urinary bladder or of TBM cells. These epitopes are detected by an antibody raised against the α subunit of Na,K-ATPase. The identification and the biochemical characterization of these antigenic determinants could probably lead to a better understanding, in molecular terms, of the early response. This question is further discussed in the context of the sodium pump-channel hypothesis (see below).

EFFECTS OF ALDOSTERONE AT THE BASOLATERAL MEMBRANE.

Much of the evidence that aldosterone affects the basolateral membrane deals with the action of the hormone on Na,K-ATPase, the biochemical equivalent of the sodium pump (25). An increasing number of papers have been published during the last 16 years (26-33) showing that aldosterone has a rapid and specific effect in restoring to its normal level the enzyme activity of the distal part of the nephron (namely cortical collecting tubules = CCT) of the adrenalectomized mammal (rodent or rabbit). In the amphibian model, however, the effect of aldosterone on enzyme activity were either not detectable (34) or minimal (35). This lack of effect in the amphibian model (toad bladder) might be due to three distinct factors : first the rather short period of observation (3-5 hours) which might have prevented the detection of long term effects; second, the impossibility to adrenalectomize the animal and third, the difficulty to measure the enzyme activity in cell homogenates of this species (high level of ouabain resistant ATPase) or the omission to take into account the enzyme recovery when the enzyme activity is measured in a cell fraction specifically enriched in plasma membranes. Despite these methodological problems one could reconcile these divergent results in proposing that aldosterone has distinct effects on Na,K-ATPase depending on whether the tissue has been totally deprived of hormone (adrenalectomy), thus undergoing partial dedifferentiation, or whether the tissue is continuously exposed to aldosterone (even at low concentration), allowing the tissue to maintain a better degree of differentiation. In order to study the biogenesis and the cell surface expression of Na,K-ATPase, we purified the holoenzyme in order to prepare polyclonal antibodies specific for each subunit of the enzyme that is the α subunit (Mr 96 Kd, also termed "catalytic" subunit) and the β subunit (Mr 60 Kd, a glycoprotein) (25). Both antisera reacted with antigenic determinants (epitopes) exposed at the cell surface (ectodomain) and with cytoplasmic and/or intramembranous epitopes (endodomain). These antisera can thus be used to selectively label the surface exposed sites of each subunit of the living cell and also to immunoprecipitate radioiodinated or biosynthetically labeled enzymes in the intact cell or in cell free reconstitution systems. Such probes should allow us to test the different sites at which aldosterone could act : (i) at the transcriptional level by increasing the *de novo* synthesis of sodium pumps (see site 5, Fig. 2)

(ii) at the post translational level by inducing (and/or repressing) regulatory proteins that control the expression of the sodium pump at the cell surface (see site 4, Fig. 2) and (iii) by a combination of these two effects in a timely and coordinated sequence of events.

Effects of aldosterone on Na,K-ATPase synthesis.

We have measured the effect of aldosterone on the relative rate of synthesis of each subunit of Na,K-ATPase in the toad urinary bladder (36). The results are summarized in Table I. At maximal concentration of aldosterone (80 nM), the synthesis of the α subunit increased 2.85 fold over the control. The β subunit (or its coreglycosylated 42 Kd precursor) was always coinduced with the α subunit but to a slightly smaller extent (2.40 fold). In a number of experimental situations (stimulation by oxytocin (37), inhibition by Actinomycin D (25) or by sodium butyrate (5)), the synthesis of α and β subunits are stoechiometrically induced or deinduced, suggesting that the transcription and the translation of both α and β mRNAs are tightly coupled. As shown in Table 1, however, there is a trend ($p < 0.1$, n = 16) toward a slightly larger effect on the α than on the β subunit. This trend is reproduced in a number of experimental conditions (+ 18%, $p < 0.025$, n = 28 paired observations). This interesting finding is discussed below in the context of the possible existence of two intracellular pools of 96 Kd polypeptides.

We checked that the Na,K-ATPase fulfilled the criteria for an AIP according to the classic model of steroid induction (18). (i) The *de novo* synthesis of Na,K-ATPase was selectively blocked by Actinomycin D (25) and by sodium butyrate (5) (ii) the induction was also inhibited by spironolactone (36) a competitive antagonist for the mineralocorticoid receptor (iii) finally, the dose-response showed that the effect of aldosterone occured at low hormone concentrations (0.4 to 7 nM) corresponding to the saturation of Type I binding sites (38). Finally amiloride did not block the induction of Na,K-ATPase synthesis suggesting that the effect of aldosterone is a primary steroid effect that is independent of any transepithelial sodium transport (36). Having established that Na,K-ATPase was indeed an AIP, does not imply that the *de novo* synthesis of pumps is required for the establishment of the early or late sodium transport response. It is clearly important to determine the time course of induction and the time course of expression at the cell surface. To do so, we have used three experimental models (Table II) which represent different states of epithelial differentiation and of aldosterone responsiveness. Our reference model is the toad urinary bladder which can be considered as expressing the complete physiological response.

TABLE I

EFFECT OF ALDOSTERONE ON THE RELATIVE RATE OF SYNTHESIS OF α AND β SUBUNIT OF Na,K-ATPase.

Fractional change (test/control) + SE

η = 16 pairs	α subunit (96 Kd)	β subunit (42 Kd)*	Difference $\alpha - \beta$	Paired t test
Control	1.00	1.00		
Aldosterone	2.85 ± 0.35	2.40 ± 0.24	0.45 ± 0.23	0.05 < p < 0.1
Pared t test Aldo vs Control	p < 0.001	p < 0.001		

Effects of aldosterone on the rate of biosynthesis of the α subunit and the β subunit of Na,K-ATPase. Homologous pools of urinary bladder tissue were incubated for 18h in the presence or absence of aldosterone. Tissue labeling (35S-methionine pulse of 30 min.), immunoprecipitation of α and β subunits and their resolution on SDS-PAGE were as described previously (36).

*
i.e. the coreglycosylated precursor.

In this model both aldosterone-dependent sodium and hydrogen transport as well as the induction of Na,K-ATPase can be studied. It is, however, a complex model system (5 cell types) responding to a large variety of hormones (39) and finally, it is not suitable for the study of the expression of antigenic determinants at the cell surface, which can be more conveniently studied in cell culture. In this respect, the TBM cell line is a continuous amphibian cell line (2 cell types), derived from the toad bladder, which expresses only sodium transport (4) and responsiveness to aldosterone. It thus mimics the response observed in the tissue of origin (see Figure 1). The A6 cell line (a single cell type) is derived from an unknown site of Xenopus laevis kidney (40). It displays a much lower sodium transport baseline (Table II) or in some culture conditions (42) no detectable sodium transport at all in spite of a high resistance (5'000 $\Omega \cdot cm^2$). A6 cells are, however, responsive to aldosterone although the early response (fall in resistance) is either greatly diminished or abolished (42). TBM and A6 cell lines are ideal model systems to study the cell surface expression of Na,K-ATPase with our immunological probes. In addition, the results obtained can be correlated

TABLE II

EXPERIMENTAL MODELS FOR STUDYING ALDOSTERONE-DEPENDENT ION TRANSPORT ACROSS EPITHELIAL CELLS.

Experimental model	Sodium transport * Transepithelial				Hydrogen transport ** Transepithelial		
	baseline		stimulated ***		baseline		stimulated
	PD	SCC	early	late	RPD	RSCC	
	mV	$\mu A/cm^2$			-mV	$-\mu A/cm^2$	
Toad urinary bladder (52)	50	10-20	+++	+++	0-5	0-5	++
TBM cell line (4)	10-20	5-8	0/+	+++	0	0	0
A6 cell line (40,42)	0-10	0-1.5	0/+ ?	++	0	0	0

* *Sodium transport was measured by the short circuit current method (SCC) in absence (baseline) or in presence of aldosterone (stimulated)*

** *Hydrogen transport was measured by the reversed short circuit current method (RSCC) that is in presence of amiloride at a supra maximal concentration which abolishes sodium transport completely. None of these 3 experimental models displays transepithelial potassium transport.*

*** *Early vs late response : see Fig.1 and Text for explanations.*

with the sodium transport responses.

In the toad bladder, the time-course of induction showed that the increased rate of synthesis of Na,K-ATPase was a late event (36) that did not occur until 6h after aldosterone addition. Thus the induction cannot be responsible (36, 25) for the early mineralocorticoid response. The induction could, however, play a role in maintaining a high capacity to transport sodium. The effects of sodium butyrate which blocks simultaneously the induction of Na,K-ATPase and the late response on sodium transport is consistent with this idea (5). Thus, in a highly differentiated tissue from an animal which has not been adrenalectomized, Na,K-ATPase would qualify as a late AIP. This might not be true in all cases. For instance, both in the A6 (42), and TBM cell lines (unpublished observations) preliminary experiments show that the time course of induction could be much faster. In A6 it may be important to synthesize new pumps before sodium transport can be established.

Effects of aldosterone on cell surface expression on Na,K-ATPase.

We have investigated the cell surface expression of each subunit of Na,K-ATPase in TBM cells. Selective labeling of the apical membrane is possible (43) as the result of the establishment of a tight monolayer, and total labeling (apical + basolateral) can be performed on detached monolayers. Intact cells can be labeled at 4^0C (to prevent internalization) by a glucose-oxydase/lactoperoxydase procedure. The overall pattern of membrane protein labeling was assessed by 2D gel analyses. Quite distinct patterns emerged when apically labelled cell extracts were compared to totally labelled cell extracts, suggesting that the latter corresponded mainly to surface labeling of basolateral membranes. The cell surface expression of each subunit of Na,K-ATPase was studied in confluent monolayers in the presence or absence of aldosterone. The β subunit (Mr : 60 Kd) was recovered from cells that were radioiodinated on their entire surface ("total" labeling) but not from cells that were labeled only from their apical surface. This indicates that the β subunit is probably restricted to the basolateral surface of the cell. The α subunit (96 Kd polypeptide) was recovered from the basolateral cell surface. A very large proportion of the specifically immunoprecipitated radioactivity was not resolved by the SDS-PAGE system, suggesting the existence of aggregates or high Mr entities. These high Mr species, but not the 96 Kd polypeptide, were also immunoprecipitated with anti α subunit antibodies from cells exclusively labeled at their apical membrane. Aldosterone progressively (between 6 and 24 h) increased the surface expression of the β subunit (3 fold at 24h) as well as that of the basolateral 96 Kd polypeptide and the apical high Mr species. These results indicate that the α subunit, like the β subunit, is restricted to the basolateral membrane. This is in agreement with what is known about the asymetrical distribution of the sodium pump in epithelial cells. In addition, aldosterone-inducible epitopes, recognized by antibodies directed against the endodomain of the 96 Kd polypeptide were also associated with the apical membrane of these cells.

In a second type of experiment (44), epitopes recognized by the anti α subunit antibodies were detected immunocytochemically on 0.5 μm frozen <u>sections</u> both on the apical and the basolateral surface, whereas the β subunit antibodies reacted with epitopes restricted to the basolateral membrane. Upon addition of aldosterone, the apical labeling of some cells increased rapidly (within 90 min) with the anti α antibody; from 3 to 6h, the increased apical labeling extended to the entire surface of the monolayer. At 6h after aldosterone addition, intracellular organelles were labeled and, at 24h, the basolateral labeling increased drastically for both α and β epitopes. These data suggest that there are two pools of molecules recognized by antibodies raised against a 96 Kd polypeptide, the α subunit of Na,K-ATPase. The first pool is associated with the β subunit and is restricted to the basolateral membrane where it should play the role of a sodium pump. The second pool appears to be associated with the apical surface where it may play a role in forming a sodium channel. Along these lines, we have obtained evidence for the existence of two cellular pools of molecules recognized by anti α subunit antibodies in A6 cells. One large pool was recovered in a 100'000 x g microsomal pellet in association with the β subunit. One small pool, which did not sediment at 100'000 x g, contained a 98 Kd protein (which comigrated with the mature α subunit of Xenopus laevis) was apparently not associated with the mature β subunit. After a 7 minute pulse, the SDS-PAGE patterns of each pool following *in vitro* proteolysis where quite distinct suggesting either that we were dealing with 2 distinct 98 Kd polypeptide or that the same 98 Kd polypeptide takes on a different configuration in the membrane in the presence or absence of the β subunit. At this point one should remember that we observed a slight imbalance between α and β subunit aldosterone-dependent synthesis (see Table I). This observation could also be consistent with the concept of two distinct pools for the 96 Kd polypeptides in the toad bladder system. We interpret our data in the context of the sodium pump-channel hypothesis that is discussed below.

THE SODIUM PUMP-CHANNEL HYPOTHESIS.

This working model (25) states that the genetic information needed to code for a sodium pump might not be dissimilar from that required for a sodium channel. In other words, the information needed to synthesize a peptide with the unique property of being selective for sodium could be used in differents cells, organelles or different plasma membrane domains of the same cell, for instance, an epithelial cell. In our experimental models, the amiloride-sensitive sodium channel (Figure 2) and the α subunit of the sodium pump at the basolateral membrane would share partial structural homology that would be recognized by antibodies directed against the α subunit of the Na,K-ATPase. At least two possibilities could be envisaged. The sodium channel and the α subunit of the sodium pump are encoded by different genes which belong to the same gene family. Alternatively, the apical sodium channel and the basolateral α subunit could be encoded by a single gene that is processed differently at the post-transcriptional level (different splicing) or at the post-translational level (different covalent modifications of the gene product). Of course, the sequencing of the α subunit in the various cellular pools and the cloning of the Na,K-ATPase gene should ultimately allow us to distinguish between these possibilities.

EFFECTS OF ALDOSTERONE ON THE TIGHT JUNCTION APPARATUS.

There are no data available, to our knowledge, on this potential site of action. One logical prediction would be that aldosterone increases the electrical resistance or decreases the overall leakiness of the junctional apparatus. Together with the specific effect of aldosterone on the trancellular transport of Na^+, it would allow the creation of higher hydro-osmotic gradients by diminishing the backleak of Na^+ across the tight junctions. Cell lines might be of special interest in studying this aspect of the question (42).

EFFECTS OF ALDOSTERONE ON MITOCHONDRIAL PROTEINS INVOLVED IN ATP SYNTHESIS.

A number of observations in the rat kidney (13, 45, 46) and in the toad bladder (47) indicate that citrate synthase fulfills the criteria for an early AIP, i.e. specifically induced during the first three hours of aldosterone action. Increased ATP (48) production could be important if it was a limiting factor for the Na,K-ATPase activity and/or if the permeability of the apical membrane would be regulated by ATP/ADP ratio as postulated (1, 18, 55). However, the hypothesis that citrate synthase is an AIP playing an important role in determining the sodium transport response to aldosterone has recently been challenged by Handler and his colleagues (49, 50). They were unable to demonstrate any significant effect of aldosterone on citrate synthase activity in TB6, TBM and A6 cell lines which normally respond to aldosterone by a marked increase of sodium transport. In the same study, however, there was a small but significant effect of aldosterone on citrate synthase activity in the toad urinary bladder, as shown previously by other authors (47). One interpretation of their results could be that induction of citrate synthase was more related to the aldosterone-dependent hydrogen transport than to aldosterone-dependent sodium transport. We have recently reexamined this question in our three experimental models (Table II) and have measured citrate synthase activity (enzyme assay) and citrate synthase synthesis (biosynthesis assay) in tissues which had been previously checked for sodium transport responses (toad bladder, TBM, A6) and for hydrogen transport (toad bladder). After 3 hours of exposure to aldosterone in the toad bladder (Table III), one could observe more than a 2 fold increase in sodium transport ($p < 0.001$, $n = 10$ pairs); in addition, hydrogen transport, which was not observed in the control situation, was readily induced in the presence of aldosterone ($p < 0.01$, $n = 10$ pairs). Even in the presence of this marked physiological responses, citrate synthase activity was not significantly increased (+ 7%, $p > 0.3$ $n = 10$). In the TBM cell line measurements of the immunoreactive pool of citrate synthase (after 18 hours of aldosterone treatment) showed no evidence for a significant change in the chemical pool of this enzyme during the period of incubation. These results suggest that the effect of aldosterone on citrate synthase activity or synthesis is small. Induction of citrate synthase synthesis might not be a prerequisite for the sodium or the hydrogen transport responses. It could represent a "permissive" type of response. Which would be consistent with the data obtained in the kidney and the heart of adrenalectomized mammals (46, 51).

CONCLUSIONS.

Aldosterone induces and represses a large number of proteins (AIPs and ARPs) which determine its specific hormonal domain. This typical pleiotropic response is responsible for at least three distinct physiological responses : increased sodium reabsorption and increased potassium and hydrogen secretion. The understanding in molecular terms of such a complex physiological response is a difficult task which appears beyond the reach of current methodological approaches. By looking at specific and potential sites of regulation (the sodium channel in the apical membrane, the sodium pump in the basolateral membrane) it may be possible to understand the regulation of transepithelial sodium transport in better but certainly not in definitive terms.

TABLE III

EFFECTS OF ALDOSTERONE ON SODIUM TRANSPORT, HYDROGEN TRANSPORT AND CITRATE SYNTHASE ACTIVITY IN THE TOAD URINARY BLADDER

$n = 10$ pairs of hemibladders	Na^+ transport * SCC	H^+ transport ** SCC	Citrate synthase activity ***
	μA/hemibladders ± SEM	μA/hemibladders ± SEM	units/mg protein ± SEM
Control	+ 217 ± 36	+ 35 ± 10	0.226 ± 0.007
Aldosterone	+ 537 ± 27	− 42 ± 16	0.242 ± 0.016
Δ A − C	+ 319 ± 31	− 77 ± 23	0.015 ± 0.013
paired t test	$p < 0.001$	$p < 0.02$	$p > 0.3$

Paired hemibladders were incubated for 3h with or without aldosterone (80 nM) as described previously (5,38)

* Na^+ transport was measured by the short-circuit current method as described (5,38)

** H^+ transport was measured in presence of amiloride (50 μM) as described (53)

*** Citrate synthase activity was measured in cell homogenates according to a published procedure (54).

AKNOWLEDGEMENTS

We gratefully acknowledge Ms M. Moget and Ms N. Benmazari for their unvaluable and excellent assistance in preparing this manuscript.

We thank J. Atkinson and R. Solari for their comments and suggestions.

The work presented in this paper was supported by Swiss National Science Foundation Grants no 3.646-0.80, 3.419-0.83 (BCR) and 3.413-0.83 (JPK).

REFERENCES

1. Marver D (1980) Vitam Horm 38:55-117

2. Fanestil DD, Park CS (1981) Annu Rev Physiol 43:637-649

3. Crabbé J (1977) In: Pasqualini JR (ed) Receptors and Mechanism of Action of Steroid Hormones, part 2. Dekker, New York, pp 513-568

4. Handler JS, Steele RE, Sahib MK, Wade JB, Preston AS, Lawson NL, Johnson JP (1979) Proc Natl Acad Sci USA 76:4151-4155

5. Truscello A, Geering K, Gäggeler HP, Rossier BC (1983) J Biol Chem 258: 3388-3395

6. Spooner PM, Edelman IS (1975) Biochim Biophys Acta 406:304-314

7. Ivarie RD, Morris JA, Eberhardt NL (1980) In: Greep RO (ed) Recent Progress in Hormone Research, vol. 36. Academic Press, New York, pp. 195-239

8. Benjamin WB, Singer I (1974) Science 186:269-272

9. Scott WN, Sapirstein VS (1975) Proc Natl Acad Sci USA 72:4056-4060

10. Scott WN, Reich IM, Brown JA Jr, Yang CPH (1978) J Membr Biol 40(special issue):213-220

11. Scott WN, Reich IM, Goodman DBP (1979) J Biol Chem 254:4957-4959

12. Law PY, Edelman IS (1978) J Membr Biol 41:15-40

13. Law PY, Edelman IS (1978) J Membr Biol 41:41-64

14. O'Farrell PH (1975) J Biol Chem 250:4007-4021

15. Geheb M, Huber G, Hercker E, Cox M (1981) J Biol Chem 256:11716-11723

16. Truscello A, Geering K, Gaeggeler HP, Rossier BC (1983) Experientia 39:675

17. Edelman IS, Bogoroch R, Poeter GA (1963) Proc Natl Acad Sci USA 50:1169-1177

18. Edelman IS (1978) In: Hoffman JF (ed) Membrane Transport Processes, Vol. 1. Raven Press, New York, pp 125-140

19. Cuthbert AW, Shum WK (1975) Proc R Soc London 189:543-575

20. Garty H, Edelman IS (1983) J Gen Physiol 81:785-803

21. Palmer LG, Edelman IS (1981) Ann NY Acad Sci 372:1-14

22. Palmer LG, Li JHY, Lindemann B, Edelman IS (1982) J Membr Biol 64:91-102

23. Sariban-Sohraby S, Burg MB, Turner RJ (1983) Am J Physiol 245:C167-C171

24. Sariban-Sohraby S, Burg M, Wiesman WP, Chiang PK, Johnson JP (1984) Science 225:745-746

25. Rossier BC (1984) Current Topics Membr Transp 20:125-145

26. Chignell CF, Titus E (1966) J Biol Chem 241:5083-5089

27. Jørgensen PL (1972) J Steroid Biochem 3:181-191

28. Charney AN, Silva P, Besarab A, Epstein FH (1974) Am J Physiol 227:345-350

29. Schmidt U, Schmid J, Schmid H, Dubach UC (1975) J Clin Invest 55:655-660

30. Garg LC, Knepper MA, Burg MB (1981) Am J Physiol 240:F536-F544

31. Horster M, Schmid H, Schmidt U (1980) Pfluegers Arch 384:203-206

32. El Mernissi G, Doucet A (1982) INSERM Symp 21:269-276

33. Petty KJ, Kokko JP, Marver D (1981) J Clin Invest 68:1514-1521

34. Hill JH, Cortas N, Walser M (1973) J Clin Invest 52:185-189

35. Park CS, Edelman IS (1984) Am J Physiol 246:F509-F516

36. Geering K, Girardet M, Bron C, Kraehenbühl JP, Rossier BC (1982) J Biol Chem 257:10338-10343

37. Rossier BC (1984) In: Proceedings of the 7th International Congress of Endocrinology, July 1-7, 1984, Quebec, in the press

38. Geering K, Claire M, Gäggeler HP, Rossier BC (1984) Am J Physiol: Cell Physiol, in the press

39. Rossier BC, Geering K, Kraehenbuhl JP (1979) In: Hormonal Control of Epithelial Transport, Les Colloques de l'INSERM Vol. 85. INSERM, Paris, pp 185-194

40. Perkins FM, Handler JS (1981) Am J Physiol 241:C154-C159

41. Kipnowski J, Park CS, Fanestil DD (1983) Am J Physiol 245:F726-F734

42. Paccolat MP, Geering K, Gaeggeler HP, Rossier BC (1984) Kidney Int 25:987

43. Girardet M, Truscello A, Geering K, Rossier BC (1983) Experientia 39:664

44. Kraehenbuhl JP, Bonnard C, Geering K, Girardet M, Rossier BC (1983) J Cell Biol 97:310a

45. Kinne R, Kirsten E (1968) Pfluegers Arch 300:244-254

46. Marver D, Schwartz MJ (1980) Proc Natl Acad Sci USA 77:3672-3676

47. Kirsten E, Kirsten R, Leaf A, Sharp GWG (1968) Pfluegers Arch 300:213-225

48. Cortas N, Abras E, Arnaout M, Mooradian A, Muakasah S (1984) J Clin Invest 73:46-52

49. Handler JS, Preston AS, Perkins FM, Matsumura M, Johnson JP, Watlington CO (1981) Ann NY Acad Sci 372:442-454

50. Johnson JP, Green SW (1981) Biochim Biophys Acta 647:293-296

51. Marver D (1984) Am J Physiol 246:E452-E457

52. Crabbé J (1963) The Sodium-Retaining Action of Aldosterone (thesis). Ed. Arscia, Presses académiques européennes, Bruxelles

53. Rossier BC, Gäggeler HP, Brunner DB, Keller I, Rossier M (1979) Am J Physiol 236:C125-C131

54. Srere PA, Brazil H, Gonen L (1963) Acta Chem Scand 17:S129-S134

55. Garty H, Edelman IS, Lindemann B (1983) J Membr Biol 74:15-24
56. Lewis SA, Wills NK (1981) Ann NY Acad Sci 372:56-63
57. Ivarie RD, Baxter JD, Morris JA (1981) J Biol Chem 256:4520-4528

RESUME

L'aldostérone augmente la réabsorption de sodium et la sécrétion de potassium et d'hydrogène dans les épithelia à haute résistance électrique. Les modèles amphibiens (vessie urinaire du crapaud, cellules en cultures TBM ou A6) permettent d'étudier le mécanisme d'action de l'aldostérone, en particulier son effet natriférique. Ce dernier se caractérise par une période de latence d'environ 45 minutes, suivie d'une réponse précoce qui dure environ $2\frac{1}{2}$ heures et enfin une réponse tardive qui peut s'étendre sur 24 heures. La réponse natriférique est totalement inhibée par les inhibiteurs de la transcription (Actinomycine D). Elle dépend donc strictement de l'induction ou de la répression de nombreuses protéines. La réponse minéralocorticoïde est donc typiquement pléiotropique. Le rôle physiologique des protéines induites ou réprimées est discuté dans le cadre d'un modèle de cellule épithéliale où 3 sites importants peuvent intervenir dans le transport de sodium à travers l'épithelium : 1) le canal au sodium, sensible à l'amiloride, situé dans la membrane apicale, 2) la pompe à sodium (Na,K-ATPase), sensible à l'ouabaïne, située dans la membrane basolatérale et 3) l'appareil jonctionnel, qui joue un rôle important en prévenant la rediffusion du sodium vers le compartiment urinaire. Les protéines contrôlées par l'aldostérone pourraient faire partie intégrante de ces trois sites ou bien en contrôler leur expression ou leur fonctionnement à la surface de la cellule. L'aldostérone contrôle l'expression et/ou le fonctionnement du canal au sodium apical. Cet effet semble surtout impliqué dans la réponse précoce. L'aldostérone induit la synthèse et augmente l'expression de la Na,K-ATPase à la membrane basolatérale. Cet effet pourrait être surtout important pour la réponse tardive. Les résultats présentés sont discutés dans le cadre d'une hypothèse de travail ("sodium pump-channel hypothesis") qui prédit que le canal au sodium possède certaines homologies structurelles avec la sous-unité catalytique de la Na,K-ATPase.

Hormones and Cell Regulation, Volume 9
INSERM European Symposium
J.E. Dumont, B. Hamprecht and J. Nunez editors
© 1985 Elsevier Science Publishers B.V. (Biomedical Division)

ISOLATION AND RECONSTITUTION OF VOLTAGE SENSITIVE SODIUM CHANNELS
FROM MAMMALIAN SKELETAL MUSCLE

JACQUELINE C. TANAKA AND ROBERT L. BARCHI
Institute of Neurological Sciences and the Departments of
Neurology, and Biochemistry and Biophysics, University of
Pennsylvania School of Medicine, Philadelphia, Pennsylvania 19104

INTRODUCTION

Ion channels form an important class of intrinsic membrane
proteins involved in the control of a wide variety of cell
functions. New techniques such as patch clamping have now
demonstrated the general role of ion channels in excitation,
secretion, motility and information transfer. These ion channels
provide a hydrophyilic pathway through the membrane; the
particulars of the ion movement through the pathway uniquely
define the channel type. Characteristic properties of ion
channels include 1) selectivity, or the ability to discriminate
among ions of similar size and charge; 2) gating, or the capacity
to open and close the conductance pathway in response to changes
in voltage or in the concentration of a ligand; 3) specificity,
or the ability to respond only to a unique physical or chemical
agent; and 4) peak conductance, or the maximal rate of ion
movement through the open channel.

One of the most widely distributed and best studied of these
ion channels is the voltage dependent sodium channel, responsible
in most excitable membranes for the generation of the rising
phase of an action potential (1). Early work on the sodium
channel has catalyzed progress in both theoretical and
technological aspects of ion channels, and has led to an
appreciation of the relationship between the macroscopic membrane
current produced by large populations of ion channels and the
discrete opening and closing events that occur at the level of
individual channel molecules (2,3). One of the primary goals of
current research on ion channels is to explain their unique
functional properties at the level of the single channel in terms
of a detailed understanding of the structure of the channel
protein.

Although biophysical studies of ion channels can be traced
back for half a century (4), their detailed biochemical analysis
has been a more recent undertaking. During the last ten years
dramatic advances have been made in elucidating the structure of

the nicotinic acetylcholine receptor (5); studies on the voltage
sensitive sodium channel are now gaining momentum. Within the
past six years the sodium channel has been purified to near
homogeneity from rat brain (6), rat skeletal muscle (7), rabbit
skeletal muscle (8) and Electrophorus electric organ (9).

Since the sodium channel has no natural ligands,
identification of this protein for biochemical studies has
depended on the availability of a group of neurotoxins that exert
their lethal effect by binding specifically to the channel
protein. Some of these toxins can be radiolabelled to high
specific activity, and of these toxins the most commonly used
during sodium channel purification are tetrodotoxin and saxitoxin
(10). Other neurotoxins can be used to manipulate the
conformational state of the channel; they include the alkaloid
toxins veratridine and batrachotoxin, which promote channel
activation (11).

Neurotoxin binding can serve as a label for the sodium
channel protein but does not provide information about the
preservation of channel function. In order to document that an
isolated protein is indeed the sodium channel, assays of its
functional properties are needed. These functional assays seek
to compare key aspects of channel behavior of the isolated
protein, such as the conductance, selectivity and the voltage
dependence of gating, with observations on this channel in the
native membrane. To demonstrate these properties, the purified
protein must be reinserted into a planar bilayer or unilamellar
vesicle so that ion movement can be measured either as an
electrical current or isotopic flux. This article will consider
the biochemical characteristics of sodium channels purified from
mammalian muscle and will review studies with the reconstituted
channel that address its functional properties.

BIOCHEMICAL PROPERTIES OF THE SODIUM CHANNEL FROM MAMMALIAN
SKELETAL MUSCLE

The voltage dependent sodium channel has been purified to
>90% purity from four different sources (6-9). The three
mammalian preparations include rat brain, rat skeletal muscle and
rabbit skeletal muscle; the fourth preparation is the electric
organ from the eel. The channels from each of these sources
appear quite similar in regard to their physical properties.

The size of the sodium channel from rat (12) and rabbit muscle (8) in lipid and detergent has been estimated by chromatography on a Sepharose Cl-6B column. An apparent Stokes radius of 8.6 nm was obtained for both these mammalian skeletal muscle channels. This agrees well with similar measurements done on rat brain (13) and eel electric organ (9,14-15), which yielded values of 8.0 and 8.5 respectively. Sedimentation measurements with the solubilized channel on sucrose gradients formed in either H_2O or D_2O were used to estimate the molecular weight of the protein. Estimates of the molecular weight of the sodium channels from the three mammalian sources show good agreement, with calculated molecular weights ranging between 285,000 and 316,000. Minimum size estimates based on the irradiation inactivation of toxin binding have been made for the eel channel (16); these ranged between 230,000 and 260,000. The accuracy of all these estimates is limited by the various assumptions necessary for their derivation from the experimental data; this is especially true of the sedimentation measurements made with the solubilized channel. Even with these limitations, the agreement between reported overall size for these channels is good.

Each of the purified sodium channel preparations has been found to contain a large glycoprotein subunit of >200,000 M_r. Although the eel sodium channel appears to contain only this large polypeptide (17), the three mammalian preparations each contain one or more smaller subunits of about 38,000 M_r (6,7,8).

The subunit composition of purified rabbit skeletal muscle sodium channel as resolved by SDS-PAGE is shown in Fig. 1. A diffuse band is seen which migrates in the region of 260 Kd. This band is heavily glycosylated and demonstrates anomalous migratory characteristics on SDS-PAGE as do the large subunits from rat brain and skeletal muscle and from the eel electric organ.

In the past, the large glycoprotein subunit of the purified sodium channel from rat sarcolemma appeared to migrate as a wider band between 160-200 Kd on gradient SDS-PAGE (18). The discrepancy in size between this rat skeletal muscle subunit and the large subunits from rabbit muscle and other sources can now be attributed to proteolytic nicking of the rat channel that occurs very early during the preparation of the muscle membranes in spite of the presence of a variety of protease inhibitors. Immunoblot analysis with monoclonal antibodies against the purified rat sodium channel was used to document this proteolysis (19).   In immunoblots from freshly homogenized rat muscle, monoclonal antibodies against the large subunit detected only a single broad band at 260 Kd. In similar homogenates held at 4°C for longer periods, a diffuse immunoreaction band appeared between 160 and 220 Kd concomittant with a decrease in the intensity of the 260 Kd band. When purified sarcolemmal membranes were used to prepare the transfers, often only the 160-220 K band was seen with little or no remaining immunoreactivity at 260 K. This interrelated pattern of bands suggests that limited cleavage of the 260 Kd subunit by one or more proteases occurs early in the membrane isolation. The sensitivity of this interconversion to EGTA but not to other protease inhibitors suggests that a Ca^{++}-activated muscle protease is a likely candidate.

Mol. Wt.

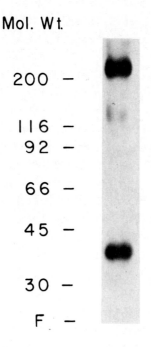

Figure 1.   SDS-PAGE of the purified sodium channel from rabbit T-tubular membranes. A large glycoprotein subunit at 260,000 d and a single smaller subunit at 38,000 are seen.

Although the purified sodium channel from eel electroplax (8) contains only the 260 Kd component, all mammalian preparations have at least one additional smaller band on gel electrophoresis. Rat and rabbit muscle have a band at 38,000 M_r

(8,18); the rat brain sodium channel has two bands in this region
(6). In the rabbit preparation this small subunit codistributes
with the 260,000 M_r subunit and with both [3H]-saxitoxin binding
activity and with functional channel activity after
reconstitution. The significance of these small subunits to
normal sodium channel function remains to be clarified. Are the
smaller bands seen with mammalian preparations necessary for the
function of the channel? Since all of the purified channels have
now been shown to gate cation movement we cannot assume that an
essential piece of the eel channel has somehow been lost during
the purification. Likewise, with respect to the mammalian
channel we cannot yet state that the channel function requires
the smaller components.

If the sodium channel, at least from the eel, is a single
subunit our classical notions concerning the mechanism of
allosteric cooperativity must be challenged (20,21). Allosteric
behavior in the sodium channel is perhaps best demonstrated by
the coupling of alkaloid toxin activaton of the channel to the
binding of Leiurus q. polypeptide toxin at a separate site. A
classical explanation for this type of cooperativity would relate
changes in the coupling energies between subunits to changes in
the ligand receptor binding energy. If the sodium channel, which
demonstrates these well characterized cooperative behaviors,
contains a single subunit another mechanism must be postulated.
Perhaps the concept of protein domains (22) will become
particularly important in such large membrane proteins. Rather
than changes in interaction between two coupled subunits after
ligand binding, there may be changes in the energetics of the
coupling of two domains within a single polypeptide chain.

FUNCTIONAL CHARACTERIZATION OF THE RECONSTITUTED SODIUM
CHANNEL
Reconstitution and Pharmacological Activation of Purified
Sodium Channels in Phospholipid Vesicles
For biochemical analysis, the sodium channel protein is
identified during purification by its high-affinity binding of
radiolabelled tetrodotoxin or saxitoxin. Unfortunately, this
binding provides no information about the functional state of the
channel. Since the principal role of this protein is the control
of transmembrane cation permeability, the purified protein must

be reconstituted into either a planar lipid bilayer or a lipid
vesicle in order to assess its functional capacity. For the
purified sodium channel from rat and rabbit muscle,
reconstitution has been accomplished by removal of the NP-40
detergent with Biobeads SM2 in the presence of excess
phospholipid (8,23,24). In order to simplify analysis of flux
data, reconstitution conditions were chosen to produce an average
incorporation of less than one channel per vesicle. Using this
approach vesicles formed in phosphatidylcholine were large and
unilamellar; these vesicles exhibited low leakage of monovalent
cations. Freeze fracture electron microscopy and Sepharose 2B
column chromatography of the vesicles showed a bimodal size
distribution with about 60% of the lipid in vesicles between
1000-2500A in diameter and the remainder in vesicles of less than
300 A diameter (23). In freeze-fracture images, intramembranous
particles of ~95A were seen in vesicles reconstituted with
purified rat sodium channel; the particles were absent from
vesicles formed in an identical buffer without the purified
protein.

Based on measurements of ^{3}H-saxitoxin binding after
reconstitution, about 35% of the sodium channels present in the
starting material could be identified in the intact vesicles.
When the vesicles were ruptured with detergent, the total number
of saxitoxin binding sites increased to about 70% of the starting
material suggesting that channels were inserted randomly during
vesicle formation. The 30% loss in binding was due to a loss of
protein on the Biobeads during reconstitution since the specific
activity of the vesicle fraction was usually the same as that of
the starting micellar fraction. The K_d for saxitoxin binding to
the reconstituted channel (23) was not significantly different
from that measured in sarcolemma or purified micelles but the
thermal stability of the binding site, markedly reduced in
detergent solution, was restored to near that seen in native
membranes.

Specific influx of ^{22}Na$^+$ could be produced in vesicles
containing the reconstituted channel by pre-incubation with
batrachotoxin (5x10^{-6} M) or veratridine (5x10^{-4} M) (Fig. 2).
Vesicles incubated under the same conditions in the absence of
alkaloid toxins showed no increase in ^{22}Na$^+$ influx over control.
As expected if the channels are randomly oriented, about 50% of

the alkaloid-activated influx could be blocked by saxitoxin or
tetrodotoxin in the external solution. All the toxin-activated
flux was inhibited when saxitoxin was also present on the inside
of the vesicles, confirming that the alkaloid toxins could
activate all channels regardless of their orientation but that
saxitoxin and tetrodotoxin blocked only those channels whose
external faces were exposed to the aqueous compartment containing
these polar toxins. Selective addition of tetrodotoxin or
saxitoxin thus allows the creation of vesicle populations with
functionally active channels oriented in either an inward or
outward direction.

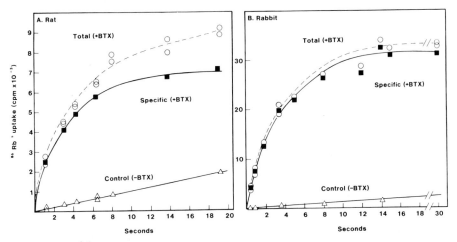

Figure 2. ^{86}Rb$^+$ influx into phosphatidylcholine vesicles
containing the purified sodium channel from rat (A) or rabbit (B)
skeletal muscle. After reconstitution, vesicles were activated
with 5x10^{-6}M batrachotoxin and fluxes measured at 22°C.

Batrachotoxin and veratridine activate the channel with K_d's
similar to those measured with sodium channels in situ (23). The
apparent K_d for batrachotoxin and veratridine activation of the
rat sodium channel were 1.5x10^{-6}M and 3.5x10^{-6}M respectively.
The K_i for saxitoxin inhibition of the stimulated flux was
5.0x10^{-9}M (24), comparable to the K_i for the inhibition of
membrane sodium currents as well as the K_d for ^{3}H-saxitoxin
binding to the native channel (23).

In order to evaluate the correlation between the capacity
for functional activity after reconstitution and the level of
high-affinity [^{3}H]-saxitoxin binding, the distribution of both

properties was examined in individual fractions of a sucrose
gradient containing partially purified rabbit muscle sodium
channel protein (8). Separate reconstitutions were done on each
fraction and the batrachotoxin-activated space and the total
intravesicular volume was measured in each. The activated flux,
plotted either as the ratio of batrachotoxin uptake to the
control or as the ratio of activated space to total accessible
space, paralleled exactly the distribution of [^{3}H]-saxitoxin
binding, and corresponded to the distribution of the 260,000 M_r
and 38,000 M_r subunits on this gradient.

Cation Selectivity Characteristics

One of the distinguishing characteristics of a voltage gated
ion channel is its ability to select for a particular ion among
others of similar charge; this selectivity is, in fact, the basis
for naming most of these familiar channels. The selectivity of
the voltage gated sodium channel has been well documented in a
variety of excitable membranes using electrophysiological
techniques and provides another benchmark for a functional
channel (25-30). If the purified sodium channel retains this
property then the rates of equilibration of different cations
into vesicles containing the reconstituted channel should reflect
this differential permeability of the channel.

Due to the small included volume of vesicles formed during
reconstitution and the expected high rate of ion movement through
a single sodium channel (1×10^6 ions/sec) very rapid equilibration
of the intravesicular space would be expected for permeant
cations. We have used a quenched flow system to resolve these
rapid kinetics in vesicles containing batrachotoxin-activated
channels (24). The dead time of our system, due mainly to the
quenching of the influx on a Dowex resin, was ~90 msec. Vesicles
were activated with batrachotoxin (5×10^{-6} M) and the rate of
vesicle equilibration was determined for Na^+, K^+, Rb^+ and Cs^+
(Fig. 3). The half-time for Cs^+ uptake was quite slow (10 s) and
could be resolved even with manual techniques. The rates
increased dramatically with Rb^+ and K^+, yielding halftimes of
about 2.5 s and 350 ms respectively for both the rat and rabbit
channel at 22°C. Measurements between 90 ms and 200 ms clearly
resolved the early linear phase of K^+ uptake at this temperature
(Fig. 3A), but demonstrated that Na^+ uptake was virtually
complete even at the earliest timepoint that could be accurately

measured. From this information, an upper limit of 50 msec could
be placed on the halftime for Na⁺ equilibration under these
conditions. Using this upper limit, selectivity ratios were
calculated for each cation and are given in Table 1. Values
shown for both rat and rabbit reconstituted preparations agree
well with selectivity measurements on the intact muscle membrane.

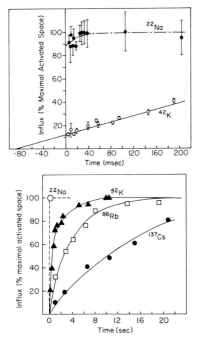

Fig. 3 Cation selectivity in
the purified, reconstituted
sodium channel from rat
muscle. A. Quenched flow
measurements of the early
linear phase of ⁴²K⁺ uptake
( ) and ²²Na⁺ uptake ( ) at
22°C after activation by
batrachotoxin. B. Uptake
curves for 4 monovalent
cations plotted on the same
time scale. In each case, only
specific uptake is shown.

Table 1.   Cation selectivity in the purified reconstituted
            sodium channel

	Na⁺	K⁺	Rb⁺	Cs⁺
Rabbit	1	0.13	0.02	0.008
Rat	1	0.14	0.02	0.005

Values represent the ratio of the half-time for equilibration of
a given cation to that for Na⁺, based on measurements made at
22°C, after activation of the reconstituted channel with
batrachotoxin (5x10⁻⁶M) at 36°C for 45 min.

The kinetics of veratridine-activated influx were several
orders of magnitude slower than those determined following
activation with batrachotoxin (8,24). In addition, selectivity

among monovalent cations, while still present, appeared to be
much less than with channel activated by batrachotoxin. The
large differences between the vesicle equilibration times for
batrachotoxin and veratridine is probably due to differences in
the rate limiting step for ion flux. The temperature dependence
of cation influx was measured for vesicles activated by both
alkaloids and a five-fold difference was found (24). The
apparent activation energy for influx into batrachotoxin-
activated vesicles was 7 Kcal/mol whereas the activation energy
for veratridine-treated vesicles was 31 kcal/mol. Our
interpretation of the kinetics and temperature dependencies of
the cation influx with batrachotoxin-modified channels is that
these channels are essentially open all the time and the
effective rate-limiting step is the passage of an ion through the
open channel. With veratridine, the kinetics may reflect a more
complicated combination of the probability of channel opening and
the actual open time. Under these conditions the ratios of
apparent influx may not correlate with actual ratios of channel
conductance for a given pair of cations.

Voltage Dependent Activation

Resolution of the kinetics of the voltage dependent gating
of purified sodium channels requires electrical measurements
using patch electrodes or planar lipid bilayers. However,
evidence for the presence of voltage dependent behavior can be
obtained with flux measurements in the presence of batrachotoxin.
Voltage clamp measurements on sodium channels from neuroblastoma
cells have shown that batrachotoxin-activated sodium channels
retain their voltage dependent activation properties although
time-dependent inactivation is virtually eliminated (31). In
batrachotoxin treated cells channels open at more hyperpolarized
potentials than untreated channels but the shape of the curve
relating the membrane potential to sodium conductance remains
unchanged and is merely translated about 50 mV in a more
hyperpolarizing direction. At $V_m > -100$ mV, for example, sodium
channels are closed despite the presence of batrachotoxin (31).

Exploiting this voltage dependent property, we examined
fluxes in vesicles containing batrachotoxin-activated sodium
channels in which a K^+ diffusion potential had been established
across the membrane. In order to create a homogenous channel
population, inward facing sodium channels were blocked with

internal saxitoxin. A K$^+$ ion gradient was established by
trapping high [K$^+$] within vesicles during reconstitution and
subsequently adjusting the concentration of K$^+$ outside. The
membrane potential was then altered by rapidly changing the
external K$^+$ in the presence of valinomycin, a K$^+$ selective
ionophore (Fig. 4A). Since batrachotoxin blocks channel
inactivation, a curve relating steady-state activation of the
batrachotoxin-modified channel to voltage was produced. The
potential across the vesicles was estimated from the Nernst
relationship of the potassium concentrations; this estimate of
the V_m is not absolute but should be relatively accurate. Under
these conditions, the purified reconstituted channel clearly
retained voltage-dependent gating properties (Fig. 4B). At very
hyperpolarizing potentials the batrachotoxin-stimulated ^{22}Na
influx was shut off; specific influx increased abruptly as the
membrane potential was reduced.

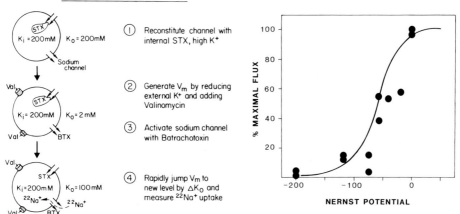

Fig. 4 Voltage-dependent activation of the purified rabbit muscle
sodium channel after batrachotoxin activation. The experimental
paradigm is shown on the left.

A major goal in the study of reconstituted sodium channels
is the resolution of individual channel kinetics. Patch clamp
technology has permitted the resolution of these kinetics on
individual sodium channels in their native membrane environment
(2,3). The constraints imposed on the lipid environment of the
reconstituted protein for patch clamping are different from those
imposed during ion flux measurements. Alternative approaches to

both the detergent solubilization and vesicle reconstitution were
therefore developed for patch recording.

Preliminary patch electrode recordings of individual channel
transitions were made following solubilization and purification
of the sodium channel in a zwitterionic detergent, CHAPS (32).
Vesicles were formed from mixtures of lipid containing 60/40
phosphatidyl ethanolamine/phosphatidyl serine; large patchable
multilamellar vesicles were formed after freezing and thawing of
these vesicles. Upon reconstitution several classes of channel-
like activity were seen repeatedly with inside-out patches (33).
One type of channel activity was voltage and toxin insensitive
with a typical conductance of 15 pS in of 100 mM NaCl. A second
class of channel events was more interesting physiologically.
These channels rarely flickered in the absence of batrachotoxin
but exhibited continuous opening flickers within seconds of the
application of this toxin and the frequency and duration of
channel opening was clearly voltage dependent. Again at very
hyperpolarized potentials the channel was mostly closed. With
increasing depolarization the channel open time increased until
only brief closing transitions were seen. The conductance of the
toxin activated single channel activity in 100 mM NaCl at 22°C
was 15 pS. This value agrees well with single channel
conductance values reported for native sodium channels (34) and
thereby suggests the conservation of another sodium channel
characteristic.

SUMMARY

A biochemical picture of the voltage gated sodium ion
channel is now emerging. The overall protein is very large with
a molecular weight in excess of 250,000 daltons. The major
functioning unit of this channel is a single large glycoprotein
that may itself be as large as 250 Kd. The mammalian sodium
channels studied to date have one or more smaller subunits but
there is currently no conclusive evidence for or against the
requirement of these subunits in the functioning of the channel.
Reconstitution studies have correlated several sodium channel
properties of the purified protein with the in situ channel,
including specificity for neurotoxins and cation selectivity.
Preliminary observations have shown that at least part of the
voltage dependent properties are also retained. In the future we

hope to use purified, reconstituted sodium channel protein to understand how the channel accomplishes the complex gating of its quantized conductance changes.

ACKNOWLEDGEMENTS

Much of the work presented in this review has been done in collaboration with other investigators in this laboratory; the authors thank Jan Casadei, Susan Kraner, Richard Roberts, and Dr. Roy Furman for their valuable contributions. This work was supported in part by NIH grants NS-18013 and NS-08075, and by a grant from the Muscular Dystrophy Association of America. We would also like to thank Lois Murphy and Sheri Irons for their technical assistance and Nancy Goodman for the preparation of the manuscript.

REFERENCES

1.   Hodgkin A, Huxley A (1952) J Physiol 117:500-544

2.   Neher E, Sakmann B (1976) Nature 260:799-802

3.   Hamill OP, Marty A, Neher E, Sakmann B, Sigworth FJ (1981) Pflugers Arch 391:85-100

4.   Hodgkin AL (1937) J Physiol 90:183-210

5.   Stroud RM (1983) Neurosci Commentaries 1:124-139

6.   Hartshorne RP, Catterall WA (1981) Proc Natl Acad Sci USA 78:4620-4624

7.   Barchi RL, Cohen SA, Murphy LE (1980) Proc Natl Acad Sci USA 78:1306-1310

8.   Kraner SD, Tanaka JC, Matesic DR, Barchi RL (1984) Submitted for publication.

9.   Agnew WS, Levinson SR, Brabson JS, Raftery MA (1978) Proc Natl Acad Sci USA 75:2606-2610

10. Ritchie JM, Rogart R (1977) Rev Physiol Biochem Pharmacol 79:2-45

11. Catterall WA (1980) Ann Rev Pharmacol Toxicol 20:15-43

12. Barchi RD, Murphy LE (1981) J Neurochem 36:2097-2100

13. Hartshorne R, Coppersmith J, Catterall W (1980) J Biol Chem 255:10,572-10,575

14. Benzer TI, Raftery MA (1973) Biochem Biophys Res Commun 51:939-944

15. Norman R, Schmid A, Lombet A, Barhanin J, Lazdusnki M (1983) Proc Natl Acad Sci USA 80:4164-4168

16. Levinson SR, Ellory JC (1973) Nature (London) New Biol 245:122-123

17. Miller JA, Agnew WS, Levinson SR (1983) Biochemistry 22:462-470

18. Barchi R (1983) J Neurochem 40:1377-1385

19. Casadei J, Gordon R, Lampson L, Schotland D, Barchi R (1984) PNAS, submitted for publication

20. Monad J, Wyman J, Changeux J-P (1965) J Mal Biol 12:88-118

21. Kashland DE, Nemethy G, Filmer D (1966) Biochemistry 5:365-386

22. Janin J, Wodak F (1983) Prog Biophys Mal Biol 42:21-78

23. Weigele JB, Barchi RL (1982) Proc Natl Acad Sci USA 79:3651-3655

24. Tanaka JC, Eccleston JF, Barchi RD (1983) J Biol Chem 258:7519-7526

25. Hille B (1972) J Gen Physiol 59:633-658

26. Campbell DT (1976) J Gen Physiol 67:295-307

27. Pappone PA (1980) J Physiol 306:377-410

28. Huang LYM, Ehrenstein G, Catterall WA (1978) Biophys J 23:219-231

29. Khodorow BI, Revenko SV (1979) Neuroscience 4:1315-1330

30. Frelin C, Vigne P, Lazdunski M (1981) Eur J Biochem 119:437-442

31. Huang LYM, Moran N, Ehrenstein G (1982) Proc Natl Acad Sci USA 79:2082-2085

32. Hjelmeland LM (1980) Proc Natl Acad Sci USA 77:6368-6370

33. Barchi RL, Tanaka JC, Furman RE (1984) J Cell Biochem in press

34. Sigworth FJ, Neher E (1980) Nature 287:447-449

RESUME

Une image biochimique du canal sodium, voltage-dépendant, émerge actuellement. La protéine globale est très grande, avec un poids moléculaire dépassant 250.000 daltons. L'unité fonctionnelle principale de ce canal est une seule grande glycoprotéine qui peut elle-même être aussi grande que 250 kd. Les canaux sodium de mammifères, étudiés à ce jour, ont une ou plusieurs sous-unités plus petites, mais il n'y a actuellement aucune évidence concluante, pour ou contre un rôle obligatoire de ces sous-unités dans le fonctionnement du canal. Des études de reconstitution ont corrélé différentes propriétés du canal sodium obtenues à partir de la protéine purifiée avec le canal in situ, y compris la spécificité pour les neurotoxines et la sélectivité pour les cations. Des observations préliminaires ont montré que au moins une partie des propriétés de voltage dépendance sont également conservées. Dans l'avenir, nous espérons utiliser des protéines du canal sodium, purifiées et reconstituées pour comprendre comment s'opèrent les modulations complexes d'ouverture du canal permettant des changements discrets de conductance.

Hormones and Cell Regulation, Volume 9
INSERM European Symposium
J.E. Dumont, B. Hamprecht and J. Nunez editors
© 1985 Elsevier Science Publishers B.V. (Biomedical Division)

MECHANISM OF β-ADRENERGIC MODULATION OF CARDIAC CALCIUM CURRENT IN ISOLATED MYOCYTES FROM ADULT MAMMALIAN HEARTS

GUSTAVO BRUM* and WOLFGANG OSTERRIEDER**

II. Physiologisches Institut der Universität des Saarlandes, Homburg/Saar, F.R.G.

INTRODUCTION

The current flow through two distinct ion channels in the cell membrane generates excitation of the working myocardium, i.e. the "fast" Na inward current and the "slow" inward Ca (I_{Ca}) current. Both currents are voltage-dependent, but only the latter appears to be subject to regulatory processes. Of great importance for the modulation of Ca conductance is the sympathetic innervation, which adapts heart function to the requirements of exercise. The increase in Ca current across the cell membrane elicited by the neurotransmitter adrenaline [1] is intimately involved in the increase in the force and rate of myocardial contractility (for review, see references [2] - [6]).

Many of the effects of catecholamines are mediated by an increase in the cytosolic concentration of cyclic AMP [7]. For example, the adrenaline-induced enhancement of contractile force in the heart is preceeded by an increase in cAMP [8]. Electrophysiological studies have shown that external application of membrane permeable analogues of cAMP to Purkinje fibers [9] and injection of cAMP into Purkinje fibers [10] or cardiac myocytes [10, 11] increases the Ca inward current which flows during the plateau phase of the cardiac action potential. Several authors [13 - 16] have suggested that I_{Ca} is increased by the activation of a cAMP-dependent protein kinase, resulting in phosphorylation of the Ca channel (or parts of the channel) in the sarcolemma. It was proposed that non-phosphorylated Ca channels cannot conduct current and that, as a consequence of Ca channel phosphorylation, more channels might open in the presence of β-adrenergic stimulation [1].

* Present address: Departamento de Biofísica, Facultad de Medicina, Montevideo, Uruguay

**Present address: F. Hoffmann-La Roche & Co., Ltd., Pharmaceutical Research Department, Basle, Switzerland

The present article describes several experiments aimed at elucidating the mechanism of the β-adrenoceptor induced increase in cardiac Ca current. In order to investigate the phosphorylation hypothesis, purified catalytic and regulatory subunits of cAMP-dependent protein kinase were pressure-injected into isolated guinea-pig ventricular myocytes and the electrophysiological response of the cell membrane was recorded. By application of the patch-clamp technique [17], the effects of β-adrenergic stimulation on the Ca current could be studied at the level of single Ca channels. Since all experiments were carried out using isolated cardiac cells, some properties of this new experimental model will briefly be described.

Part of these results has recently been published [12, 18, 19].

MATERIALS AND METHODS

Solutions

The composition of the Tyrode's solution was (in mM): NaCl 135, KCl 5.4, $CaCl_2$ 1.8, $MgCl_2$ 1, glucose 10, HEPES 5; pH was adjusted to 7.4. The "storage solution" (see also RESULTS) contained (in mM): taurine 10, oxalic acid 10, glutamic acid 70, KCl 25, K_2HPO_4 10, glucose 11, HEPES 10 and EGTA 0.5. The pH was adjusted to 7.3 with KOH. Adrenaline (Hoechst), isoprenaline (Sigma) and forskolin (Hoechst) were prepared freshly as stock solutions and diluted with Tyrode's solution.

Preparation of single ventricular cells

The method to isolate single cells is described in refs. [12], [20] and [21]. Briefly, hearts were enzymatically dissociated by retrograde perfusion with collagenase (Sigma, type I, or Worthington, type II) in a Langendorff apparatus.

Preparation of catalytic and regulatory subunits

Homogeneous preparations of catalytic and regulatory subunits of cyclic AMP-dependent protein kinase II were prepared by Drs. Flockerzi and Hofmann (Heidelberg, F.R.G.) as previously described [22, 23]. For injection, they were dialysed against a 150 mM KCl buffer in 2 mM HEPES (pH 7.1).

Injection

The single cells were kept on the glass bottom of a small perspex chamber which was continuously perfused with Tyrode's solution at

36 °C. Intracellular glass microelectrodes were used for stimulation of the cell and for pressure injection of the subunits. Details of the injection technique are reported in refs. [11] and [19].

Action potential recording, voltage clamp and patch clamp

Single cells were stimulated by injection of current through a conventional intracellular microelectrode (filled with 2 M KCl); the transmembrane potential was recorded with the same electrode. In order to voltage clamp the cell membrane, the two microelectrode voltage-clamp technique [24] was applied. The current through individual Ca channels was recorded by the patch clamp method [17]. The patch pipettes were produced from Pyrex glass, coated with Sylgard 182 (Dow Corning) and fire-polished to a final diameter of 1 - 1.5 μm. The patch pipettes were filled with a solution containing (in mM): $BaCl_2$ 90, $MgCl_2$ 5, NaCl 4, HEPES 5 (pH 7.3). The elevated Ba content enabled the conductance of the Ca channel to be increased [25] and potassium outward currents to be suppressed (cf. [26]).

Data processing

Original data were displayed on a storage oscilloscope (Tektronix) and stored on FM tape (Hewlett Packard). Data were evaluated following transfer to a computer (Nicolet, MED 80).

RESULTS

Isolated ventricular cells

The major problem in the isolation of single ventricular cells is the "Ca-paradox" [27]. After perfusion of whole hearts with Ca-free Tyrode's solution, myocardial tissue undergoes an irreversible contracture upon reperfusion with Ca-containing solutions (> 0.1 mM Ca), resulting in cell death. The situation is similar when cells are enzymatically dissociated, because the collagenase solution must be kept (nominally) Ca-free. In our experiments, the yield of Ca-tolerant cells could be increased (to about 30 - 60%) by incubating the dispersed cells in a "storage solution" before restoring physiological Ca concentrations [20, 28]. The storage solution contains mainly potassium ions and a number of substrates to replete and maintain intracellular energy stores. This treatment also improves the electrophysiological properties of these cells [29]. Viable, Ca-tolerant cells (as shown in Fig. 1) are rod-shaped and show clear cross-striations. These cells have a morphology similar

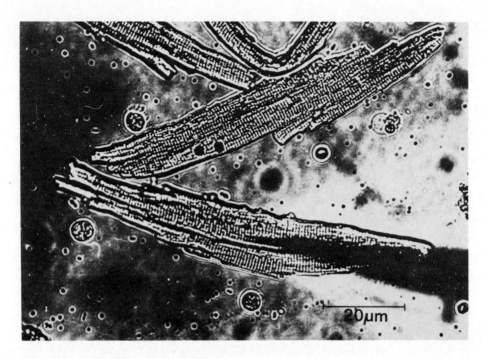

Fig. 1. Isolated ventricular cells from adult guinea-pig heart. This photomicrograph of cells in the recording chamber was taken during perfusion with Tyrode's solution. The cell in the lower part of the picture has been impaled by a conventional microelectrode (black wedge).

to that of undissociated cells [30], being typically about 100 μm long, 20 - 40 μm wide and less than 10 μm thick.

The isolated cells offer several advantages over multicellular preparations. For instance, there are no cell-to-cell connections and there is no series resistance due to an interstitial space, allowing good potential control of the cell membrane in the voltage clamp [31]. When compounds are injected into a cell, they are able to freely diffuse within the cytosolic space. The cell surfaces appear very "clean" (no connective tissue) and, thus, the patch-clamp technique [17] can be applied to record the activity of single ion channels in the cell membrane. For these reasons isolated ventricular cells were especially suited to study the mechanism of β-adrenergic increase in Ca current.

245

Electrical response of isolated cardiac myocytes to stimulation of
the adenylate cyclase system

    In a first series of experiments, the responses of isolated cells
to stimulation of the adenylate cyclase system were studied. By injection of depolarizing current into a cell, using an intracellular
glass microelectrode, action potentials can be elicited which are
quite similar in configuration to those recorded from cardiac tissue
(see also [32]). Upon application of adrenaline to the bath, a dose-
dependent increase in the amplitude and duration of the action potential was recorded (Fig. 2 A). Since the long plateau phase of the
ventricular action potential is mainly determined by the amplitude
of the Ca current, these effects are indicative of an increase in
I_{Ca}. The enhancement of I_{Ca} was directly demonstrated in voltage
clamp experiments (data not shown).

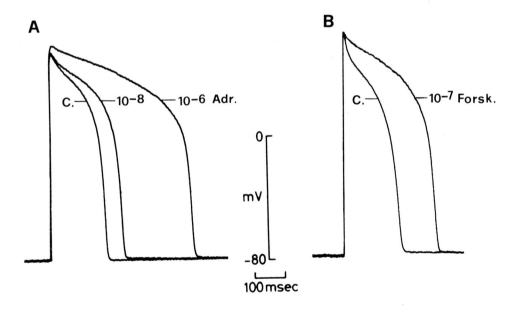

Fig. 2.   Effects of adrenaline and forskolin on guinea-pig ventricular cells.
A) The superimposed action potentials were recorded before (C.) and
during perfusion with 10^{-8} and 10^{-6} M adrenaline (in steady state).
B) Action potentials before (C.) and during perfusion with 10^{-7} M
forskolin (steady state).

    The binding of adrenaline to the β-adrenergic receptor stimulates
adenylate cyclase [7], resulting in increased cytosolic concentra-

tions of cAMP. Forskolin is a diterpene alkaloid from the Indian plant Coleus forskohlii, which exerts its effects on a variety of eukaryotic cells by a direct activation of adenylate cyclase [33], thereby markedly increasing intracellular levels of cAMP. In the heart, forskolin increased contractile force and rate [34, 35]. In the experiment shown in Fig. 2 B, the electrical response of a cell to forskolin was examined. Forskolin, at a concentration of 10^{-7} M, had essentially the same effects on the action potential of guinea-pig ventricular cells as adrenaline, namely an increase in both amplitude and duration (Fig. 2 B).

Pressure injection of the catalytic subunit of cAMP-dependent protein kinase

The cyclic AMP-dependent protein kinase has been isolated to a high degree of purity from bovine heart [36, 37]. The inactive holoenzyme consists of two catalytic (C) and two regulatory (R) subunits. cAMP activates the enzyme by causing dissociation of the inactive holoenzyme (R_2C_2), according to the scheme [38, 39]:

$$R_2C_2 + 4 \ cAMP \rightleftharpoons R_2(cAMP)_4 + 2 \ C.$$

The reaction yields the free catalytic subunit which is able to phosphorylate substrate proteins. When injected into a myocyte, the catalytic subunit should produce adrenaline-like alterations of the action potential or of the Ca current.

Injection of the catalytic subunit of cAMP-dependent protein kinase elevated the plateau and prolonged the action potential of ventricular cells (Fig. 3 A). These effects result from an increased Ca current as was demonstrated in voltage clamp experiments (Fig. 3 B). The effects were already detectable 3 sec (the stimulus interval) after the onset of injection. After termination of the injection, the effects persisted for a long time [18, 19]. The relationship between I_{Ca} and membrane potential was not changed except for the increased amplitude of I_{Ca} [19]. In multiple injection experiments the first injection produced the largest effect, whereas a third or fourth identical injection was ineffective, suggesting that the Ca conductance of the cell membrane had attained a saturation level. When a high concentration (e.g. 5 x 10^{-7} M) of adrenaline was administered to a cell in which injection of catalytic subunit had produced a maximal effect, the configuration of the action potential remained totally unaltered. This finding indicates that the same

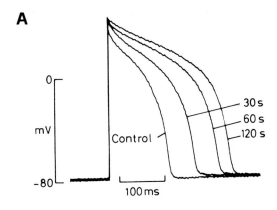

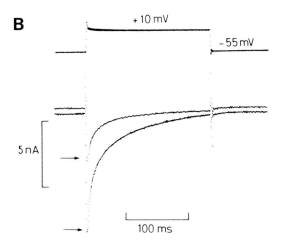

Fig. 3. Injection of the catalytic subunit of cAMP-dependent protein
kinase into guinea-pig ventricular cells.
A) Injection of the catalytic subunit increased amplitude and duration of the action potential. The cell was stimulated regularly at
0.33 Hz. The injection pressure was increased from 0 to 5 atm within
1 min and then maintained for 1 min. The control action potential
and action potentials recorded 30, 60 and 120 s after the onset of
injection are superimposed.
B) Increase in the amplitude of I_{Ca} by injection of catalytic subunit (voltage clamp experiment). Superimposed voltage traces (top)
and current traces (bottom) are shown. The cell membrane was depolarized from a -55 mV holding potential to +10 mV, the depolarization lasted 200 ms. Arrows indicate the peak amplitude of the Ca
current before and after injection.
(From ref. 12, with permission)

population of Ca channels is modulated by adrenaline and the catalytic subunit. These results strongly support the hypothesis that adrenaline, injected cAMP (data not shown) and injected catalytic subunit of cAMP-dependent protein kinase increase I_{Ca} by the same basic mechanism.

Injection of the regulatory subunit of cAMP-dependent protein kinase

It was attempted to establish whether the catalytic subunit of cAMP-dependent protein kinase also contributes to the regulation of Ca current in the absence of β-adrenergic stimulation. For this purpose the regulatory subunit was injected into myocytes. The rational for these experiments was that since the only known function of the regulatory subunit [40] is to bind the catalytic subunit, thereby inhibiting its phosphotransferase activity, injection of regulatory subunit should exert an effect upon the myocardial action potential if the catalytic subunit contributes to Ca conductance processes. Indeed, injection of the regulatory subunit into myocytes caused a decrease in the plateau height as well as duration of the action potential (Fig. 4). To substantiate an effect of the regulatory sub-

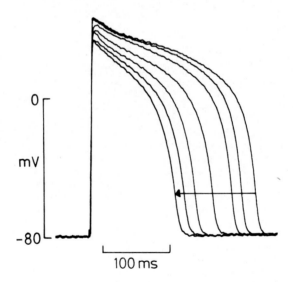

Fig. 4 Effects of injection of the regulatory subunit of cAMP-dependent protein kinase on the ventricular action potential. Pressure injection of cyclic AMP-free regulatory subunit at 3 atm for 30 s shortened the action potential and decreased its amplitude. The arrow indicates the direction of change. (From ref. 18, with permission).

unit on the Ca current, the resting membrane potential of some cells
was decreased to about -50 mV by increasing the K concentration in
the Tyrode's solution to 22 mM. As a result, the fast Na inward current is inactivated, and the upstroke velocity and amplitude of the
Ca-mediated action potential can be considered a crude estimate of
the Ca current [41]. Under these conditions, both parameters were
depressed suggesting a decrease in Ca current [18]. The decrease in
Ca current has meanwhile been shown directly in voltage clamp experiments. After intracellular dialysis of cells with the regulatory
subunit using a perfused patch clamp pipette, reductions in the
amplitude of I_{Ca} to about 30 % of the control value were observed
(M. Kameyama and W. Trautwein, personal communication).

The results of these studies indeed suggest that phosphorylation
of the Ca channel is an essential mechanism which occurs even in
the absence of β-adrenoceptor stimulation. It also argues for a
physiological role of both subunits of cAMP-dependent protein kinase
in regulating conductance through myocardial Ca channels.

β-adrenergic stimulation and gating of the Ca channel

In the past, investigations on the slow inward Ca current of the
heart were seriously hampered by inadequate potential control of
the cell membrane with the voltage clamp, due to the complex morphology of the heart and due to interference by other overlapping
current systems [42]. With the advent of the patch clamp technique
[17] it has become possible to record the activity of single Ca
channels in the cell membrane and to analyze their opening and closing behaviour with great precision [25,43,44,45]. Recordings from
adult guinea-pig myocytes [25] and from cultured neonatal rat heart
cells [43] in the "cell-attached" configuration have yielded similar
results. In response to voltage steps, channel openings occur singly
or in bursts of closely spaced unitary current pulses separated by
longer shut intervals. Channel openings become more probable with
increased amplitude of deporalizing test pulses; concomitantly, the
current amplitude declines in an ohmic manner. With 90-96 mM $BaCl_2$
in the patch pipette, the single-channel slope conductance has been
found to range from 15 to 25 pS [25,43,44], which corresponds to a
single current amplitude of about 1.5 pA at 0 mV membrane potential
and allows a good resolution of this current. With 50 mM Ba or Ca in
the pipette, the conductance decreases to 9-10 pS, and with still

smaller concentrations of divalent cations, the unitary currents
are too small to be resolved from the electronic background noise.

The mean open times (t_o) of myocardial Ca channels are exponentially distributed, with the time constant ranging from 0.5 to 1.5 ms
[19,25,43]. In contrast, the mean shut times are composed of the sum
of two exponentials: a fast exponential with a time constant (τ_s)
of 0.2 ms and a slow one with a time constant (τ_{ag}) ranging from
3-20 ms [19,25,43,46]. These kinetic data are similar to those reported for Ca channels in chromaffin cells [47] and snail neurons
[48]. The most simple kinetic model which can explain one open time
constant and two shut time constants is a sequence of two closed
states (C_1, C_2) and one open state (O), according to the following
scheme:

$$C_1 \underset{k_2}{\overset{k_1}{\rightleftharpoons}} C_2 \underset{k_4}{\overset{k_3}{\rightleftharpoons}} O$$

It is assumed that the short closing events are due to dwell times
of the Ca channel in the state C_2, placed between those of the
longer closed state C_1 and the opening pathway. k_1 to k_4 are the
respective rate constants for these transitions.

Generally, the net current (I) through N identical ion channels
can be expressed as $I = N \cdot i \cdot p(t)$, where i is the single channel
current and $p(t)$ the probability of the channel being in the open
state (which may be time-dependent). In single channel current
measurements it should be possible to decide whether the β-adrenergic increase in Ca current is due to an increase in the number
of channels, in the single channel current or in the probability of
the channel to be in the open state. An experiment aimed at answer-
ing this question is shown in Fig. 5. In this experiment, Ca channel
activity was elicited by 80 mV voltage jumps from the cell's resting
potential. Throughout the experiment no superpositions of unitary
currents were observed, indicating that the membrane patch did not
contain more than one Ca channel. In the control condition during
39 of 119 test pulses "blanks" were observed, i.e. no channel activity was seen. In panel A, six representative control patch currents
are illustrated. In panel B, six original current traces are shown
from the same patch 4 min after perfusion of the bath with isoprenaline (10^{-6} M). In the presence of the catecholamine, "blanks"

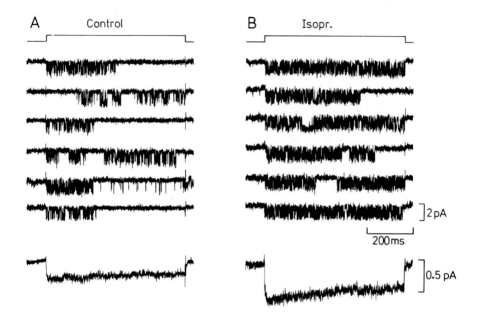

Fig. 5. Effects of isoprenaline on single Ca channel activity in response to some depolarizing pulses at a rate of O.5 Hz. The duration
of the pulses was 600 ms (indicated schematically at the top).
Part A: control; part B: bath application of 10^{-6} M isoprenaline.
The upper 6 traces in each panel are original recordings, while the
lowest traces are average currents (computed from 119 and 126 patch
currents, respectively).

were observed during only 18 of 126 test pulses. The lowest traces

in each panel are the respective averaged currents. Isoprenaline

increased the average current (representing I_{Ca}) by a factor of

about 2.7. β-adrenergic increases in I_{Ca} in multicellular mammalian

heart tissue have been reported to be on the same order of magnitude

[1,49,50]. It is obvious that the increase in average current was

caused neither by an enlargened single channel current nor by an

increase in the number of channels opening in the patch. It seemed,

however, that in the presence of isoprenaline the channel was in

its open state for longer periods of time.

   In order to determine the effect of β-adrenergic stimulation on

the gating of the Ca channel, open and closed time histograms were

computed. An example of the computation is shown in Fig. 6. The

procedure to measure open and shut times is described in ref. [25].

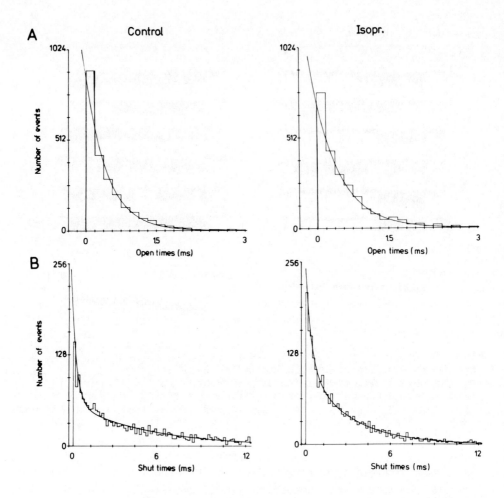

Fig. 6. Time histograms of open (A) and shut times (B) of the Ca channel. At the left the data calculated for control, at the right for isoprenaline (10^{-6} M). The fits to the open times data are single exponentials with time constants of 0.36 ms (control) and 0.46 ms (isoprenaline). The shut times data were fitted with double exponentials, the respective time constants were 0.33 ms and 5.10 ms for the control and 0.46 ms and 2.81 ms for isoprenaline.

Briefly, a threshold level was set at the half-maximal single channel current amplitude and open or shut times were measured as the time interval between two crossings of this level. As mentioned above, the open time histogram could be fitted with a single exponential (τ_o = 0.36 ms), whereas the closed time histogram was the

sum of two exponentials (τ_s = 0.33 ms and τ_{ag} = 5.10 ms). In presence of isoprenaline in the bath the open time constant was larger
(τ_o = 0.46 ms) and the time constant for long closure was shorter
(τ_{ag} = 2.81 ms). In a total of 6 experiments, where either adrenaline or isoprenaline (10^{-6} M) was applied, the open time constants
increased by about 15 % and the long shut times were reduced by
approximately 35 %, whereas the short shut times were not consistently affected [19]. From the time histogram data, the rate
constants k_1 to k_4 in the kinetic model described above can be
calculated [51]. The values obtained in the control situation were
(means of 6 experiments $\pm$ SD): k_1 = 1,178 $\pm$ 599, k_2 = 1,696 $\pm$ 892,
k_3 = 1,606 $\pm$ 812 and k_4 = 1,482 $\pm$ 781 s^{-1}. These were changed by
β-adrenoceptor stimulation to k_1 = 1,994 $\pm$ 1,098, k_2 = 1,877 $\pm$ 754,
k_3 = 2,125 $\pm$ 1,250 and k_4 = 1,143 $\pm$ 609 s^{-1} [19]. This implies that
for both reaction steps in the kinetic model, the equilibrium is
shifted towards the open state of the Ca channel in the presence of
isoprenaline or adrenaline. Therefore, the Ca current is increased
by longer and more frequent sojourns in its conducting state, equivalent to an increase in its probability to be open. Similar results were recently reported for the Ca channel in neonatal rat
heart cells in tissue culture [46].

DISCUSSION

   In the present study it could experimentally be demonstrated that
the Ca current in the heart is modulated by cAMP-dependent protein
kinase. The Ca current was increased when the intracellular concentration of the catalytic subunit was increased by pressure injection into guinea-pig ventricular myocytes. Injection of the regulatory subunit resulted in a decrease of the Ca conductance, presumably by binding to and inhibiting the phosphotransferase activity
of the catalytic subunit. The effect of the regulatory subunit was
observed in the absence of β-adrenergic stimulation, underlining
the importance of protein phosphorylation for the function of the
Ca channel. The increase in Ca current by protein kinase was indistinguishable from that produced by stimulation of adenylate
cyclase (either by adrenaline or forskolin) and cAMP-injection (data
not shown, see [19]), in line with the view of a common basic mechanism of increase in I_{Ca}. This result is also in accordance with

the findings that both adrenaline and cAMP increase the activity of protein kinase (for review, see refs. [16], [52]).

The nature of the membrane proteins associated with the Ca channel, which are supposed to be phosphorylated, remains unclear. Investigations on membrane fractions of adrenaline-perfused hearts showed a rapid phosphorylation of a protein with a 27,000 molecular weight [53]. Other groups reported the in-vitro phosphorylation of two membrane proteins with lower molecular weights of 22,000-23,000 and 10,000 [54-58]. With certainity, these proteins do not represent the Ca channel which has been reported to be of a much higher molecular weight (178,000 [59]), but they may be part of an oligomeric structure of the channel.

Phosphorylation of the Ca channel results in an increased probability of the channel to be in the open state upon depolarization. By this mechanism, the total Ca influx into the cell during β-adrenergic stimulation will be increased in mammalian heart [19,46]. In our studies, there was no indication of an increase in the number of channels. In those membrane patches which occasionally contained many Ca channels, overlappings of more single channel currents were observed during β-adrenoceptor stimulation [19]. This has, however, to be expected if the open probability of each individual channel is increased. Indeed, non-stationary current fluctuation analysis [60] provided evidence that the increase in overlappings was caused by an increased open probability [19]. It can, however, not be excluded that in other cell types the Ca current is also regulated by affecting N, the number of channels. Such a mechanism has been proposed for the frog heart where the increase of I_{Ca} by catecholamines is much larger [61,62].

ACKNOWLEDGEMENTS

This work was supported by the Deutsche Forschungsgemeinschaft. Dr. G. Brum was a stipendiary of the German Academic Exchange Service (DAAD). We are grateful to Drs. V. Flockerzi and F. Hofmann, University of Heidelberg, for the preparation of enzymes and to Dr. M. Holck for comments on the manuscript.

REFERENCES

1. Reuter H, Scholz H (1977) J Physiol 264:49-62

2. Reuter H (1979) Ann Rev Physiol 41:413-424

3. Chapman RA (1979) Prog Biophys Molec Biol 35:1-52

4. Fozzard HA (1980) In: Zipes DP, Bailey JC, Elharrar V (eds)
    The Slow Inward Current and Cardiac Arrhythmias. Martinus
    Nijhoff Publishers, The Hague/Boston/London, pp 173-203

5. Reuter H (1980) In: Zipes DP, Bailey JC, Elharrar V (eds)
    The Slow Inward Current and Cardiac Arrhythmias. Martinus
    Nijhoff Publishers, The Hague/Boston/London, pp 205-219

6. McDonald TF (1982) Ann Rev Physiol 44:425-434

7. Robison GA, Butcher RW, Sutherland EW (1971) Cyclic AMP. Acade-
    mic Press, New York

8. Robison GA, Butcher RW, Øye I, Morgan HE, Sutherland EW (1965)
    Mol Pharmacol 1:168-177

9. Tsien RW, Giles W, Greengard P (1972) Nature New Biol 240:181-183

10. Tsien RW (1973) Nature New Biol 245:120-123

11. Trautwein W, Taniguchi J, Noma A (1982) Pflügers Arch 392:307-314

12. Brum G, Flockerzi V, Hofmann F, Osterrieder W, Trautwein W
    (1983) Pflügers Arch 398:147-154

13. Greengard P (1976) Nature 260:101-108

14. Sperelakis N, Schneider JA (1976) Am J Cardiol 37:1079-1085

15. Niedergerke R, Page S (1977) Proc Roy Soc Lond B 197:333-362

16. Tsien RW (1977) Adv Cycl Nucl Res 8:363-420

17. Hamill OP, Marty A, Neher E, Sakmann B, Sigworth FJ (1981)
    Pflügers Arch 391:85-100

18. Osterrieder W, Brum G, Hescheler J, Trautwein W, Flockerzi V,
    Trautwein W (1982) Nature 298:576-578

19. Brum G, Osterrieder W, Trautwein W (1984) Pflügers Arch 401:
    111-118

20. Isenberg G, Klöckner U (1982) Pflügers Arch 395:6-18

21. Trube G (1983) In: Sakmann B, Neher E (eds) Enzymatic Dispersion
    of Heart and other Tissues. Plenum Publishing Corporation, New
    York, pp 69-76

22. Beavo JA, Bechtel PJ, Krebs EG (1974) Methods Enzymol 38:299-308

23. Hofmann F (1980) J Biol Chem 255:1559-1564

24. Deck KA, Kern R, Trautwein W (1964) Pflügers Arch 280:50-62

25. Cavalié A, Ochi R, Pelzer D, Trautwein W (1983) Pflügers Arch
    398:284-297

26. Osterrieder W, Yang Q.-f., Trautwein W (1982) Pflügers Arch
    394:78-84

27. Zimmermann ANA, Hülsmann WC (1966) Nature 211:646-647

28. Taniguchi J, Kokubun S, Noma A, Irisawa H (1981) Jpn J Physiol 31:547-558

29. Pelzer D, Trube F, Piper HM (1984) Pflügers Arch 400:197-199

30. Dow JW, Harding NGL, Powell T (1981) J Physiol 331:599-635

31. Brown AM, Lee KS, Powell T (1981) J Physiol 318:455-477

32. Isenberg G, Klöckner U (1982) Pflügers Arch 395:19-29

33. Seamon KB, Daly JW (1981) J Cycl Nucl Res 7:201-224

34. Lindner E, Dohadwalla AN, Bhattacharya BK (1978) Drug Res 28: 284-289

35. Lindner E, Metzger H (1983) Drug Res 33:1436-1441

36. Rubin CS, Ehrlichmann J, Rosen OM (1972) J Biol Chem 247:36-44

37. Hofmann F, Beavo JA, Bechtel PJ, Krebs EG (1975) J Biol Chem 250:7795-7780

38. Smith SB, LaPorte D, Builder S, Beavo JA, Krebs EG (1979) Federation Proc 38:465

39. Corbin JD, Sugden PH, West L, Flockhart DA, Lincoln TM, McCarthy D (1978) J Biol Chem 253:3997-4003

40. Krebs EG, Beavo JA (1979) Ann Rev Biochem 48:429-451

41. Pappano AJ (1970) Circ Res 27:379-390

42. Reuter H (1984) Ann Rev Physiol 46:473-484

43. Reuter H, Stevens CF, Tsien RW, Yellen G (1982) Nature 297:501-504

44. Reuter H (1983) Nature 301:569-574

45. Reuter H (1984) In: Opie LH (ed.) Calcium Antagonists and Cardiovascular Disease. Raven Press, New York, pp 43-51

46. Cachelin AB, de Peyer JE, Kokubun S, Reuter H (1983) Nature 304:462-464

47. Fenwick EM, Marty A, Neher E (1982) J Physiol 331:599-635

48. Lux HD, Nagy K (1981) Pflügers Arch 391:252-254

49. Noma A, Kotake H, Irisawa H (1980) Pflügers Arch 388:1-9

50. Brown H, DiFrancesco D (1980) J Physiol 308:331-351

51. Colquhoun D, Hawkes AG (1981) Proc Roy Soc Lond B 211:205-235

52. Tada M, Katz AM (1982) Ann Rev Physiol 44:401-423

53. Walsh DA, Perkins JP, Krebs EG (1968) J Biol Chem 243:3763-3765

54. Will H, Levchenko TS, Levitsky DO, Smirnov VN, Wollenberger A (1978) Biochem Biophys Acta 543:175-193

55. StLouis PJ, Sulakhe PV (1979) Arch Biochem Biophys Acta 198: 227-240

56. Lamers JMJ, Stinis JT (1980) Biochem Biophys Acta 624:227-240

57. Rinaldi ML, Capony J-P, Demaille JG (1982) J Mol Cell Cardiol 14:279-289

58. Manalan AS, Jones LR (1982) J Biol Chem 257:1005Z
59. Venter JC, Fraser CM, Schaber JS, Jung CY, Triggle DJ (1983) J Biol Chem 15:9344-9348
60. Sigworth FJ (1980) J Physiol 307:97-129
61. Bean BP, Nowycky MC, Tsien RW (1984) Nature 307:371-375
62. Tsien RW, Bean BP, Hess P, Nowycky MC (1983) In: Cold Spring Harbor Symposia on Quantitative Biology, Vol. 18, pp 201-212

RESUME

L'augmentation de la conductance au Ca^{++} de la membrane de la cellule cardiaque, induite par les β adrénergiques, a été étudiée dans des myocytes isolés à partir de coeur de cochon d'Inde. L'adrénaline, ajoutée au bain, augmente l'amplitude et la durée du potentiel d'action transmembranaire indiquant une augmentation du courant Ca^{++} vers l'intérieur. Ces effets ont été reproduits par la forskoline, un activateur de l'adénylate cyclase. L'injection, sous pression, dans les cellules de la sous-unité catalytique de la protéine kinase AMP cyclique dépendante, augmente la conductance au Ca^{++}; l'injection de la sous-unité régulatrice a l'effet opposé. On en conclut que l'adrénaline, la forskoline et une protéine kinase AMP cyclique dépendante, modulent le courant Ca^{2+} vers l'intérieur, par le même mécanisme de base, probablement une phosphorylation de protéine membranaire.

Le courant à travers des canaux calcium individuels a été mesuré par la technique du "patch-clamp". Lorsqu'il n'y a qu'un seul canal calcium dans une membrane, l'application d'agonistes β adrénergiques, résulte en une augmentation du courant moyen (calculé à partir de nombreux enregistrements consécutifs) qui ne peut être la conséquence ni d'une augmentation de courant d'un canal unique, ni de l'apparition de canaux supplémentaires dans le "patch". Le calcul des histogrammes de temps d'ouverture et de fermeture ont montré que le canal restait plus longtemps dans son état ouvert et moins longtemps dans son état fermé. Par conséquent, le mécanisme d'augmentation du courant calcium induit par les β adrénergiques est une augmentation de probabilités pour le canal Ca^{2+} individuel, à être ouvert sous dépolarisation.

Hormones and Cell Regulation, Volume 9
INSERM European Symposium
J.E. Dumont, B. Hamprecht and J. Nunez editors
© 1985 Elsevier Science Publishers B.V. (Biomedical Division)

THE Na+/H+ EXCHANGE SYSTEM. PROPERTIES AND ROLE IN CELL FUNCTION

CHRISTIAN FRELIN, PAUL VIGNE and MICHEL LAZDUNSKI

Centre de Biochimie du CNRS, Université de Nice, Faculté des Sciences, Parc Valrose, 06034 Nice Cedex (France)

INTRODUCTION

A membrane mechanism that exchanges Na+ for H+ has recently been characterized in several eukaryotic cells. It is inhibited by amiloride, a widely used diuretic drug. This exchanger is a major mechanism for regulating the internal pH in eukaryotic cells (1). It has furthermore been implicated in other important physiological processes such as the reabsorption of Na+ by the renal proximal tubule, the regulation of the cell volume in lymphocytes, the fertilization of sea urchin eggs, and it seems to mediate some of the action of growth factors on fibroblasts.

This chapter summarizes the properties of this important membrane cation transporting system and analyses some of its functions. It is stressed that the same exchange system can fulfil different functions in different cell types depending on (i) its intrinsic catalytic properties, (ii) the presence of other membrane structures that also contribute to the maintenance of ion gradients across the membrane and (iii) the presence of extracellular effectors that modify its catalytic properties.

OCCURENCE OF THE Na+/H+ EXCHANGE SYSTEM

Table 1 lists the cell types in which a Na+/H+ exchange system has been characterized in details and stresses its wide distribution. The only cell type that is apparently devoid of Na+/H+ exchanger is the anucleated erythrocyte.

THE PHARMACOLOGICAL PROPERTIES OF THE Na+/H+ EXCHANGE SYSTEM

High affinity inhibitors of ion transport systems are essential to analyse both their physiological function annd their molecular properties. Well-known examples are ouabain for the (Na+,K+)ATPase and tetrodotoxin for the voltage-dependent Na+ channels. For the Na+/H+ exchange system, amiloride is such an inhibitor.

Half-maximum effect for amiloride action on the Na+/H+ exchanger, measured in $^{22}Na^+$ uptake experiments or in H+ efflux experiments are in the range of 3 μM to 1 mM. The reason for such a large range is that both external Na+ and H+ antagonize the action of amiloride in a competitive manner (3, 9, 20, 21). The K_i for Na+ at the amiloride binding site is about 20 mM (3, 9) and the pK for H+ inhibition of amiloride action is 7.3 to 7.5 (20). When experiments are performed in a low Na+ medium and at

Table 1 : Occurence of the Na+/H+ exchange system

- Skeletal muscle cells   - mouse (2)
                                               - chicken (3)
                                               - frog (4)

- Cardiac cells   - chicken (5, 6)
                                             - rat (6)

- Fibroblasts   - mouse 3T3 cells (7, 8)
                                             - Chinese hamster lung fibroblasts (9)
                                               - human foreskin fibroblasts (10)
                                               - A 431 (11)

- Lymphocytes (12)

- Neuronal cells   - mouse neuroblastoma cells (13)
                                             - rat brain synaptosomes (14)

- Kidney cells   - MDCK cells (15)
                                             - brush border membranes (16)
                                               - renal proximal tubule (17)

- PC 12 cells (18)

- Necturus gall-bladder (19)

references to published work are given in parentheses

alkaline external pH's so that neither external Na^+ nor H^+ prevent the binding of amiloride to its receptor site, the K_d for amiloride is observed between 3 and 5 μM (2, 9, 21).

Because of its low affinity for the exchanger under physiological conditions of external Na^+, and because amiloride is a weak base (pKa = 8.5) which accumulates into intracellular acidic compartments (22, 23), the use of amiloride for analyzing the functions of the Na^+/H^+ exchanger is not without problems.

A structure-activity relationship for several amiloride derivatives with selected modifications on each of the functional groups of the molecule was undertaken with the hope to find more potent inhibitors. An unsubstituted guanidino group is essential for the activity of the molecule since any substituted derivative of amiloride (e.g. benzamil) is less potent than amiloride (24). Substitutions on the 5-amino group with alkyl or alkenyl groups produce compounds that are more active than amiloride. As a rule N-5 disubstituted derivatives of amiloride are more potent than the N-5 monosubstituted derivatives (24). The best inhibitor found is ethylisopropyl-amiloride with a K_d for the Na^+/H^+ exchange system of 30 to 50 nM under low Na^+ conditions (25). The pharmacological properties of the Na^+/H^+ exchange system are identical in chick and rat skeletal muscle cells, chick and rat cardiac cells, 3T3 fibroblasts, CCl39 fibroblasts, MDCK cells, renal brush border membranes and rat brain synaptosomes (6, 14, 24, 26, 27).

Although N-5 disubstituted derivatives of amiloride are more potent than amiloride for inhibiting the Na^+/H^+ exchanger of brush border membranes of renal proximal tubules (27), their diuretic properties are identical to those of amiloride (28). The reason for this apparent discrepancy is that N-5 disubstituted derivatives of amiloride are readily transformed into amiloride when injected to animals.

Amiloride is also known to inhibit electrogenic Na^+ channels of frog skins and toad urinary bladders. This action can also contribute to its diuretic properties. Structure activity relationship for selected amiloride analogs in relation to the inhibition of Na^+ channels has also been performed and produced different results. Benzamil, which is almost inactive for inhibiting the Na^+/H^+ exchange system, is the most potent molecule found for inhibiting Na^+ channels. Conversely N-5 disubstituted derivatives for amiloride that are highly potent on the Na^+/H^+ exchanger are inactive on Na^+ channels (29). High affinity binding sites for [^3H]benzamil have been identified in the kidney (30).

Benzamil and dichloroderivatives of benzamil have recently been shown to inhibit the Na^+/Ca^{2+} exchange system of cardiac cells (31). Amiloride and N-5 disubstituted derivatives of amiloride have no action on the Na^+/Ca^{2+} exchanger of cardiac cells (6).

PROPERTIES OF THE Na^+/H^+ EXCHANGE SYSTEM

1. The Na^+/H^+ exchange system catalyses the electroneutral and reversible exchange of Na^+ for H^+ (1, 3, 12, 13). A 1:1 stoichiometry means that its activity is independent of the membrane potential and that it does not contribute to the value of the membrane

potential. In cardiac cells, however, amiloride produces a partial depolarization of the membrane and the stimulation of the exchanger produces a membrane hyperpolarization (6, 32). These effects are due to a tight coupling between the (Na^+,K^+)ATPase, which is electrogenic and the Na^+/H^+ exchanger which is electroneutral (33).

2. External Na^+ stimulates H^+ efflux and $^{22}Na^+$ uptake. The apparent K_m for Na^+ is comprised between 5 mM and 60 mM in different cell types (3, 9, 11-13, 15, 16, 19). Such a large range of values does not mean the existence of several types of Na^+/H^+ exchanger. They are only apparent values since, in most cells, the internal pH depends on the Na^+ gradient across the membrane, so that, when the external Na^+ concentration varies, the internal pH also varies.

3. External Li^+ also stimulates H^+ efflux but not choline, K^+, Rb^+ or Cs^+. The effect of Li^+ is blocked by amiloride, indicating that Li^+ can substitute for Na^+. The apparent K_m for Li^+ is lower than the K_m for Na^+, and the maximum rate of H^+ efflux that is induced by Li^+ is only half that produced by Na^+ (9, 13, 19, 34).

4. Na^+ uptake and H^+ efflux by the Na^+/H^+ exchanger are inhibited by external H^+ with a pK comprised between 7.0 and 7.4 and a Hill coefficient of 1 (3, 6, 9, 12, 14, 20). The nature of external H^+ inhibition of the Na^+/H^+ exchanger is not yet clear. Non-competitive type of inhibition, competitive inhibition and a mixed type of inhibition have been described in different cell types (3, 9, 20).

THE INTERNAL pH DEPENDENCE OF THE Na^+/H^+ EXCHANGE SYSTEM

The pHi dependence of the Na^+/H^+ exchange system does not follow simple Michaelis Menten kinetics as first pointed out by Aronson et al. (35) for renal brush border membranes. This finding has now been confirmed in other cell types such as CCl39 fibroblasts (36), chick skeletal muscle cells (37) and rat thymocytes (12, 38). The most interesting observation is that the internal pH dependence of the Na^+/H^+ exchanger is different in different cell types and that extracellular messages can modify this dependence. Two extreme situations have been found. In chick skeletal muscle cells and cardiac cells the internal pH dependence of the Na^+/H^+ exchanger is very steep and full activation is observed within less than 1 pH unit. Hill coefficients are higher than 3. The half-maximum effect for internal H^+ activation of the exchanger is observed at pH 7.3 to 7.4, which is close to the physiological pHi value. This means that for these cells, the exchanger is already half-maximally activated under physiological conditions of pHi (37).

In 3T3 fibroblasts, brush border membrane vesicles, rat thymocytes and human foreskin fibroblasts, the internal pH dependence is less steep and the apparent pK for internal H^+ is shifted to more acidic values as compared to chick skeletal muscle cells. As a consequence the activity of the exchanger is negligible at physiological pHi values (7.0 to 7.3) and an intracellular acidification is necessary to reveal its activity (7, 10, 12, 34, 38, 39).

Internal Na^+ ions most likely also interact with the internal H^+ binding sites on the exchanger, so that at high $[Na^+]_i$, the internal pH dependence of the exchanger may be shifted to more acidic pH values (12).

THE ROLE OF THE Na+/H+ EXCHANGER IN RELATION TO THE CONTROL OF THE STEADY STATE pHi

For an electroneutral, fully reversible and symetric Na+/H+ exchanger, like the carboxylic ionophore monensin, thermodynamic equilibrium will be attained when

$$\frac{[H^+]_i}{[H^+]_o} = \frac{[Na^+]_i}{[Na^+]_o}$$

Since the Na+ gradient across the membrane is actively maintained by the (Na+,K+)-ATPase so that $[Na^+]_i$ is 10 to 20 mM when $[Na^+]_o$ is 140 mM, such an exchanger could drive pHi up to almost 1 pH unit above pHo. Under physiological conditions, the pHi of eukaryotic cells in only 0.1 to 0.4 pH units more acidic than the external pH (1). The difference could mean either that the other membrane systems contribute to the steady-state pHi value or that the behaviour of the exchanger is not assymetric so that its activity is turned off before reaching high alkaline pHi values.

In chick skeletal muscle cells, the half-maximum activation of the Na+/H+ exchanger is observed at pHi values close to the physiological pHi; the activity of the exchanger is already high under physiological conditions and its function is to regulate the steady-state pHi level. Changing the Na+ gradient across the membrane modifies the pHi and these variations in pHi can be suppressed by amiloride derivatives (37).

An intracellular acidification can still produce a 2-fold activation of the exchanger and because of the very steep dependence on the pHi, a second function is to allow cells to recover from an intracellular acidosis. Actually this is how the Na+/H+ exchanger was first discovered in mouse muscle fibers (2) and later studies have shown that this is a very general phenomenon.

THE ROLE OF THE Na+/H+ EXCHANGER IN RELATION TO THE CONTROL OF THE INTERNAL Na+ CONCENTRATION

In chick cardiac cells, pHi dependence of the Na+/H+ exchanger is similar to that of chick skeletal muscle cells. However in cardiac cells, the Na+/H+ exchange system does not contribute to the regulation of the physiological steady-state pHi value (39). Other, yet unidentified mechanisms should be involved in pHi regulation and the main function of the exchanger is to adjust the rate of Na+ uptake to the H+ gradient. As a consequence, amiloride, and its more potent derivatives, are able to antagonize the effect of digitalis compounds on the heart (6).

INFLUENCE OF EXTRACELLULAR SIGNALS ON THE ACTIVITY OF THE Na+/H+ EXCHANGER

In a number of cell types, the activity of the Na+/H+ exchanger is not simply determined by the membrane Na+ and H+ gradients; it can also be modified by specific extracellular messages. Best studied examples are the following.

1. Effect of insulin on frog skeletal muscle :

In frog skeletal muscle, insulin activates a Na+ uptake component which is associated with an increased rate of H+ efflux and a cell alkalinisation (4). These early effects of insulin are blocked by amiloride as expected if insulin activates the Na+/H+ exchanger. Further studies, carried out on chick skeletal muscle cells, rat muscle cells and human foreskin fibroblasts have failed to show a similar effect of insulin on avian and mammalian cells (10, 37). However in CCl39 fibroblasts and human foreskin fibroblasts, insulin potentiates the action of growth factors on the Na+/H+ exchanger (10, 36, 41).

2. Effect of growth factors on fibroblasts :

In quiescent, growth-arrested fibroblasts, the addition of serum or of purified growth factors (EGF, PDGF and α-thrombin) stimulates a ^{22}Na+ uptake component, H+ efflux and a cell alkalinisation (ref. 42 for a review). There early events are blocked by amiloride and its more potent derivatives (42) and it is believed that the activation of the Na+/H+ exchanger is important for mediating the action of growth factors on proliferation (7-11, 41, 42). Serum and mitogens do not modify the external Na+ or H+ dependence of the Na+/H+ exchange system or its interaction with amiloride (7). Rather they increase the sensitivity of the Na+/H+ exchanger to the pHi (7, 10, 36) and Paris and Pouysségur (36) have clearly shown in CCl39 fibroblasts that growth factors shift the internal pH dependence of the Na+/H+ exchanger to more alkaline values.

3. The regulation of the cell volume :

In lymphocytes two membrane mechanisms contribute to the regulatory volume increase : the Na+/H+ exchange system and the Cl-/HCO3- exchanger which operate in a coupled fashion to allow Na+ and Cl- to enter the cells (42). Cell shrinkage produces a shift in the internal pH dependence of the exchanger to more alkaline pH's (39). As a consequence, the Na+/H+ exchanger which is inactive at physiological pHi values becomes activated and pumps Na+ inside the cells. Chloride ions accompany Na+ by the coupled operation of the Cl-/HCO3- exchange system (44). The same phenomenon also applies to Chinese hamster ovary cells (45).

4. Fertilization of sea urchin eggs :

Unfertilized sea urchin eggs regulated their pHi independently of the external Na+ concentration. Upon fertilization, the activity of the Na+/H+ exchange system is unmasked, Na+ enters the cells, thereby activating the (Na+,K+)ATPase and H+ efflux is promoted. A cell alkalinisation follows, which seems essential for the early development of the sea urchin embryo (46, 47). The mechanism of activation of the Na+/H+ exchanger that is triggered by fertilization is not known.

5. The mechanism of activation of the Na+/H+ exchange system :

One of the main question to be solved is how different stimuli such as growth factors, cell shrinkage or fertilization can all produce an activation of the Na+/H+ exchange system. The finding that different growth factors, and also the regulatory volume increase involve an increased sensitivity of the Na^+/H^+ exchanger to the pHi (7, 10, 36, 39) strongly suggests that these different stimuli act _via_ a common mechanism which is to switch the internal pH dependence of the Na^+/H^+ exchanger to more alkaline values. It has been proposed that a phosphorylation of the Na^+/H^+ exchanger might be responsible for its change in catalytic properties (40, 41). This hypothesis has received recent support by the finding that phorbol esters which stimulate protein kinase C increase the activity of the Na^+/H^+ exchanger in sea urchin eggs (48), murine pre B-lymphocytes (49), human leukemic cells (50), 3T3 cells (51) and in rat thymocytes (52). Grinstein et al. (52) further showed that phorbol esters also shift the internal pH dependence of the Na^+/H^+ exchanger to more alkaline values. Growth factors like EGF, PDGF and insulin can also potentially produce the same effect since their respective membrane receptor have also a protein kinase activity (53-55). An alternative hypothesis may be that the growth factor activated protein kinases stimulate the breakdown of inositol phospholipids (56) with the release of two intracellular messengers. One would be diacylglycerol, a known activator of protein kinase C. The second one would be inositol 1, 4, 5 triphosphate which acts by mobilizing intracellular Ca^{2+} stores (57). An increase in cytosolic Ca^{2+} after exposure of human fibroblasts to growth factors has recently been shown (58). More detailed evidences for a role of phospholipase activity in the activation by serum of the Na^+/H^+ exchanger in human fibroblasts has also been presented (59).

IDENTIFICATION OF THE AMILORIDE BINDING SITES

The identification of the protein molecule that allows the exchange of Na^+ for H^+ across the membrane is now essential to test the hypothesis that the activation of the Na^+/H^+ exchange system is produced by a phosphorylation by a protein kinase.

A first step in this direction has been made by synthesizing a highly radioactive derivative of amiloride than can be used to titrate amiloride binding sites. High affinity [^{3}H]ethylpropylamiloride binding sites (K_d = 45 nM) have been characterized in brush border membranes from rabbit renal cortex (37). [^{3}H]EPA binding sites have the same pharmacological properties as the Na^+/H^+ exchange system of brush border membrane vesicles, as defined by $^{22}Na^+$ flux experiments. Such a compound will probably proved to be very useful for characterizing the Na/H^+ exchanger in other cell types.

REFERENCES

1. Roos A, Boron WF (1981) Physiol Rev 61:296

2. Aickin C, Thomas RC (1977) J Physiol 273:395

3. Vigne P, Frelin C, Lazdunski M (1982) J Biol Chem 257:9394

4. Moore RD (1981) Biophys J 33:203

5. Piwnica-Worms D, Lieberman M (1983) Am J Physiol 244:C422

6. Frelin C, Vigne P, Lazdunski M (1984) J Biol Chem 259:8880

7. Frelin C, Vigne P, Lazdunski M (1983) J Biol Chem 258:6272

8. Cassel D, Rothenberg P, Zhuang YX, Deuel TF, Glaser L (1983) Proc Natl Acad Sci USA 80:6224

9. Paris S, Pouysségur J. (1983) J Biol Chem 258:3503

10. Moolenaar WH, Tsien RY, Van Der Saag PT, De Laat SW (1983) Nature 304:645

11. Rothenberg P, Glaser L, Schlessinger P, Cassel D (1983) J Biol Chem 258:12644

12. Grinstein S, Cohen S, Rothstein A (1984) J Gen Physiol 83:341

13. Moolenaar WH, Boonstra J, Van Der Saag PT, De Laat SW (1981) J Biol Chem 256:12883

14. Sauvaigo S, Vigne P, Frelin C, Lazdunski M (1984) Brain Res 301:371

15. Rindler M J , Saier MH (1981) J Biol Chem 256:10820

16. Kinsella JL, Aronson PS (1980) Am J Physiol 238:F461

17. Boron WF, Boulpaep EL (1983) J Gen Physiol 81:29

18. Boonstra J, Moolenaar WH, Harrison PM, Moed P, Van Der Saag PT, De Laat SW (1983) J Cell Biol 97:92

19. Weinman SA, Reuss L (1982) J Gen Physiol 80:299

20. Aronson PS, Suhm MA, Nee J (1983) J Biol Chem 258:6767

21. Rindler MJ, Taub M, Saier MH (1979) J Biol Chem 254:11431

22. Smith RC, Macara IG, Levenson R, Houseman D, Cantley L (1982) J Biol Chem 257:773

23. Benos DJ, Reyes J, Schoemaker DG (1983) Biochim Biophys Acta 734:99

24. Vigne P, Frelin C, Cragoe EJ, Lazdunski M (1984) Mol Pharmacol 25:131

25. Vigne P, Frelin C, Cragoe EJ, Lazdunski M (1983) Biochem Biophys Res Commun 116:86

26. L'Allemain G, Franchi A, Cragoe EJ, Pouységur J (1984) J Biol Chem 259:4313

27. Vigne P, Frelin C, Audinot M, Borsotto M, Cragoe EJ, Lazdunski M (1984) EMBO J in press

28. Cragoe EJ (1983) In: Cragoe EJ (ed) Diuretics : Chemistry, Pharmacology and Medicine. John Wiley and Sons Inc, pp 229-240

29. Cuthbert AW, Fanelli GM (1978) Br J Pharmacol 63:139

30. Cuthbert AW, Edwardson JM (1981) Biochem Pharmacol 30:1175

31. Siegl PKS, Cragoe EJ, Trumble MJ, Kaczorowski GJ (1984) Proc Natl Acad Sci USA 81:3238

32. Piwnica-Worms D, Jacob R, Horres CR, Lieberman M (1982) J Gen Physiol 80:20a

33. Jacob R, Piwnica-Worms D, Horres CR, Lieberman M (1984) J Gen Physiol 83:47

34. Kinsella JL, Aronson PS (1981) Am J Physiol 241:C220

35. Aronson PS, Nee J, Suhm MA (1982) Nature 299:161

36. Paris S, Pouysségur J (1984) J Biol Chem in press

37. Vigne P, Frelin C, Lazdunski M (1984) EMBO J 3:1865

38. Vaughan Jones RD, Lederer WJ, Eisner DA (1983) Nature 301:522

39. Grinstein S, Goetz JD, Rothstein A (1984) J Gen Physiol in press

40. Moolenaar WH, Tertoolen LGJ, De Laat SW (1984) J Biol Chem 259:7563

41. Pouysségur J, Chambard JC, Franchi A, Paris S, Van Obberghen Schilling E (1982) Proc Natl Acad Sci USA 79:3935

42. Rozengurt E (1984) Hormones and Cell Regulation 8:17

43. Grinstein S, Clarke CA, Rothstein A (1983) J Gen Physiol 82:619

44. Grinstein S, Rothstein A, Sarkadi B, Gelfand EW (1984) Am J Physiol 246:C204

45. Sarkadi B, Attisano L, Grinstein S, Buchwald M, Rothstein A (1984) Biochim Biophys Acta 774:159

46. Johnson CH, Epel E (1981) J Cell Biol 89:284

47. Payan P, Girard JP, Ciapa B (1983) Dev Biol 100:29

48. Swann K, Whitaker M (1984) J Physiol 353:86P

49. Rosoff PM, Stein LF, Cantley LC (1984) J Biol Chem 259:7056

50. Besterman JM, Cuatrecasas P (1984) J Cell Biol 99:340

51. Burns CP, Rozengurt E (1983) Biochem Biophys Res Commun 116:931

52. Grinstein S, Cohen S, Goetz JD, Rothstein A, Gelfand EW (1984) Proc Natl Acad Sci USA in press

53. Cohen S, Ushiro H, Stoscheck C, Chinkers M (1982) J Biol Chem 257:1523

54. Heldin CH, Ronnstrand L (1983) J Biol Chem 258:10054

55. Van Obberghen E, Rossi B, Kowalski A, Gazzano H, Ponzio G (1983) Proc Natl Acad Sci USA 80:945

56. Sugimoto Y, Whitman M, Cantley L, Erikson RL (1984) Proc Natl Acad Sci USA 81:2117

57. Streb H, Irvine RF, Berridge MJ, Schulz I (1983) Nature 306:67

58. Moolenaar WH, Tertoolen LGJ, De Laat SW (1984) J Biol Chem 259:8066

59. Vicentini LM, Miller RJ, Villereal ML (1984) J Biol Chem 259:6912

RESUME

Un mécanisme membranaire catalysant l'échange électroneutre de Na+ et de H+ a récemment été décrit chez la plupart des cellules eucaryotes. Cet échangeur est inhibé par l'amiloride, un diurétique puissant, et il a été impliqué dans nombre de processus physiologiques importants tels que la régulation du pH intracellulaire, la réabsorption de Na+ par le tubule proximal rénal, la régulation du volume intracellulaire dans les lymphocytes et la fertilisation de l'oeuf d'oursin. De plus cet échangeur semble être associé étroitement à l'action des facteurs de croissance.

Les propriétés pharmacologiques de l'échangeur Na+/H+ sont définies ainsi que ses interactions avec Na+, Li+ et H+ extracellulaires. Elles sont identiques pour tous les types cellulaires étudiés. Cependant l'échangeur Na+/H+ peut remplir des fonctions différentes dans différents types cellulaires selon 1) ses propriétés d'interaction avec les protons intracellulaires, 2) la présence d'autres structures membranaires qui interviennent dans le maintien des gradients transmembranaires de H+ et de Na+ et 3) la présence d'effecteurs extracellulaires qui peuvent modifier ses propriétés catalytiques et en particulier l'interaction avec les protons intracellulaires.

Hormones and Cell Regulation, Volume 9
INSERM European Symposium
J.E. Dumont, B. Hamprecht and J. Nunez editors
© 1985 Elsevier Science Publishers B.V. (Biomedical Division)

SIGNAL TRANSDUCTION BY POLYPEPTIDE GROWTH FACTOR RECEPTORS

S.W. de LAAT, W.H. MOOLENAAR, L.H.K. DEFIZE, J. BOONSTRA[*] and P.T. van der SAAG
Hubrecht Laboratory, Uppsalalaan 8, 3584 CT Utrecht and [*]Department of Molecular Cell Biology, University of Utrecht, Padualaan 8, 3584 CH Utrecht, The Netherlands

INTRODUCTION

Polypeptide growth factors play a prominent role in the control of cellular growth and differentiation. In general, growth factors are capable of directing growth and differentiation of target cells at low concentrations, usually in the nanomolar range. Their action is initiated by binding to specific receptor molecules localized in the plasma membrane, so that the plasma membrane is their initial site of action on the cell. Interestingly, various growth factors show a striking similarity in their plasma membrane-mediated responses, suggesting that common molecular pathways are utilized in reaching diverse biological end points (1,2).

Epidermal growth factor (EGF) and platelet-derived growth factor (PDGF) are among the most intensively studied polypeptide growth factors. The general importance of the growth regulatory function of EGF and PDGF and their respective receptors is emphasized by the recent findings that: 1) the cytoplasmic domain of the EGF-receptor shows great homology with the V-erb B oncogene product (3,4), and 2) the human C-sis proto-oncogene is a structural gene for PDGF (5-8). Recently a number of review articles have appeared describing the molecular, biochemical and physical properties of these growth factors and their receptors (1,9,10), and we refer hereto for further details.

Growth factor-receptor interaction initiates a cascade of biochemical and physiological responses in cells, which ultimately lead to changes in cellular behaviour, such as initiation of proliferation or differentiation. However, the causal relationships between growth factor-induced receptor mediated responses and the ultimate changes in genetic expression in the target cell are still unclear. In the search for specific mitogenic signals in growth factor action, attention has been focused upon the earliest detectable responses, which include a rapid stimulation of tyrosine-phosphokinase activity (11,12) the kinase being an intrinsic part of the receptor molecule (13,14), phospholipid breakdown (15,16) and ionic changes (1,2).

In this paper we will restrict ourselves to a description of the early growth factor-induced ionic changes and their possible underlying mechanisms, and we will describe recent experiments on the interrelationships between the various early growth factor responses by the use of monoclonal anti-EGF-receptor antibodies.

CAUSE AND EFFECT IN IONIC SIGNAL TRANSDUCTION

Activation of Na^+, H^+ exchange by growth factors

Ionic responses have been included in the mitogenic signal transduction events since the original observation of Rozengurt and Heppel (17) that serum, EGF, insulin and prostaglandins E_1 and $F_{2\alpha}$ stimulated the activity of the Na^+, K^+-ATPase (Na^+, K^+ pump) in quiescent mouse 3T3 cells. Since then increasing evidence has demonstrated that various growth factors activate the Na^+, K^+ pump in a variety of cells (1,2). EGF was for example shown to stimulate Na^+, K^+ pump activity in quiescent human fibroblasts (18), neuroblastoma cells (19), pheochromocytoma cells (20), and other cell types (17,21). In general, the activity of the Na^+, K^+ pump increased gradually in time, a maximum being reached within 30 minutes of addition of EGF. These findings raised questions concerning the underlying mechanism of activation of the Na^+, K^+ pump. Under normal conditions, the Na^+, K^+ pump is regulated by the intracellular Na^+ concentration, as shown by its stimulation upon addition of the Na^+-ionophore monensin (22,23). Accordingly, studies from several laboratories demonstrated that the activation of the Na^+, K^+ pump activity by growth factors is a consequence of an increased availability of intracellular Na^+, due to the rapid stimulation of Na^+ influx (24). These observations have been made for various growth factors (including EGF, insulin, PDGF, NGF) in a variety of cells, such as hepatocytes, normal and viral transformed murine, hamster and human fibroblasts, murine neuroblastoma cells and rat pheochromocytoma cells (1,2). It was concluded that enhanced Na^+ influx is the primary ionic response to growth factors, whereas the stimulation of the Na^+, K^+ pump is secondary, and serves merely for the extrusion of the excess Na^+.

The nature of the Na^+ transport system was subsequently identified as a Na^+, H^+ exchange system. Strong evidence in favor of the involvement of Na^+, H^+ exchange by comparison of electrophysiological and tracer-flux studies in mouse neuroblastoma (22,25), human fibroblasts (18) and rat pheochromocytoma cells (20). It was shown that mitogen-induced Na^+ influx was mediated largely by an electroneutral pathway, distinct from electrodiffusional permeability. Furthermore, the transport system appeared to be sensitive to amiloride, a known inhibitor of Na^+,H^+ exchange. The presence of Na^+,H^+ exchange in mammalian cells was demonstrated by the observation that addition of weak acids to the cells, at a constant external pH, caused an increase in amiloride-sensitive Na^+ influx, whereas addition of Na^+ to Na^+-deprived cells resulted in an increase of amiloride sensitive H^+ efflux (25). Together, these results provided evidence for the involvement of an electroneutral, amiloride-sensitive Na^+,H^+ exchange system in growth factor action. As shown in Table 1, activation of Na^+,H^+ exchange appears to be a common effect of various growth factors in a variety of cells.

Table 1. Activation of amiloride-sensitive Na^+ Influx by Various Mitogens

STIMULUS	CELL TYPE	REFERENCES
Serum	Mouse embryo fibroblasts	24
	Mouse neuroblastoma	
	(N1E-115) cells	22
	Human fibroblasts	18,26
EGF	Mouse neuroblastoma	
	(N1E-115) cells	19,27
	Rat pheochromocytoma	
	(PC12) cells	20
	Human fibroblasts	18,26
	Human epidermoid carcinoma	
	A431 cells	28
PDGF	Mouse fibroblasts (NR6)	29
	Human fibroblasts	Moolenaar, unpubl.
NGF	Mouse neuroblastoma	
	(N1E-115) cells	27
	Rat pheochromocytoma	
	(PC12) cells	20
EGF + glycagon + insulin	Rat hepatocytes	30
Thrombin	Hamster fibroblasts	31
Lys-brady-kinin	Human fibroblasts	32

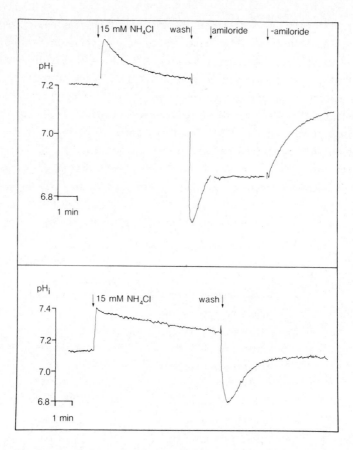

Fig. 1 Continuous recordings of pH_i recovery from an acid load induced by a pre-pulse with 15 mM NH_4Cl in human fibroblasts (upper panel) and N1E-115 neuroblastoma cells (lower panel). In fibroblasts pH_i was recorded fluorimetrically, using bis(carboxyethyl)-carboxyfluorescein (BCECF) as an intracellularly trapped pH_i-indicator. This experiment shows in addition the reversible inhibition of pH_i-recovery by 1 mM amiloride (see ref. 34,35). In neuroblastoma cells pH_i was monitored electrophysiologically, using ultrafine pH sensitive microelectrodes.

The observed activation of Na^+,H^+ exchange by growth factors included the possibility of intracellular pH (pH_i) as a cytoplasmic signal in growth factor action. Recent improvements in pH_i monitoring techniques, especially the use of intracellularly trapped fluorescent pH indicators (33,34) and ultrafine pH microelectrodes that allow for the impalement of single or fused cultured cells, allowed the study of the role of Na^+,H^+ exchange in the regulation of pH_i in general, and in the action of growth factors in particular. The role of Na^+,H^+

exchange in regulation of pH_i was demonstrated by monitoring pH_i recovery from a rapid decrease of pH_i, induced by an NH_4^+ prepulse (34,35). As shown in Fig. 1, both pH_i monitoring methods, intracellularly fluorescent pH indicators and ultrafine pH microelectrodes, are suitable for continuous monitoring of rapid pH_i changes. The observed pH_i recovery followed an exponential time course and is mediated entirely by Na^+,H^+ exchange. This conclusion is based upon the following observations: 1) pH_i recovery was accompanied by Na^+ influx and net H^+ efflux, 2) pH_i recovery and accompanying Na^+ and H^+ fluxes were sensitive to amiloride and 3) the rate of pH_i recovery depends upon the external Na^+ concentration (half-maximal rate at ~ 30 mM Na^+). From experiments such as shown in Fig. 1, it has been concluded that amiloride-sensitive Na^+,H^+ exchange is an efficient regulator of pH_i (34,35).

The exponential recovery of pH_i after an NH_4-prepulse is an important feature, since it implies that the rate of Na^+,H^+ exchange, proportional to $d(pH_i)/dt$, is linearly dependent on pH_i. Therefore, the value of the resting pH_i depends upon the balance between H^+ influx and intracellular H^+ production on one hand and H^+ efflux via the Na^+,H^+ exchanger on the other. It should be noted however, that under normal ionic conditions no effects of amiloride were observed on basal Na^+ influx, although the system is not in thermodynamic equilibrium. That means that the Na^+,H^+ exchanger is relatively inactive under those conditions. Evidence has been presented that the pH_i sensitivity of the Na^+,H^+ exchanger reflects an allosteric activation by internal H^+ at the site which is distinct from the internal transport site (36) and growth factors apparently act by increasing the internal H^+ sensitivity of the exchanger (34). In contrast, the external H^+ may compete with both Na^+ and amiloride for binding at the same external transport site.

Activation of Na^+,H^+ exchange by growth factors might consequently result in a rise of pH_i. Using the conventional weak acid distribution method, initial indications were found that mitogenic stimulation of quiescent fibroblasts indeed resulted in an amiloride-sensitive rise of pH_i (34,35). Direct evidence for growth factor-induced increase of pH_i was obtained using trapped fluorescent pH_i indicators for continuous pH_i recording during mitogenic stimulation (29,35,37). A typical example of pH_i shift after addition of EGF or PDGF to human fibroblasts is shown in Fig. 2.

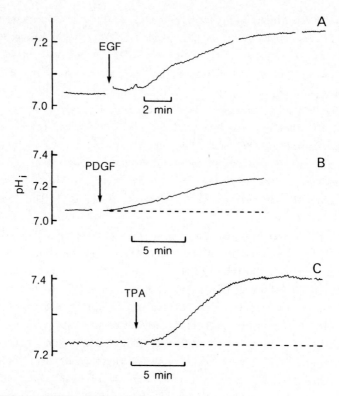

Fig. 2 Effect of EGF, PDGF and TPA on pH_i.
Cells were loaded with the fluorescent pH_i indicator BCECF, and fluorescence was calibrated as described previously (34,35). Response of quiescent human fibroblast to A: EGF (50 ng/ml); B: PDGF (40 ng/ml) and C: TPA (50 ng/ml). Data obtained from Moolenaar et al. (34,42).

The pH_i shift is detectable within 30 seconds after growth factor addition and is completed by 10-15 minutes. The involvement of Na^+, H^+ exchange in this growth-factor induced pH_i shift is demonstrated by the observations that the pH_i shift was inhibitable by amiloride and depended upon the presence of a transmembrane Na^+ gradient. In the continuous presence of growth factors the elevated pH_i is maintained for at least 2 hours, whereas pH_i slowly returns to its value after washout of growth factors. Thus activation of the Na^+, H^+ exchanger is a rapid event, whereas deactivation is a relatively slow process.

Evidence for the mechanism of activation of the Na^+,H^+ exchanger was obtained from studies of the effects of the tumor promoting phorbol ester 12-0-tetra-decanoyl phorbol-13-acetate (TPA). TPA has been demonstrated to bind to and to activate the phospholipid-dependent protein kinase C. Under normal conditions, this enzyme is activated by endogenous diacylglycerol derived from the prior breakdown of inositol phospholipids (40,41). TPA was shown to mimic growth factors with respect to a rise of pH_i after addition to human fibroblasts and other cell types (42-44). In HF cells TPA caused a rise of pH_i by 0.15 pH units within 10 minutes of addition (Fig. 2). The effect of TPA on pH_i was amiloride-sensitive and dependent upon extracellular Na^+, demonstrating the involvement of Na^+,H^+ exchange. In addition, a non-tumor promoting phorbol ester, which has been demonstrated to have no effect on protein kinase C, was unable to raise pH_i (42). Furthermore, synthetic diacylglycerol, a potent activator of protein kinase C, has been demonstrated to raise pH_i in human fibroblasts, Hela cells and mouse NlE-115 neuroblastoma cells (42). These findings strongly suggest that protein kinase C is responsible for modifying the pH_i sensitivity of the Na^+,H^+ exchanger. The simplest explanation would be that protein kinase C phosphorylates the Na^+,H^+ exchanger directly, thereby increasing its sensitivity for cytoplasmic H^+.

Ca^{2+} mobilization by growth factors

The intracellular free Ca^{2+} has been recognized for a long time as an important parameter in the regulation of numerous cellular processes. However, relatively little is known about the role of intracellular free Ca^{2+} in the action of growth factors, mainly due to the lack of reliable methods to determine cytoplasmic, free Ca^{2+}. The fluorescent Ca^{2+} indicator quin-2 has, however, recently provided a means of determining these values. It has been established, for example, that a rapid 2-fold rise in $[Ca^{2+}]_i$ occurs following lectin-stimulation of lymphocytes (45). Using the same technique, the effect of EGF and PDGF on $[Ca^{2+}]_i$ were recorded in quiescent human fibroblasts (46). As demonstrated in Fig. 3, EGF and PDGF caused 20-30% increase in quin-2 fluorescence without a detectable lag period and completed within 20-40 seconds. The increase in quin-2 fluorescence represents a 2-3 fold increase in $[Ca^{2+}]_i$. After reaching the maximal value, the fluorescence declines slowly to a new steady state level about 30% above the normal resting level of 100-150 nM (46).

Since the growth-factor induced rise of $[Ca^{2+}]_i$ was independent upon the presence of extracellular Ca^{2+}, it was concluded that the rise of $[Ca^{2+}]_i$ is not caused by changed Ca^{2+} influx, and therefore Ca^{2+} is released from intracellular sources (mitochondria, plasmamembrane, endoplasmic reticulum).

The response of $[Ca^{2+}]_i$ to growth factors in the fastest response thus far described, suggesting that Ca^{2+} acts as a primary trigger in the action of growth factors. A similar conclusion has been reached for the action of mitogenic lectin in lymphocytes (45). It is of particular interest to note, that TPA, in contrast to its effects on pH_i, did not affect $[Ca^{2+}]_i$ (Fig. 3) indicating that TPA acts at a point distal from Ca^{2+} mobilization. These observations demonstrate that activation of the Na^+, H^+ exchanger and Ca^{2+} mobilization occur in parallel but independently of each other.

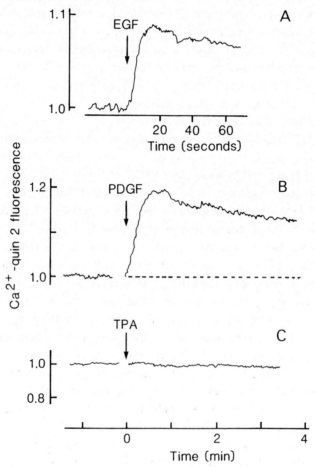

Fig. 3 Effect of EGF, PDGF and TPA on $[Ca^{2+}]_i$.
Response of human fibroblasts, loaded with the fluorescent Ca^{2+} indicator quin-2, to A: EGF (50 ng/ml), B: PDGF (40 ng/ml) and C: TPA (50 ng/ml) respectively. Quin-2 loading of quiescent HF cells and fluorescence monitoring were performed as described in detail elsewhere (46). Data obtained form Moolenaar et al. (42,46).

Altogether, the data are consistent with the proposal that growth factor-receptor interaction causes an immediate breakdown of inositol phospholipids and in particular the formation of inositol 1,4,5-triphosphate constitutes a key signal for internal Ca^{2+} release (47). Furthermore, the subsequent formation of diacylglycerol and the increased $[Ca^{2+}]_i$ stimulates protein kinase C activity, resulting in an activation of the Na^+,H^+ exchanger.

DISSOCIATION OF EGF-RECEPTOR MEDIATED RESPONSES

As stated before, cause and effect among the various cellular responses evolved by the interaction of growth factors with their receptors is far from clear. Such knowledge is however, crucial to understand the mechanism of growth factor action, in particular in relating the early receptor-mediated responses to the ultimate effect on cellular proliferation. In this respect several approaches have been applied, such as a temporal analysis of the early responses and the study of effects of experimental activation of molecules which are likely to be involved in the signal transduction machinery, as described in the previous section. Another approach involved the dissociation of the multitude of growth factor-induced responses by partial activation of the receptor molecule. An early example of this approach has been presented by Schechter et al. (48), who compared the biological activity of native EGF with that of cyanogen-bromide cleaved EGF (CnBr-EGF). Cnbr-EGF was unable to stimulate DNA-synthesis in quiescent human fibroblasts, and this was related to the inability to cause clustering of EGF-receptors (EGF-R) within the plasma membrane. Experimentally induced clustering of the CnBr-EGF/EGF-receptor complexes, however, resulted in mitogenic responses, demonstrating the importance of the spatial reorganization of the receptors in evoking the ultimate biological growth factor response. Subsequently, it was demonstrated that, at similar receptor occupancy, CnBr-EGF was as potent as EGF in enhancing tyrosine phosphorylation of plasma membrane proteins in A431 cells (49). Therefore, stimulation of tyrosine-phosphokinase activity was independent of receptor clustering, and, importantly, was either irrelevant or a necessary, but insufficient signal for the mitogenic response.

An important tool in the dissociation of EGF-receptor responses has been obtained by the availability of monoclonal antibodies directed against different antigenic determinants of the EGF-R. Such antibodies have been raised by several laboratories (50-54), including ours, and we have exploited these to obtain further knowlegde of the causes and effects of the various EGF-R-mediated responses, such as tyrosine phoshokinase activity, EGF-R autophosphorylation, morphological changes and mitogenic responses.

Production of monoclonal antibodies against EGF-receptor

Monoclonal antibodies directed against EGF-R were raised by immunization of Balb/c mice with a plasma membrane preparation of human epidermoid carcinoma A431 cells. Hybridoma cultures were obtained as described (55) and subjected to a number of assays with increasing specificity (56). Using this approach, three out of 200 hybridoma cultures, designated as 2E9, 2D11 and 2G5 respectively, were selected for subcloning by limiting dilution and further characterization. The antibodies produced by these hybridoma cultures were characterized as IgG's, and were able to precipitate a functional EGF-R tyrosine phosphokinase from solubilized A431 plasma membranes. However, binding studies revealed that only 2E9 was able to compete with EGF for binding to EGF-R and vice versa (Fig. 4), in contrast to 2D11 and 2G5, where no effects on EGF binding, vice versa, were observed. ^{125}I-2E9 binding was concentration dependent and Scatchard analysis revealed a similar number of 2E9 binding sites as compared to EGF (i.e. 2.3×10^6 receptors/cell) in A431 cells. Scatchard analysis of 2D11 antibody binding revealed a number of sites 4-5 times higher than those of EGF or 2E9, suggesting the presence of antigenic sites for 2D11 on other membrane constituents or multiple antigenic sites on EGF-R.

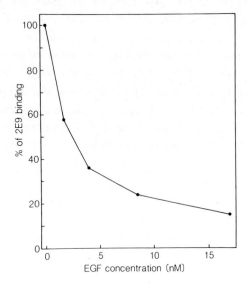

Fig. 4   Effect of EGF on 2E9 binding on A431 cells. A431 cells were incubated at room temperature for 30 minutes with increasing concentrations of EGF, followed by incubation for 60 minutes with ^{125}I-2E9. Data obtained from Defize et al. (56).

The nature of the antigenic determinants of 2E9, 2D11 andd 2G5 respectively, was subsequently identified by immunoprecipitation of solubilized A431 cells, labeled with ^{35}S-methionine in the presence of tunicamycin. In these cell preparations, 2E9 was able to precipitate a single protein with M_r of 130,000, equal to that of unglycosylated EGF-R (57), demonstrating that 2E9 is directed against a peptide determinant of EGF-R located close to or at the binding domain for EGF. In contrast, no membrane constituents were precipitated in these cell preparation using either 2D11 or 2G5, demonstrating that these antibodies are directed against sugar moieties. This indication has been confirmed by using cells labeled with ^{35}S-methionine in the absence of tunicamycine. In these cell preparations, both 2D11 and 2G5 were able to precipitate multiple proteins, consistent with the contention that both antibodies cross-react with various membrane constituents. Further support has been obtained by a haemagglutination test, using human erythrocytes of all four blood groups. Recently it has been demonstrated that at least part of the sugar determinants carried by EGF-R, are structurally similar to the blood group A antigen (58). Both 2D11 and 2G5 were able to agglutinate exclusively blood group A antigen carrying erythrocytes, in contrast to 2E9. Altogether these findings demonstrate that 2E9 is directed against a peptide determinant of EGF-R, while 2D11 and 2G5 are directed against sugar moieties.

Visualization of EGF-R

Pertinent to an understanding of the mechanism of growth factor action is the knowledge about the localization of both ligand and receptor during the endocytotic pathway. An improvement to this end is the application of cryo-ultramicrotomy in combination with immuno-gold labeling, using mono- or polyclonal antibodies directed against ligand or receptor. This method permits a post-sectioning labeling and therefore enables localization of both ligand and receptor at any time during the endocytotic pathway and furthermore, the availability of gold particles of different defined sized allows labeling of both ligand and receptor simultaneously (59,60).

We have applied this method to localize EGF-receptor in cryo-sections of A431 cells, using the monoclonal anti-EGF-receptor antibody 2E9. After a consecutive labeling with 2E9, rabbit-anti-mouse antibody and protein A-colloidal gold complex, a high number of gold particles was observed on the cell surface (Fig. 5A). It was possible to identify coated pits, heavily labeled with gold particles (61). Intracellularly, gold particles were observed in large vesicles (Fig. 5A) these vesicles probably arising from fusion of membrane folds with the plasma membrane. Furthermore, a high number of gold particles was observed on intracellular vesicular membrane structures (Fig. 5B). These structures resemble most likely the compartement previously called CURL (60), and are thought to be

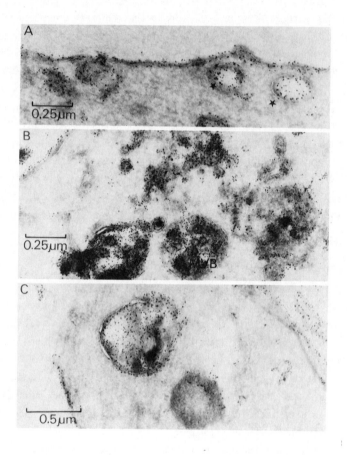

Fig. 5 Localization of EGF-receptor in cryo-sections of A431 cells.
A431 cells were cryo-sectioned and labeled with monoclonal anti EGF-receptor antibody, rabbit anti-mouse antibody and protein A-gold complex as described in detail elsewhere (Boonstra et al., 61)
A: surface labeling of EGF-receptor note the presence of large vesicles
B: intracellular labeling of EGF-receptor, note presence of multivesicular bodies
    and intracellular membrane structures (CURL)
C: labeling of EGF-receptor in lysozomes

involved in the sorting-out process of ligand and receptor. In addition, multivesicular bodies (MVB) (Fig. 5B) and structures resembling lysosomes were labeled (Fig. 5C) indicating the degradation of EGF receptors. These results demonstrate that this method provides the possibility to identify EGF-receptors in A431 cells, and, using the proper antibodies, enables an analysis of the temporal relationship between the localization of both ligand and receptor and the cellular responses of growth factor receptor interaction. Further experiments along this line are presently being carried out.

EGF-R functioning as assayed by monoclonal antibodies

An important function in growth factor action has been assigned to the activation of tyrosine phosphokinase (11,12), the kinase being an intrinsic part of the receptor molecule (13,14). Recently, exogenous, tyrosine containing peptides have been used as a simple substrate to demonstrate growth factor induced activation of tyrosine phosphokinase in cellular membranes (62). As shown in Fig. 6, EGF and 2E9 stimulated the phosphorylation of the tyrosine-containing peptide angiotensin I, while no stimulation of phosphorylation was observed with 2D11 or 2G5. These findings were partially confirmed by endogenous protein phosphorylation

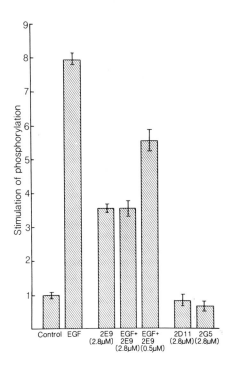

Fig. 6 Phosphorylation of Angiotensin I by A431 plasma membranes. The phosphorylation reaction was performed essentially as described (60), Angiotensin I was used at a concentration of 2mM and EGF at a concentration of 0.28 µM. Data obtained from Defize et al. (56).

282

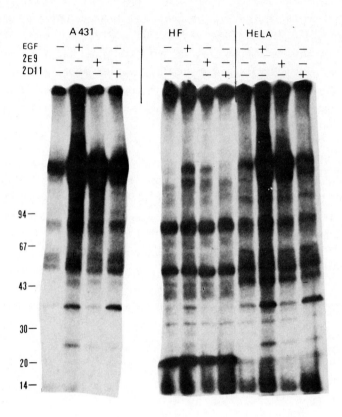

Fig. 7 Effect of anti-EGF-R antibodies on protein phosphorylation of NP-40 solubilized membranes. Protein phosphorylation has been measured as described in detail elsewhere; (56) using EGF (50 ng/ml) or 2E9 (10 g/ml) or 2D11 (10 g/ml) respectively. Data obtained from Defize et al. (56).

studies in plasma membrane preparations of A431 cells, human fibroblasts (HF) and HeLa cells (Fig. 7). Phosphorylation of EGF-R (M_r 170,000) was stimulated by EGF, 2E9, and 2D11 in all three cell types; 2D11 being least effective, in particular in HF cells. In addition various other membrane proteins are being phosphorylated under these conditions but interesting differences are found for the various effectors. In particular, the phosphorylation of a 36KD protein by addition of 2D11 to A431 and HeLa plasmamembrane is similar as found for EGF, while 2E9 addition has only a minor effect. It should also be noted that phosphorylation of a 24KD band is exclusive for EGF. 2G5 had no effects on endogenous protein phosphorylation. These results demonstrate that stimulation of the intrinsic tyrosine phosphokinase of the EGF-R, as well as its substrate specificity are dependent upon the nature of the anti-EGF-R antibodies.

Preliminary experiments have indicated that the stimulation of the tyrosine phosphokinase activity of the EGF-R is not directly related to the ionic signal transduction, as described above, since 2E9 and 2D11 were ineffective in raising pH_i and $[Ca^{2+}]_i$.

Another characteristic response of EGF on A431 cells involves morphological changes. Within a few minutes after EGF-addition an extensive membrane ruffling and rounding-up of cells can be observed, especially in Ca^{2+}-free medium (63). These effects of EGF on A431 cell morphology, could be mimicked by 2D11, but not by 2E9 and 2G5.

Finally, the mitogenic properties of the antibodies were examined on quiescent HF cels. According to Schreiber et al. (64), clustering of EGF-R is a sufficient signal for induction of mitogenesis. However, none of the antibodies were able to induce DNA synthesis in HF cells, even in the presence of a second cross linking rabbit-anti-mouse antibody.

CONCLUSIONS

The differential responses upon binding of EGF and the respective anti EGF-R antibodies to the EGF-R have been summarized in Table 2. It will be clear that monitoring the effects of monoclonal antibodies directed against different antigenic determinants of the receptors is an effective tool in analyzing cause and effect in EGF-R functioning. Taking into account that extrapolations have to be made from in vitro experiments to the situation in intact cells, our results provide evidence for the following conlusions:

1) Cross-linking of EGF-R and/or stimulation of EGF-R tyrosine phosphokinase activity are insufficient to induce DNA synthesis or morphological changes.

2) Substrate specificity and activity of EGF-R tyrosine phosphokinase are dependent on the nature of the activator-receptor interaction, and are thus probably determined by the spatial configuration of the EGF-R, in which sugar moieties may play a role.

3) Sofar, our data suggest that the stimulation of EGF-R tyrosine phosphokinase activity is not directly related to the activation of Na^+/H^+ exchange or Ca^{2+} mobilization. It seems more likely that the primary ionic changes are linked to the indirect stimulation of protein kinase C.

4) Growth factor-induced mitogenic stimulation of cells requires a multitude of receptor-mediated parallel responses, rather than a sequence of serial events.

Table 2 Properties of anti EGF-R monoclonals compared with EGF

	EGF	2E9	2D11	2G5
Precipitation of EGF-R		+	+	+
EGF binding competition	+	+	−	−
Tunicamycin sensitivity		−	+	+
Blood-group A specificity		−	+	+
Number of binding sites on A431 cells	2.10^6	2.10^6	10^7	ND
TyrPK activity towards exogenous substrate	+	+	−	−
Stimulation of EGF-R autophosphorylation	+	+	+	−
Cytoplasmic alkalinization	+	−	−	ND
Ca^{2+} mobilization	+	−	−	ND
Induction of morphological changes	+	−	+	−
Induction of DNA-synthesis in HF cells	+	−	−	−

ND = Not Determined

ACKNOWLEDGEMENT

We thank Ms Marie-Jeanne Mens for preparing the manuscript. Research reported in this paper, carried out at the Hubrecht Laboratory, was supported by the Koningin Wilhelmina Fonds (Netherlands Cancer Foundation).

REFERENCES

1. de Laat SW, Boonstra J, Moolenaar WH, Mummery CL, van der Saag PT and Zoelen EJJ (1983)In: Development in Mammals Vol. 5, Ed. M.H. Johnson, Elsevier Science Publishers, pp. 33-106

2. de Laat SW, Boonstra J, Moolenaar WH and Mummery CL (1984) In: Growth and Maturation Factors Vol. 3, Ed. G. Guroff, John Wiley & Sons, Inc. New York. In press

3. Downward J, Yarden Y, Mayes E, Scrace G, Totty N, Stockwell P, Ullrich A, Schlessinger J and Waterfield MD (1984) Nature 307: 521-527

4. Ullrich A, Coussens L, Hayflick JS, Dull, TJ, Gray A, Tam AW, Lee J, Yarden Y, Libermann TA, Schlessinger J, Downward J, Mayes ELW, Whittle N, Waterfield MD and Seeburg PH (1984) Nature 309: 418-425

5. Doolittle RF, Hunkapiller MW, Hood LE, Devare SG, Robbins KC, Aaronson SA and Antionades HN (1983) Science 221: 275-277

6. Waterfield MD, Scrace GT, Whittle N, Stroobant P, Johhson A, Wasterson A, Westermark B, Heldin CH, Huang JS and Deuel TF (1983) Nature 304: 35-39

7. Robbins KC, Antoniades HN, Devare SG, Hunkapiller MW and Aaronson SA (1983) Nature 305: 605-608

8. Chiu IM, Reddy EP, Givol D, Robbins KC, Tronick SR and Aaronson SA (1984) Cell 37: 123-129

9. James R and Bradshaw RA (1984) Ann Rev Biochem 53: 259-292

10. Carpenter G and Cohen S (1979) Ann Rev Biochem 48: 193-216

11. Hunter T and Cooper JA (1981) Cell 24: 741-752

12. Ek B and Heldin CH (1982) J Biol Chem 257: 10486-10492

13. Cohen S, Ushiro H, Stoschek C and Chinkers M (1982) J Biol Chem 257: 1523-1531

14. Buhrow SA, Cohen S and Staros JV (1982) J Biol Chem 257: 4019-4022

15. Sawyer ST and Cohen S (1981) Biochemistry 20: 6280-6286

16. Habenicht AJR, Glomset JA, King WC, Nist C , Mitchell CD and Ross R (1981) J Biol Chem 256: 12329-12335

17. Rozengurt E and Heppel LA (1975) Proc Natl Acad Sci USA 72: 4492-4495

18. Moolenaar WH, Yarden Y, de Laat SW and Schlessinger J (1982) J Biol Chem 257: 8502-8506

19. Mummery CL, van der Saag PT and de Laat SW (1983) J Cell Biochem 21: 63-75

20. Boonstra J, Moolenaar WH, Harrison PH, Moed P, van der Saag PT and de Laat SW (1983) J Cell Biol 97: 92-98

21. Fehlman M, Canivet B and Freychet P (1981) Biochem Biophys Res Comm 100: 254-260

22. Moolenaar WH, Mummery CL, van der Saag PT and de Laat SW (1981) Cell 23: 789-798

23. Boonstra J, van der Saag PT, Moolenaar WH and de Laat SW (1981) Exp Cell Res 131: 452-455

286

24. Smith JB and Rozengurt E (1978) Proc Natl Acad Sci USA 75: 5560-5564

25. Moolenaar WH, Boonstra J, van der Saag PT and de Laat SW (1981) J Biol Chem 256: 12883-12887

26. Villereal ML (1981) J Cell Physiol 107: 359-369

27. Moolenaar WH, Mummery CL, van der Saag PT and de Laat SW (1982) In: Membranes in Tumour Growth, eds. T. Galeotti, A Cittadini, G Neri and S Papa, Elsevier Biomedical Press, Amsterdam pp. 413

28. Rothenberg P, Glaser L, Schlesinger P and Cassel D (1983) J Biol Chem 258: 4883-4889

29. Cassel D, Rothenberg P, Zhuang YX, Deuel TF and Glaser L (1983) Proc Natl Acad Sci USA 80: 6224-6228

30. Koch KS and Leffert HL (1979) Cell 18: 153-163

31. Pouysségur J, Chambardt JC, Franchi A, Paris S and van Obberghen-Schilling E (1982) Proc Natl Acad Sci USA 79: 3935-3939

32. Owen NE and Villereal ML (1983) Cell 32: 979-985

33. Thomas JA, Buchsbaum RN, Zimniak A and Racker E (1979) Biochemistry 18: 2210-2218

34. Moolenaar WH, Tsien RY, van der Saag PT and de Laat SW (1983) Nature 304: 645-648

35. Moolenaar WH, Tertoolen LGJ and de Laat SW (1984) J Biol Chem 259: 7563-7569

36. Aronson PS, Nee J and Suhm M (1982) Nature 299: 161-163

37. Rothenberg P, Glaser L, Schlesinger P and Cassel D (1983) J Biol Chem 258: 12644-12653

38. Castagna M, Takai Y, Kaibuchi K, Sano K, Kikkawa U and Nishizuka Y (1982) J Biol Chem 257: 7847-7851

39. Kikkawa U, Takai Y, Tanaka Y, Miyake R and Nishizuka Y (1983) J Biol Chem 258: 11442-11445

40. Nishizuka Y (1983) Trends Biochem Sci 8: 13

41. Nishizuka Y (1984) Nature 308: 693-697

42. Moolenaar WH, Tertoolen LGJ and de Laat SW (1984): submitted

43. Besterman JM and Cuatrecasas P (1984) J Cell Biol 99: 340-343

44. Rosoff PM, Stein LF and Cantley LC (1984) J Biol Chem 259: 7056-7060

45. Tsien RY, Pozzan T and Rink TJ (1982) Nature 295: 68-71

46. Moolenaar WH, Tertoolen LGJ and de Laat SW (1984) J Biol Chem 259: 8066-8069

47. Streb H, Irvine RF, Berridge MJ and Schulz I (1983) Nature 306: 67-69

48. Schechter Y, Hernaez L, Schlessinger J and Cuatrecasas P (1979) Nature 278: 835-838

49. Schreiber AB, Yarden Y and Schlessinger J (1981) Biochem Biophys Res Comm 101: 517-523

50. Schreiber AB, Lax I, Yarden Y, Eshhar and Schlessinger J (1981) Proc Natl Acad Sci USA 78: 7535-7539

51. Waterfield MD, Mayes E, Stroobant P, Bennet P, Young S, Goodfellow P, Banting G and Ozanne B (1982) J Cell Biochem 20: 115-127

52. Kawamoto T, Sato JD, Le A, Polikoff J, Sata GH and Mendelsohn J (1983) Proc Natl Acad Sci USA 80: 1337-1341

53. Richert N, Willingham MD and Pastan I (1983) J Biol Chem 258: 8902-8907

54. Gregoriou M and Rees A (1982) EMBO J 3: 929-937

55. Herzenberg L, Herzenberg C and Milstein C (1978) In: Handbook of experimental immunology Weiz DM (ed) Blackwell, Oxford U.K. pp. 25.1-25.7

56. Defize LHK, van der Saag PT and de Laat SW (1984): submitted

57. Decker S (1984) Mol Cell Biol 4: 571-575

58. Gooi H, Schlessinger J, Lax I, Yarden Y, Libermann T and Fuzi T (1983) Biosc Reports 3: 1045-1052

59. Geuze HJ, Slot JW, Strous GJAM, Lodish HF and Schwartz AL (1983) Cell 32: 277-287

60. Geuze HJ, Slot JW, Strous GJAM, Peppard J, von Figura K, Hasilik A and Schwartz AL Cell 37: 195-204

61. Boonstra J, van Maurik PAM, Defize LHK, de Laat SW, Leunissen JLM and Verkley AJ (1984): submitted

62. Pike LJ, Marquardt H, Todaro GJ, Gallis B, Casnellie JE, Bornstein P and Krebs EG (1982) J Biol Chem 257: 14628-14631

63. Haigler HT, McKanna JA and Cohen S (1979) J Cell Biol 81: 382-395

64. Schreiber AB, Libermann T, Lax I, Yarden Y and Schlessinger J (1983) J Biol Chem 258: 846-853

RESUME

   Parmi les réponses les plus rapides à l'interaction facteur de croissance-
récepteur, on trouve la stimulation de l'activité phospho-tyrosine kinase du
récepteur, le clivage des inositol-phospholipides avec la mobilisation subsé-
quente du Ca^{2+} intracellulaire $[Ca^{2+}_i]$ et la stimulation de la protéine kinase C
Ca^{++} et lipide dépendante, et l'activation de l'échange Na^+, H^+ qui donne lieu à
une augmentation du pH intracellulaire. La comparaison des effets de l'EGF, du
PDGF et de l'ester de phorbol TPA (promoteur de tumeur et activateur direct de
la protéine kinase C) sur le pH_i et la $[Ca^{2+}_i]$ met en évidence la présence d'un
lien entre le clivage des phospholipides et les premiers changements ioniques.
En particulier, nos données indiquent que la stimulation de l'activité de la
protéine kinase C par le diacylglycérol active l'échangeur Na^+, H^+ tandis que la
mobilisation du Ca^{2+}_i est initiée de manière indépendante. Une connaissance plus
approfondie de l'interrelation entre les réponses rapides a été obtenue en
étudiant l'activation du récepteur de l'EGF par des anticorps monoclonaux dirigés
contre différents déterminants de ce récepteur. On a montré que le
"cross-linking" du récepteur de l'EGF et/ou la stimulation de l'activité tyrosine
phosphokinase du récepteur de l'EGF sont insuffisants pour induire la synthèse de
DNA ou des changements morphologiques. De plus, la spécificité vis-à-vis des
substrats et l'activité de la tyrosine phosphokinase du récepteur de l'EGF sont
déterminées par la configuration spatiale du récepteur de l'EGF dans laquelle des
résidus carbohydrates pourraient jouer un rôle.

   En conclusion, nous n'avons pas observé de relation directe entre l'activité
tyrosine phosphokinase du récepteur de l'EGF et l'activation de l'échange Na^+, H^+
ou la mobilisation du Ca^{2+}_i.

CALCIUM AND SECRETION

CALCIUM ET SECRETION

Hormones and Cell Regulation, Volume 9
INSERM European Symposium
J.E. Dumont, B. Hamprecht and J. Nunez editors
© 1985 Elsevier Science Publishers B.V. (Biomedical Division)

THE ROLE OF CALCIUM-ACTIVATED POTASSIUM CHANNELS IN HORMONE SECRETION AND IN THE ACTION OF HORMONES ON EXOCRINE CELLS

O. H. PETERSEN, Y. MARUYAMA, I. FINDLAY and D.V. GALLACHER

M.R.C. Secretory Control Research Group, Department of Physiology, University of Liverpool, P.O. Box 147, Liverpool L69 3BX U.K.

INTRODUCTION

Membrane potassium (K^+) channels activated by calcium (Ca^{2+}) acting from the cytoplasmic side have been found in many different tissues including endocrine and exocrine gland cells (1). There are several types of Ca^{2+}-activated pores, but the one that is best characterized is the 'large' (1), 'maxi' (2) or 'big' (3) K^+ channel. This channel is interesting both from a biophysical point of view, because of its high unit conductance combined with extreme selectivity (2,4), as well as from a functional viewpoint since its role is so radically different in endocrine and exocrine cells (1). In this short review we summarize the basic conductance properties of this K^+ channel, discuss the Ca^{2+}-activation, describe hormonal activation of the channel in intact cells and speculate about the role of the channel in hormone secretion and action.

BASIC PROPERTIES OF LARGE Ca^{2+}-ACTIVATED K^+ CHANNEL

Conductance and ion selectivity

Fig. 1 shows a short excerpt of a single-channel current recording from an excised outside-out membrane patch taken from a rat pancreatic islet cell. The solution in the recording pipette (in contact with the inside of the plasma membrane) was a K^+-rich 'intracellular' solution with a very low ionized Ca^{2+} concentration ($[Ca^{2+}]_i$) whereas the bath (in contact with the outside of the plasma membrane) was filled with a Na^+-rich 'extracellular' solution. The membrane patch from which the trace shown in Fig. 1 was obtained contained 2 channels and the 3 different current levels corresponding to closure, one channel open and two channels open are seen.

It is also clear from Fig. 1 that the current trace is 'noisier' when the channels are open and this open-state noise corresponds to very rapid and shortlasting closures of the open channels. By applying an electrical potential (negative or positive) to the pipette interior) it is possible to record the relationship between the amplitude of the single-channel current and the membrane potential. The single-channel current-voltage (i/V) relationship is shown in Fig. 2 both for the ionic situation described in Fig. 1 (Na^+/K) as well as for an experiment in which the patch membrane was exposed to symmetrical K^+ rich intracellular solutions (K^+/K^+). With Na^+ outside and K^+ inside it is possible to observe outward (K^+) current but not inward (Na^+) current. When there is a

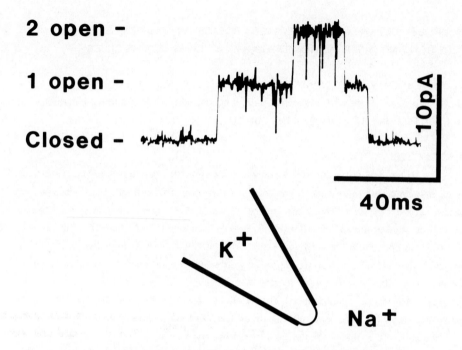

Fig. 1. Single-channel current recording from an excised outside-out membrane patch from a rat pancreatic islet cell. The 'intracellular' pipette solution (in contact with the inside of the plasma membrane) contained K^+-rich solution with no added Ca and 1 mM EGTA. The 'extracellular' bath solution (in contact with the outside of the plasma membrane) contained Na^+-rich solution. The current trace was obtained at 0 mV membrane potential. Filtering: 1 kHz (low pass). (From ref. 5).

high K^+ concentration on both sides both inward and outward current can be measured. In the latter situation the i/V relationship is linear with a slope corresponding to a single-channel conductance of 200-300 pS (in the case shown in Fig. 2 the value is 250 pS). A detailed study of the ion-selectivity has recently been carried out in the salivary glands in which it was shown that the channel is surprisingly selective as it discriminates between K^+ and Rb^+. With extracellular Rb^+ and intracellular K^+ outward (K^+) current but not inward (Rb^+) current can be evoked. With the Rb^+/K^+ gradients reversed inward (K^+) current but not outward (Rb^+) current can be observed. Nevertheless, the channel is not impermeable to Rb^+ as null potential measurements in a bi-onic K^+/Rb^+ situation indicate a Rb^+/K^+ permeability ratio of close to 1 unlike the extremely low ratios obtained for Na^+/K^+ or Li^+/K^+ (4). These surprising, but important findings can most easily be interpreted by assuming that Rb^+ binds much more strongly to sites within the channel than K^+ (4).

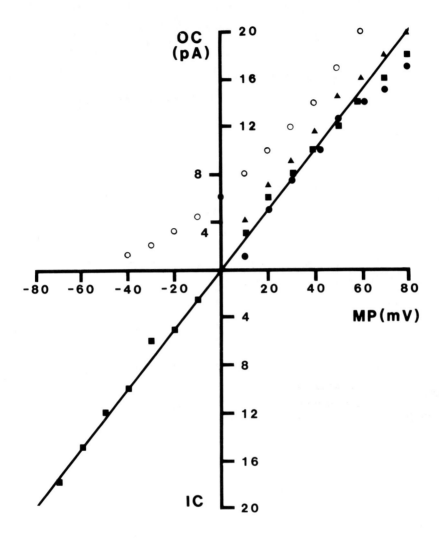

Fig. 2. Single-channel current voltage relationships from excised membrane patches in the presence of quasi-physiological cation gradients (Na$^+$/K$^+$ situation) (data from same experiment as in Fig. 1) (open circles) and in the presence of high-K$^+$ solutions on both sides (K$^+$/K$^+$ situation) (closed symbols). Circles: $[Ca^{2+}]_i$ <10^{-9}M, triangles: $[Ca^{2+}]_i$ = 1.5 x 10^{-7}M and squares $[Ca^{2+}]_i$ = 10^{-6}M. (From ref. 5)

294

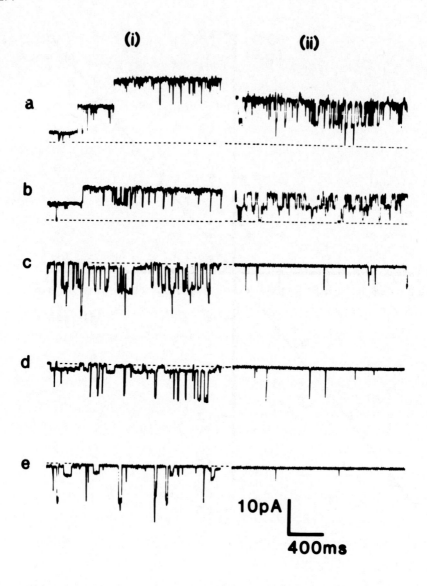

Fig. 3. Single-channel current traces from an excised inside-out mouse submandibular acinar membrane patch. Two different intracellular solutions were used in the bath; in (1) the solution contained 1 mM Ca^{2+} and 2 mM EGTA ($[Ca^{2+}]_i$ 10^{-7}M), while in (2) there was no added Ca^{2+} and EGTA (1 mM) was present ($[Ca^{2+}]_i < 10^{-9}$M). a-e show traces obtained at different membrane potentials: a, +25 mV: b, +15 mV; c, -25 mV; d, -35 mV; and e, -45 mV. (From ref. 6).

Control of opening by membrane potential and Ca^{2+}

As seen in Fig. 3 channel opening is very much dependent on the membrane potential. In the virtual absence of internal Ca^{2+} ($[Ca^{2+}]_i < 10^{-9}M$) positive membrane potentials (inside positive with respect to outside) are associated with frequent and long-lasting opening bursts, whereas at negative membrane potentials there are only very few and brief openings. When $[Ca^{2+}]_i$ is increased, in the case shown in Fig. 3 to $10^{-7}M$, there are many more and longer openings at all membrane potentials, but the effect of Ca^{2+} is particularly clear at the negative membrane potentials. The sensitivity to Ca^{2+} varies considerably from tissue to tissue, but the effect of Ca^{2+} is always to shift the curve relating open state probability of the channel to membrane potential to the left (Fig. 4). In contrast Ca^{2+} has no effect on single-channel conductance (Fig. 2).

Counting the number of channels in a cell

In order to estimate the number of Ca^{2+}-activated K^+ channels in a single cell it is necessary to combine single-channel with whole-cell current recording (8). The best results can be obtained if one channel type dominates the cell membrane. This is the case for the pig pancreatic acinar cells in which only one type of Ca^{2+}-activated K^+ channel is observed. Fig. 5 shows currents recorded from a single isolated acinar cell. It is seen that the whole plasma membrane acts as a perfect rectifier allowing outward (K^+ current) when the membrane is depolarized, but hardly any inward (Na^+) current when the membrane is hyperpolarized in agreement with the single-channel current recordings under quasi-physiological conditions (Fig. 2). At positive membrane potentials the channels are open most of the time (Figs. 3 and 4) and the total outward current through the active plasma membrane should therefore simply be the sum of the single-channel currents. At a membrane potential of +60 mV (120 mV depolarization from a holding potential of -60 mV) the outward total current is about 800 pA (Fig. 5), the corresponding single-channel current is about 20 pA (Fig. 2) and therefore a total of just 40 channels could account for the whole-cell current. A detailed quantitative analysis indicates that in the resting pig pancreatic acinar cells (with a very low $[Ca^{2+}]_i$) there are only about 50 functional K^+ channels (7). Another approach is to analyze the noise associated with the outward currents (Fig. 5) as originally described by Sigworth (9). Such measurements have recently been carried out on both lacrimal and salivary acinar cells and in these preparations there appear to be about 50-100 large K^+ channels per cell (10,11).

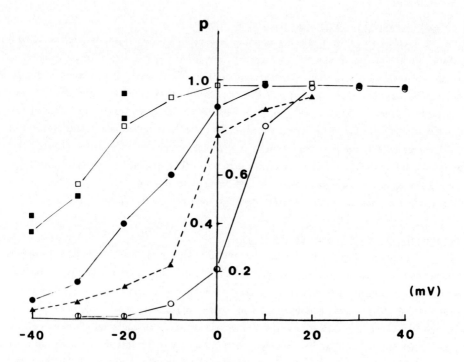

Fig. 4. Pig pancreatic acinar cell, excised inside-out patch. Curves showing the relationship between open-state probability of K^+ channel and membrane potential at three different levels of $[Ca^{2+}]_i$: open circles 10^{-8}M, closed circles 10^{-7}M and open and closed boxes 10^{-6}M. All the experiments were carried out in the presence of normal Na^+/K^+ gradients. The open-state probability was determined by examining 30-40-s long continuous stretches of recordings at each potential level. The triangles (dashed line) represent results from an experiment on a cell-attached patch. $[Ca^{2+}]_i$ in the intact cell was apparently between 10^{-7}M and 10^{-8}M. (From ref. 7).

HORMONAL ACTIVATION OF LARGE Ca^{2+}-ACTIVATED K^+ CHANNEL

Since many hormones and neurotransmitters act on target cells by increasing $[Ca^{2+}]_i$ it might be expected that it should be possible to demonstrate hormonal activation of membrane K^+ conductance in intact cells.

Fig. 6 shows the action of cholecystokinin (CCK) (2×10^{-10}M) on the outward K^+ current in a single pig pancreatic acinar cell. CCK is seen to markedly enhance the outward currents associated with depolarizing voltage jumps. In the sustained phase of stimulation the effect is acutely dependent on the presence of Ca^{2+} in the extracellular solution. The effect of CCK is reversible and can never be evoked if the pipette and therefore the cell is filled with a solution containing a high (10 mM) EGTA concentration

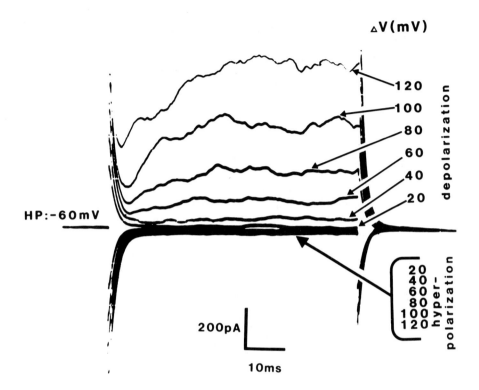

Fig. 5. Whole-cell patch-clamp current recording from single pig pancreatic acinar cell. Traces of whole-cell currents associated with depolarizing or hyperpolarizing voltage steps. The recording pipette and therefore the cell was filled with a nominally Ca-free intracellular solution containing 1 mM EGTA. The holding potential was -60 mV. The upward deflections represent outward currents associated with depolarizing steps of 20, 40, 60, 80, 100 and 120 mV. The downward deflections represent the inward currents when hyperpolarizing steps of the same magnitudes were used. (From ref. 12).

(12). The experiment shown in Fig. 6 was carried out with a pipette solution containing only 0.5 mM EGTA. Thus the importance of both intracellular and extracellular Ca^{2+} for the action of CCK has been demonstrated. These patch-clamp data fit in very well with recent microelectrode studies on segments of pig pancreatic acinar tissue showing that stimulation evokes sustained hyperpolarization in the presence of extracellular Ca^{2+}, but only a brief transient response during exposure to a Ca^{2+}-free solution (7,13).

It would appear therefore that hormonal stimulation releases intracellular Ca^{2+} most likely via the newly discovered intracellular messenger Inositol-trisphosphate (14), but that during sustained stimulation there is under normal circumstances a

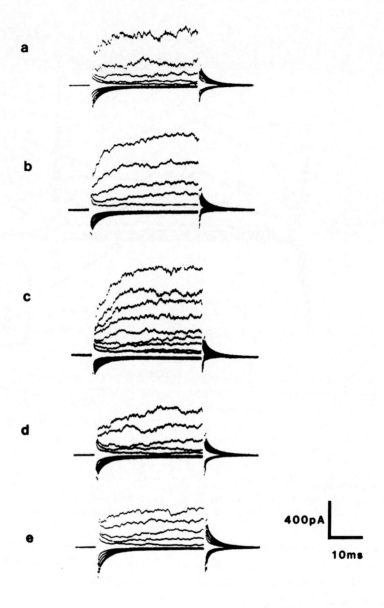

Fig. 6. Whole-cell current recordings from a single pig pancreatic acinar cell. The action of CCK. No Ca^{2+} was added to the pipette solution, which contained 0.5 mM EGTA. External solution contained 2.5 mM Ca except in d where Ca^{2+} was omitted from the solution. Holding potential was -85 mV throughout. (a) Control before CCK application. Voltage steps: ±40,60,80,100 and 120 mV. (b) 3 min after start of exposure to CCK (2×10^{-10}M). Voltage steps: ±40,60,80,100 and 120 mV. (c) 5 min after start of CCK stimulation. Voltage steps ±30,40,50,60,70,80,90,100 and 110 mV. (d) 10 sec after replacing 2.5 mM Ca^{2+} solution with nominally Ca^{2+}-free solution, but still with CCK (2×10^{-10}M). Voltage steps: ±40,60,80,100 and 120 mV. (e) 1 min after return to control (without CCK). Voltage steps: ±40,60,80,100 and 120 mV. (From ref. 12).

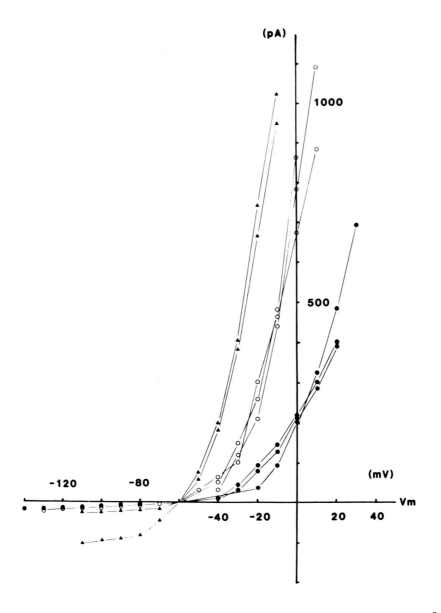

Fig. .7. Single pig pancreatic acinar cells. Whole cell currents at different $[Ca^{2+}]_i$ (strongly buffered Ca-EGTA solution) plotted as function of membrane potential. Closed circles: $[Ca^{2+}]_i = 5 \times 10^{-8}$M; open circles: $[Ca^{2+}]_i = 10^{-7}$M and closed triangles: $[Ca^{2+}]_i = 5 \times 10^{-7}$M. (From ref. 12).

continuing Ca^{2+} influx. It is possible that a Ca^{2+}-activated non-selective cation channel is involved in this part of the response (15), but as yet there is no hard evidence to support or refute this hypothesis.

In view of the importance of both intra- and extracellular Ca^{2+} in the CCK response shown in Fig. 6 there can be little doubt that CCK acts on the K^+ conductance via changes in $[Ca^{2+}]_i$. However, it would obviously be of importance to know something about the magnitude of changes in $[Ca^{2+}]_i$ required to change the K^+ conductance in an isolated cell. For this reason a series of experiments were carried out in which pipette solutions with different ionized Ca^{2+} concentrations were used. In this series highly buffered solutions (10 mM EGTA) were used so as to clamp $[Ca^{2+}]_i$ effectively at the desired levels (12). Fig. 7 shows the results of 8 experiments on 8 individual isolated pig pancreatic acinar cells. It is seen that the outward K^+ currents associated with depolarizing voltage pulses were very sensitive to small changes in $[Ca^{2+}]_i$. Thus increasing $[Ca^{2+}]_i$ from 5×10^{-8}M to 10^{-7}M very markedly enhanced the outward current and a further increase in $[Ca^{2+}]_i$ to 5×10^{-7}M produced an additional effect. The effects evoked by these very modest changes in $[Ca^{2+}]_i$ are very similar to those evoked by CCK and it is therefore possible to conclude that CCK can control the membrane K^+ conductance via rather small changes in $[Ca^{2+}]_i$ (12).

ROLE OF Ca^{2+}-ACTIVATED K^+ CHANNEL IN ENDOCRINE CELLS

In the endocrine cells the K^+ channel acts as a link between $[Ca^{2+}]_i$ and the membrane potential. In an endocrine cell, such as the pancreatic B-cell, Ca^{2+} needed in excitation-secretion coupling enters the cells through specific voltage-gated Ca^{2+} channels (16,17) which are opened by membrane depolarization and closed by hyperpolarization. Therefore, as the entry of Ca^{2+} produces an increase in $[Ca^{2+}]_i$ so the K^+-selective channels will be activated leading to membrane hyperpolarization and impedance of further Ca^{2+} influx. The interplay between Ca^{2+} influx, $[Ca^{2+}]_i$ and K^+ conductance results in the oscillating membrane potential pattern (depolarization - plateau with Ca^{2+} action potentials - hyperpolarization-gradual slow depolarization) seen in mammalian pancreatic B-cells (17). Ca^{2+}-activated K^+ channels can indeed by regarded as the crucial regulatory link between $[Ca^{2+}]_i$ and therefore insulin secretion, and Ca^{2+} influx (5). The insulin secretagogue glucose, according to one hypothesis, acts by decreasing the Ca^{2+} sensitivity of the K^+ channels, prolonging the plateau phase with the Ca^{2+} action potentials and thereby enhancing $[Ca^{2+}]_i$ (17).

The detailed recent study on the quantitative relationship between membrane potential, $[Ca^{2+}]_i$ and open state probability of the large K^+ channel in rat pancreatic islet cells shows that the regulation of the opening of the K^+ channel occurs over a very narrow range of $[Ca^{2+}]_i$ (5). Fig. 8 shows the open state probability curves at different levels of $[Ca^{2+}]_i$. Interestingly and in marked contrast to the situation in the epithelial cells (see Figs. 3,4 & 7) Ca^{2+} has no effect up to a level of about $10^{-7}M$ then as it is further increased the curves shift very markedly to the left. In order to hyperpolarize the membrane $[Ca^{2+}]_i$ would have to move into the range 8×10^{-7} to $2 \times 10^{-6}M$. Even a small decrease in $[Ca^{2+}]_i$ from that level would depolarize the membrane. It is clearly very useful for the cell to require such relatively high levels of Ca^{2+} for K^+ channel activation as otherwise it would be impossible to achieve maximal secretion since the stimulant-evoked Ca^{2+} influx would be switched off too quickly by K^+ channel-mediated hyperpolarization. A very important theme for further investigation will be the possible modulation of the Ca^{2+}-sensitivity of the large K^+ channel by various agents influencing insulin secretion.

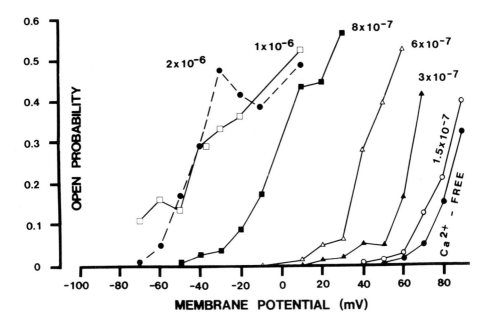

Fig. 8. Rat pancreatic islet cell, excised inside-out membrane patch. Curves relating open-state probability of Ca^{2+}-activated K^+ channel to membrane potential at different levels of $[Ca^{2+}]_i$. (From ref. 5).

302

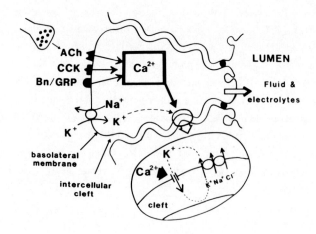

Fig. 9. Simplified scheme to account for control of electrolyte transport in baso-lateral plasma membrane of fluid secreting epithelia. The figure specifically represents part of a pig pancreatic acinus and its major secretagogues CCK, acetylcholine and bombesin/gastrin releasing peptide, but with other secretagogues and/or various additions, the model could represent a wide range of secretory epithelia. In the proposed scheme, secretagogues first increase $[Ca^{2+}]_i$ opening Ca^{2+}-activated K^+ channels in the baso-lateral membranes of the cell. K^+ released through the channels into the narrow intercellular clefts would be taken up again, primarily through a K^+-Na^+-Cl^- co-transport system. The rate of transport in the cycle of K^+ release and reuptake would be regulated at the K^+ channel by internal Ca^{2+} and would be directly linked to the rate of NaCl uptake by the cotransport system. Part of the K^+ uptake occurs via the Na^+-K^+ pump. The operation of the Na^+-K^+-Cl^- co-transport system is of course dependent on the Na^+ gradient and therefore on the operation of the Na^+-K^+ pump. (Note that there is no evidence in regard to specific localizations of Na^+-K^+ pump and co-transport system to basal and lateral membranes, respectively and the two transport proteins probably co-exist both in the lateral and basal plasma membranes). The Na^+-K^+-Cl^- co-transport is electroneutral as the most likely stoichiometry is 1 Na^+, 1K^+ and 2 Cl^-. (From ref. 1).

ROLE OF Ca^{2+}-ACTIVATED K^+ CHANNEL IN EXOCRINE CELLS

There is much evidence to suggest that the stimulant-evoked conductance changes in the exocrine acinar cells is principally linked to the primary acinar fluid secretion (17). The main problem for salt and fluid secreting cells is to move Cl^- into the cells as this process occurs against an electrochemical gradient (18). There is now considerable evidence in favour of the Cl^- uptake being mediated by a Na^+-K^+-Cl^- cotransport mechanism driven by the Na^+ gradient established by the Na^+-K^+ pump (1). This Cl^- uptake could be regulated by a fine control of the essential K^+ recirculation as schematically indicated in Fig. 9, exerted by internal Ca^{2+} acting on the K^+ channel. According to this model there are 3 essential transport proteins in the baso-lateral acinar cell membrane concerned with electrolyte transport: the Na^+-K^+-Cl^- co-transporter, the K^+ channel and the Na^+-K^+ pump. In the steady-state situation these three transporters will act as a Cl^- pump with the Na^+ and K^+ recirculating through the membrane. In order for this system to work the K^+ channel must be very sensitive to Ca^{2+} as otherwise the K^+ exit could not be stimulated at the negative resting potential. As seen in Fig. 7 an increase in $[Ca^{2+}]_i$ from 5×10^{-8} to 10^{-7}M and further to 5×10^{-7}M has a considerable effect on outward K^+ current already at a membrane potential of -50 mV. The K^+ channel in lacrimal acinar cells is even more Ca^{2+}-sensitive as it is virtually constantly open at all membrane potentials at $[Ca^{2+}]_i = 10^{-7}$M (19). The high intracellular Cl^- concentration would enable Cl^- to diffuse into the lumen perhaps via a Ca^{2+}-activated luminal Cl^- channel, although at present there is little direct evidence, and Na^+ would follow via the paracellular route. In epithelial cells the Ca^{2+}-activated K^+ channel is therefore crucial in coupling the stimulant-evoked increase in $[Ca^{2+}]_i$ to an enhanced uptake of Cl^- and thereby also explaining as indicated in Fig. 9 the stimulation of the Na^+-K^+ pump.

REFERENCES

1. Petersen OH, Maruyama Y (1984) Nature 307: 693-696

2. Latorre R, Miller C (1983) J Membr Biol 71: 11-30

3. Marty A (1983) Trends Neurosci 6: 262-265

4. Gallacher DV, Maruyama Y, Petersen OH (1984) Pflugers Archiv 401: 361-367

5. Findlay I, Dunne MJ, Petersen OH (1984) J Membr Biol In the press

6. Maruyama Y, Gallacher DV, Petersen OH (1983) Nature 302: 827-829

7. Maruyama Y, Petersen OH, Flanagan P, Pearson GT (1983) Nature 305: 228-232

8. Hamill OP, Marty A, Neher E, Sakmann B, Sigworth FJ (1981) Pflugers Archiv 391: 85-100

9. Sigworth FJ (1980) J Physiol 307: 97-129

10. Trautmann A, Marty A (1984) Proc Natl Acad Sci USA 81: 611-615

11. Maruyama Y, Nishiyama A, Izumi T, Hoshimiya N, Petersen OH In preparation

12. Maruyama Y, Petersen OH (1984) J Membr Biol 79: 293-300

13. Pearson GT, Flanagan PM, Petersen OH (1984) Amer J Physiol in the press

14. Streb H, Irvine RF, Berridge MJ, Schulz I (1983) Nature 306: 67-69

15. Petersen OH, Maruyama Y (1983) Pflugers Archiv 396: 82-84.

16. Reuter H (1983) Nature 301: 569-574

17. Petersen OH (1980) The Electrophysiology of Gland Cells. Academic Press London

18. Frizzell RA, Field M, Schultz SG (1979) Amer J Physiol 236: F1-F8

19. Findlay I (1984) J Physiol 350: 179-195

304

RESUME

La technique de "patch-clamp" extracellulaire a rendu possible l'enregistrement
direct de faibles courants passant au travers de canaux membranaires individuels,
dans des cellules de glandes exocrines et endocrines. Un canal qui présente un
intérêt tout particulier est le canal K^+, activé par le calcium. Cet article a
pour objet l'étude des propriétés fondamentales de ce canal, sa quantification et
son contrôle direct par le calcium externe et le potentiel de membrane, et
indirect par les hormones. Le canal K^+ est extrêmement sélectif, en ce sens
qu'il préfère le K^+ au Rb^+ et qu'il est activé par des concentrations très
faibles de calcium libre ionisé. On a montré, dans les cellules exocrines, que
l'activation, par les hormones, du canal K^+ est médiée par du calcium interne.
Les rôles des canaux K^+ activés par le calcium sont tout à fait différents dans
les cellules exocrines et endocrines. Dans les cellules acinaires sécrétant des
sels et du fluide, les canaux contrôlent la circulation de K^+ à travers la
membrane plasmatique qui, à son tour, contrôle le transport du chlore, tandis que
dans les cellules endocrines, les canaux K^+ relient le potentiel de membrane à la
concentration de calcium ionisé libre et, de cette façon, fournissent un contrôle
en retour ("feed-back") des mécanismes de captation du calcium.

Hormones and Cell Regulation, Volume 9
INSERM European Symposium
J.E. Dumont, B. Hamprecht and J. Nunez editors
© 1985 Elsevier Science Publishers B.V. (Biomedical Division)

STIMULUS-SECRETION COUPLING: STUDIES WITH HUMAN PLATELETS

T.J. RINK, Smith Kline & French Research Limited, The Frythe,
Welwyn, Hertfordshire, AL6 9AR, U.K. and A.W.M. SIMPSON and
T.J. HALLAM, The Physiological Laboratory, Downing Street,
Cambridge, U.K.

INTRODUCTION

The term "secretion" covers several kinds of cellular
processes including: transfer of fluid across an epithelial
layer; the release of newly synthesised, membrane permeant
compounds such as steroid hormones and prostaglandins; and the
discharge of the contents of vesicles or granules by exocytosis -
fusion of the organelle membrane with the plasma membrane
allowing secretion of the granule contents without loss of other
cellular components. Here we discuss mechanisms by which the
combination of a platelet agonist with its receptor switches on
the activation cascade that results in secretory exocytosis.

Exocytosis itself is the basis of many very different types
of cellular and secretory process, e.g. neurotransmitter release
at nerve terminals, the once-off events of the acrosome reaction
and the cortical reaction of the egg at fertilisation, and
numerous examples of endocrine and exocrine secretion.
Exocytosis is an excellent method for specifically secreting
preformed hydrophilic molecules of any size. Very high
concentrations of secreted products are often contained within
secretory granules; several different substances forming a matrix
or core containing large amounts of material in an osmotically
inactive form [1]. For instance, amine-storage (or dense)
granules contain many hundred millimolar serotonin and adenine
nucleotides as well as other constituents; if these were all in
free solution the osmotic pressure in the granules would draw in
water from the cytoplasm and cause lysis. On discharge to the
outside the granule matrix typically disperses because of the
very different conditions, e.g. pH and cationic composition, of
the extracellular fluid compared with that inside the granule.
Exocytosis usually therefore results in the release of several
substances in relatively constant proportions. For example, the

α-granules of platelets contain many types of proteins [2]
including growth factors, anti-heparin factors, and thrombo-
spondin and β-thromboglobulin [2]. Secretion from dense
granules releases ADP, ATP and serotonin. ADP is a potent
platelet stimulus and can help recruit further platelets to a
haemostatic plug. Serotonin also helps stimulate platelets and
has powerful effects on the microvasculature. ATP seems to be an
important component of the core of many types of secretory
granules and vesicles and may have a role as a secreted product.
A neurotransmitter role has been suggested for the ATP that is
released along with catecholamines from sympathetic nerve endings
[3]. One possible role for the ATP released from the dense
granules of platelets is as a stimulant to prostacyclin production by healthy endothelium surrounding a region of vessel damage
[4]. Prostacyclin is a powerful inhibitor of platelet activation
and its production could prevent unwanted extension of a haemostatic
plug.

Exocytosis has another consequence which can be just as
important and sometimes more important than the discharge of the
contents to the outside. New membrane is rapidly inserted into
the plasma membrane. This makes it important for the vesicle
membrane not to be leaky to ions or other solutes which the
plasma membrane has to exclude from the cell. It also offers the
opportunity to insert new phospholipid and new proteins or glyco-
proteins into the cell membrane. This is the method by which ADH
is thought to trigger the rapid increase in water permeability in
cells of the renal collecting system [5], and recently many other
examples have been proposed along these lines including the
insertion of proton pumps into the membrane of the parietal cells
of the gastric mucosa [6]. During platelet activation new phospholipids appear in the outer leaflet of the plasma membrane and
help activate the coagulation cascade. They may come from the
inner leaflet of exocytosed secretory granules. Clearly
exocytosis could also add new surface antigens to a cell or new
types of receptor molecule. This would be the counterpart to the
well-known endocytotic down-regulation of receptors following
agonist binding.

SECOND MESSENGERS

An enormous body of experimental work including the classic findings of Katz and his colleagues at the frog motor nerve terminal [7] and Douglas and his colleagues in the perfused adrenal gland [8] has focused on the key role of calcium ions as the trigger for exocytosis. The evidence supporting a key role for a rise in cytosolic free calcium in triggering secretory exocytosis is extensive. In essence one seeks to show: 1, that secretion is preceded by either or both an influx of calcium across the plasma membrane and a discharge of internal pools of calcium, thus raising cytosolic free calcium, $[Ca^{2+}]_i$; 2, that experimentally induced elevations of cytosolic free calcium are effective stimuli to secretion; 3, that conditions which prevent a rise in cytosolic free calcium prevent secretory response. It has not always been possible to get good evidence in each of these three classes for each type of secretion. However, the evidence is convincing that calcium is the trigger in many instances including the motor nerve terminal and the adrenal medulla.

Despite the undoubted importance of calcium, it now seems likely that the extent or rate of secretory exocytosis is not always a unique function of cytosolic free calcium at the relevant site. First, there are a number of examples of stimulated secretion in which there appears to be little or no change in cytosolic free calcium but rather a critical requirement for changes in some other second messenger [e.g. 9-15]. Sometimes calcium even seems to inhibit secretion [14]. Second, experimental activation of other pathways, while $[Ca^{2+}]_i$ remains at basal levels can stimulate secretion [eg. 10,13,15].

CALCIUM AND PLATELET SECRETION

The first convincing evidence that calcium ions played an important role in stimulus secretion coupling in platelets came from the finding that calcium ionophores, whose major action is to translocate calcium from regions of very high, e.g. mM calcium concentration, to regions of very low, e.g. sub uM concentration, were effective secretagogues [16,17]. It certainly seemed as if

a rise in cytosolic calcium was a sufficient stimulus to
secretion, but it does not of course follow that calcium ions
played a key or final role in the process; they might be serving
to generate other messages which actually do the job. It was
also found that stimulation of secretion was accompanied by
uptake of ^{45}Ca [18]. In some secretory systems there is a very
strong dependence on external calcium and removal of external
calcium abolishes secretion, e.g. the neuromuscular junction and
chromaffin cells where the dominating source of calcium is entry
across the plasma membrane through voltage dependent calcium
channels. In many cell types, including platelets, the removal
of external calcium does not suppress the secretory response to
natural stimuli. It has been traditional to interpret this
finding as indicating the presence of calcium mobilisation from
within an internal store.   The same result would be expected,
however, if messengers other than calcium were mediating the
response, a point we shall take up below. A role for calcium has
also been proposed on the basis of experiments with various types
of inhibitory drugs. Compounds such as trifluoperazine, which
are known to interfere with calcium-calmodulin, do block the
secretory response in platelets [eg. 19,20] and many other cells.
However these compounds are not very specific for calmodulin.
Phenothiazines turn out to be effective inhibitors of protein
kinase-C [for refs. see ref. 20] and may well exert their effects
at least in part through this pathway [20]; trifluoperazine had
also recently been found to be anticholinergic on chromaffin
cells [21].   Another compound which has been used in an attempt
to analyse the role of calcium in intact cells is TMB-8. As a
result of early work with this compound in smooth muscle [22] it
has been assumed by a large number of investigators that TMB-8
can be used to block the intracellular mobilisation of calcium.
When TMB-8 is found to inhibit a cellular process it is supposed
to reflect inhibition of intracellular calcium discharge.
However TMB-8 is known to have many different actions including
inhibition of cyclooxygenase, an enzyme very important in
platelets in converting liberated arachidonic acid into the
highly active agent thromboxane. Furthermore we have recently
found that TMB-8 is relatively ineffective at suppressing calcium
mobilisation in platelets, at least as judged from the quin2
signal (see below), but is reasonably effective at blocking the

response to direct activators of protein kinase-C [23]. This compound therefore turns out to be a somewhat blunt investigative tool and it may be dangerous to interpret inhibition of secretion or any other process by TMB-8 as indicative of a role for calcium.

A convincing further demonstration that calcium can be an effective stimulus for secretion in blood platelets was obtained with cells made permeable to calcium buffers by exposure to a transient intense electric field [24,25]. This technique which renders the membrane of the cells permeable to small molecules of up to at least 1000 molecular weight, and therefore including the nucleotides and calcium buffers. Increasing the concentration of calcium from 100 nM (the resting level in most cells) into the uM range promoted the discharge of serotonin from amine-storage granules. However even this very direct approach does not rule out the possibility that the elevation of calcium could have effects via the generation of other messages. One thing that could be eliminated in those experiments was a role for thrombo-xane since the addition of drugs which prevent the formation of thromboxane did not affect the secretory response to calcium. A hint that calcium is not the only factor important in stimulus secretion coupling in platelets came from a comparison of the calcium sensitivity of the release of the contents of dense granules and of lysosomes [25]. In the intact cell lower concentrations of thrombin are required to stimulate dense granule release than are needed for stimulation of secretion of the contents of lysosomes. In the permeabilised cell however the calcium sensitivity of the release from these two classes of secretory granules is virtually the same. This implies that thrombin does something extra to enhance the release of dense granule contents in the intact cell.

EXPERIMENTS WITH QUIN2

Until recently it has been difficult or impossible to measure $[Ca^{2+}]_i$ in most small cells. However Tsien's invention of 1, a fluorescent calcium indicator, quin2, and 2, a method of entrapping it in the cytosol of intact cells by permeation of the acetoxymethyl ester and hydrolysis by cytoplasmic esterases to generate the tetracarboxylic acid indicator [26-29], has allowed

measurements of $[Ca^{2+}]_i$ and function in intact platelets, and
many other cells besides. We could now find out what levels of
calcium were required for secretion to be evoked by calcium
ionophores. The use of calcium ionophores of course bypasses
normal receptor mediated processes and hopefully allows one to
determine the effect of calcium ions themselves within the cell.
Since calcium ionophores can promote formation of thromboxane, it
is useful to examine the responses in cells in which thromboxane
formation has been blocked, for example by aspirin. Under these
conditions we find that relatively high levels of calcium, in the
uM range, can be effective at stimulating discharge of dense
granules. However the "therapeutic index" of elevated calcium
appears to be relatively small since concentrations of calcium
ionophore that are effective in stimulating secretion are very
close to those at which cell lysis occurs. The important point
is that the calcium levels needed to produce the secretory
response in quin2-loaded platelets are not very different from
those observed in the original experiments with cells made
permeable to calcium buffers. Very little secretion occurs until
$[Ca^{2+}]_i$ exceeds 1 uM and a substantial secretory response
requires a $[Ca^{2+}]_i$ of many micromolar [9,24,25,29,30].

AGONIST-EVOKED RISES IN $[Ca^{2+}]_i$

If calcium is the key trigger for secretion, then we should
expect natural agonists that stimulate secretion to raise $[Ca^{2+}]_i$
into the uM range where, from the experiments discussed above,
calcium is an effective trigger for secretory exocytosis. In
quin2-loaded platelets we find that many agonists including
thrombin, arachidonic acid, vasopressin, PAF, and ADP produce a
rise in $[Ca^{2+}]_i$ within a few seconds to a level many fold above
the basal level [9,29-32]. With all these agonists the rise is
much larger in the presence of physiological, ca. 1mM, external
Ca^{2+} than in the absence of external calcium. The simplest
explanation for these findings is that much of the $[Ca^{2+}]_i$ rise
is due to triggered calcium influx, but that a component is due
to the discharge of an internal store of calcium. The mechanism
by which the agonist receptor complex stimulates calcium influx
in platelets is unknown. It does not appear to be mediated by
voltage gated calcium channels. Organic calcium channel blockers

are not very effective at inhibiting the rise in $[Ca^{2+}]_i$ and
manipulation of the membrane potential with valinomycin and
gramicidin has rather little effect on the agonist evoked change
in $[Ca^{2+}]_i$ [33]. Presumably calcium comes through receptor-
operated calcium channels, but we have little idea how the
agonist causes the channel to open. Various possibilities
include: 1, a direct effect of receptor-agonist complexation
analagous to the effect of acetyl choline binding to the acetyl
choline receptor; 2, channel opening could be mediated by GTP
binding proteins of the sort that regulate the activity of adenylate cyclase [34]; 3, hydrolysis of inositol lipids could be
important. Possibly inositol trisphosphate which is known to
liberate calcium from intracellular organelles [35,36] could also
operate a calcium channel in the surface membrane. Direct experimental evidence for any of these mechanisms is lacking in
platelets. Whatever the mechanism, the response is fast; with
our present methods we have not been able to resolve the onset of
the quin2 signal but the maximum rate of rise occurs within the
mixing time, about 1 second, after addition of agonist. Little
is known about the mechanism of the release of internal calcium.
Since calcium ionophores and natural agonists appear to release
the same pool of calcium [9,10,31,32] we believe this to be a
store inside a Ca-sequestering organelle. We have argued
elsewhere [29] that the formation of thromboxane is most unlikely
to mediate the discharge of this internal calcium store. One
possibility for which there is as yet no direct evidence in
platelets is that inositol trisphosphate could provide the link
between the surface receptor and the discharge of internal
calcium as appears to be the case in pancreatic cells, liver
cells and a number of other cell types [35,36].

Whatever the mechanisms that lead to the elevation of
$[Ca^{2+}]_i$, the important point for stimulus-secretion coupling is
that, of the agonists we have examined, only thrombin is reliably
able to elevate $[Ca^{2+}]_i$, in quin2-loaded platelets, into the uM
range [29,30]. The other agonists, even in optimal doses, rarely
raise $[Ca^{2+}]_i$ above 800 nM, yet arachidonic acid, platelet-
activating factor and vasopressin for instance are effective
secretagogues. There seem to be two main classes of explanation
for the difference in the relationship between $[Ca^{2+}]_i$ and secretion when the cells are stimulated by calcium ionophore and when

312

they are stimulated by natural agonists. 1, For some reason the quin2 may not be reporting correctly the $[Ca^{2+}]_i$ at the relevant site; such an artefact could arise either from highly localised rises in $[Ca^{2+}]_i$ within individual platelets or from inhomogeneities in the platelet population so that only some show secretion and an elevation in $[Ca^{2+}]_i$ while others are unaffected by natural agonists, and the absence of a rise in $[Ca^{2+}]_i$ in those cells reduces the summed signal from the suspension. 2, Factors other than $[Ca^{2+}]_i$ may be important in stimulating secretion, in other words there may be other intracellular messages that either sensitise the secretory process to lower levels of $[Ca^{2+}]_i$ or independently stimulate secretory exyocytosis.

It is difficult positively to rule out the kind of artefacts discussed above; but arguments against these explanations are marshalled in ref. 37. Also, a variety of experimental protocols with quin2-loaded platelets point quite strongly to an important role for one or more intracellular messengers other than calcium ions. For instance, cells heavily loaded with 2-3 mM internal quin2, and incubated in calcium free medium containing EGTA show only a small rise in $[Ca^{2+}]_i$ when stimulated with natural agonists or with calcium ionophores. This is expected because no influx is possible and the release of a finite internal store into cytosol heavily buffered with quin2 should produce only a small change in $[Ca^{2+}]_i$. Under such conditions $[Ca^{2+}]_i$ often peaks at no more than 150 nM, yet thrombin and to a lesser extent PAF [31] and vasopressin [32] can stimulate substantial secretion from amine-storage granules. A variant on this theme takes quin2-loaded cells incubated in calcium-free medium and then exposed to saturating levels of the calcium ionophore ionomycin. Typically this raises $[Ca^{2+}]_i$ to about 200 nM, presumably discharging calcium sequestered inside the dense tubular system. This small rise in $[Ca^{2+}]_i$ causes no secretion at all. Subsequent addition of a natural agonist such as thrombin [9], platelet activating factor [31] or vasopressin evokes no further change in the quin2 signal, not surprisingly since there is no source of external calcium and the internal source has been discharged already by the ionophore. Yet these agonists evoke a substantial secretion, pointing quite strongly to the generation of some other stimulus to secretion.

Although they highlight a potentially important intracellular pathway, these experiments are done under somewhat un
physiological conditions, namely the absence of external calcium
and the presence of a calcium ionophore. We have found more
physiological conditions in which secretion can be evoked with no
significant increase in quin2 fluorescence. 1, Collagen can
stimulate secretion from dense granules in aspirin-treated cells
incubated in normal medium containing 1 mM external calcium, with
a minimal rise in $[Ca^{2+}]_i$. [10,24]. 2, Low concentrations of
thrombin, e.g. 0.05 units per ml., stimulate secretion of
thromboglobulin (an α- granule protein) without elevating
$[Ca^{2+}]_i$ [38].

Support for the idea that agonists acting as surface
receptors stimulate secretion by routes additional to changes in
$[Ca^{2+}]_i$ comes from recent studies of platelets made permeable by
high voltage discharge. It was found that thrombin could greatly
enhance the calcium sensitivity of secretion, i.e. it shifted the
$[Ca^{2+}]$-secretion curve to the left, and indeed it was found that
thrombin could actually stimulate secretion from amine-storage
granules while $[Ca^{2+}]$ was buffered at 100 nM [39-41].

DIACYLGLYCEROL AND PROTEIN KINASE-C

Breakdown of inositol lipids is an early response to activation of many cell types by many different agonists [42].
Those agonists that are capable of inducing platelets to secrete
at or near basal $[Ca^{2+}]_i$, as measured by quin2, are all known to
promote hydrolysis of inositol lipids by phospholipase-C [for
refs. see ref. 29]. The immediate products of this hydrolysis
are inositol phosphates and diacylglycerol. A possible second
messenger role for inositol trisphosphate has been discussed
above. The work of Nishizuka and his colleagues shows that
diacylglycerol has a key second messenger function in activating
protein kinase-C [43]. This kinase is found in many different
cell types, in which it phosphorylates a variety of different
proteins. In platelets the chief substrate seems to be a set of
soluble proteins of molecular weight variously described to be
40-47 kilodaltons depending on the calibration system of the
particular gel used by a particular investigator. Greatly
increased phosphorylation of these proteins is often associated

with secretion from platelets [43-45] and suggests that the
activity of protein kinase-C might play an important part in
stimulus-secretion coupling. It therefore seemed possible that
the secretion we observed at very low $[Ca^{2+}]_i$ might result from
agonist evoked formation of diacylglycerol with subsequent activation of the protein kinase-C pathway. It was shown that an
exogenous diacylglycerol, oleoylacetylglycerol (OAG), could
stimulate protein kinase-C in intact platelets [46,47]. It was
also shown that the phorbol ester 12-O-tetradecanoylphorbol 13-
acetate (TPA), which was known to cause secretion and aggregation
of human platelets, (in addition to its many other effects in
many other cell types) could directly activate protein kinase-C
[48]. The obvious test therefore was to add these direct
activators of protein kinase-C to quin2-loaded platelets and to
measure $[Ca^{2+}]_i$ and the concurrent secretion. The result was
that these agents produced a delayed and sluggish but subtantial
stimulation of secretion of the contents of dense granules,
without measurably altering $[Ca^{2+}]_i$ [10, 29]. There was also a
substantial release of the contents of $\propto$-granules [10].
Interestingly the time course of the secretion from dense
granules evoked by these direct activators of C-kinase was rather
like that evoked by collagen from aspirin-treated cells. A
further test was to look for inhibitors of protein kinase-C to
see if they would block both the secretion evoked by OAG and TPA
and the secretion evoked by collagen. It turned out that among
the most effective inhibitors of protein kinase-C were the
so-called calmodulin antagonists such as trifluoperazine and
chlorpromazine. We found that at relatively low doses, 10-20 uM,
these agents did indeed suppress the secretion evoked at basal
$[Ca^{2+}]_i$ by OAG, TPA and collagen [20].

Work with platelets permeabilised by high voltage discharge
has supported the idea that the C-kinase pathway can play an
important part in stimulus secretion coupling. Addition of
OAG or TPA shifted the $[Ca^{2+}]$-secretion relation markedly to the
left and addition of these agents could evoke a significant
secretion when calcium was buffered around 100 nM [38]. It was
suggested that thrombin might act in the permeabilised cell
preparation to stimulate the hydrolysis of inositol lipids and
provide an endogenous source of diacylglycerol [38].
Direct evidence that thrombin does increase diacylglycerol levels

in permeabalised platelets has now been reported [40]. This is an important finding because it shows rather directly that the binding of an agonist to its receptor can promote the formation of diacylglycerol while calcium remains buffered at the basal level. This fits in with much other evidence, that stimulated breakdown of inositol lipids does not require an increase in $[Ca^{2+}]_i$.

The evidence outlined above argues that diacylglycerol can cause secretion at basal $[Ca^{2+}]_i$. One can then ask whether this represents a Ca-independent route or a sensitation to low $[Ca^{2+}]_i$. Only when the molecular details are worked out will we know for sure. At present, those who use permeabilised cells tend to think in terms of increased sensitivity to Ca while those who focus on quin2-loaded intact cells think in terms of, at least partly independent pathways. We found that OAG and TPA could evoke secretion at only 50nM $[Ca^{2+}]_i$. Pozzan and colleagues have gone further in showing tha TPA can stimulate secretion from polymorphs with $[Ca^{2+}]_i$ as low as 5nM [11]. TPA also caused protein phosphorylation under these conditions indicating that inside cells, if not in the test-tube, C-kinase is not necessarily Ca-dependent.

Evidence from other cell types suggest that not only the diacylgycerol/protein kinase-C pathway can support stimulus response coupling that does not rely on calcium mobilisation. In the exocrine pancreas it seems likely that that Ca and cAMP can each trigger secretory exocytosis [49]. Experiments with quin2-loaded parathyroid cells show that elevation of $[Ca^{2+}]_i$ inhibits secretion and reduced $[Ca^{2+}]_i$ increases it [14]. In these cells it looks as though cAMP may be an excitatory message for exocytosis. Ca also seems to inhibit renin secretion from the juxtaglomerular apparatus [50]. The message here is unknown but does not appear to be diacylglycerol; TPA did not stimulate renin release (May and Rink, unpublished observations) and renin release is enhanced by trifluoperazine.

CALCIUM AND DIACLGLYCEROL IN COMBINATION

It seems likely that under physiological conditions stimulation of secretion in platelets relies on contribution from both calcium and diacylglycerol. Physiological levels as opposed to maximal levels of agonists probably do not raise $[Ca^{2+}]_i$ high enough for calcium ions alone to be an effective trigger for secretory exocytosis. The formation of diacylglycerol in response to natural agonists is usually found to be short-lived [46, 51] and so the stimulation of protein kinase-C may not be long enough maintained for that to be an effective stimulus to secretion, and in any case secretion evoked by direct activation of protein kinase-C is much more sluggish than that usually evoked by natural agonists [10]. The experiments of Kaibuchi et al. suggested the existence of synergy between calcium and diacylglycerol [46,47]. They found that application of a concentration of calcium ionophore too low itself to be an effective stimulus to secretion, greatly enhanced the amount of serotonin secreted in response to OAG. We examined this effect in quin2-loaded cells and concluded that the main effect of sub-threshold elevation for $[Ca^{2+}]_i$ was to accelerate the response to subsequently added OAG or TPA [10]. The brisk secretory response now looked much more like that evoked by natural agonists which are known to cause both a modest elevation of $[Ca^{2+}]_i$ and the formation of diacylglycerol. The molecular basis for this synergy is not well worked out. One obvious possibility is that the combination of elevated $[Ca^{2+}]_i$ and enhanced formation of diacylglycerol provides a much more effective stimulus to protein kinase-C. Results with protein kinase-C isolated in a test tube are interpreted as showing that diacylglycerol makes the enzyme more sensitive to lower concentrations of Ca^{2+} [43,46]. However, the evidence in intact platelets suggests that protein kinase-C in the cell can be fully activated at basal levels of $[Ca^{2+}]_i$ [40,46]. For instance, in the very same experiments that the synergy between calcium and OAG was found for secretion, OAG alone evoked the same degree of phosphorylation for the substrate of C-kinase as did OAG in combination with calcium ionophore [46,47]. Another possible mechanism for the synergy is that some step subsequent to the phosphorylation of the substrate of C-kinase is not rate limiting and is greatly speeded by the elevation of $[Ca^{2+}]_i$. One intriguing possibility is that the main role for the modest elevation of $[Ca^{2+}]_i$ might be to evoke the cytoplasmic rearrangements of shape-change which result in bringing secretory granules close to sites of exocytosis.

The possible importance of synergy between these two intra-cellular pathways may help to explain the different abilities of different agonists to trigger secretion under various experimental conditions. For example, as mentioned above collagen produces a relatively slow and sub-maximal secretion from aspirin-treated platelets. We believe this may be because under these conditions collagen can produce the formation of diacylglycerol but is a very poor stimulus to elevated $[Ca^{2+}]_i$ [10,29]. In cells that have not been treated with cyclooxygenase blockers and are capable of generating thromboxane, collagen produced a somewhat delayed elevation of calcium to several hundred nM, an effect attributed to the calcium mobilising effect of thromboxane A_2. Under these conditions, at just the time the $[Ca^{2+}]_i$ rises, there is a brisk and substantial secretory response. We imagine that this reflects the combined actions of $[Ca^{2+}]_i$ and the diacylglycerol. With ADP it seems to be the other way round [29]. In aspirin or indomethacin-treated platelets ADP is able to elevate $[Ca^{2+}]_i$ to about 800 nM, not enough for it to stimulate secretion. And, as is well known, ADP does not stimulate secretion in the absence of formation of thromboxane. We believe that this failure to stimulate secretion reflects the very poor ability of ADP to promote hydrolysis of inositol lipids [D.E. MacIntyre, personal communication] and therefore its poor ability to generate diacylglycerol. In platelets that are capable of generating thromboxane A_2, ADP can be an effective secretagogue provided that aggregation occurs and thromboxane is indeed generated. As best we can judge from the quin2 signal, the elevation of $[Ca^{2+}]_i$ under these conditions is very little greater than in conditions where secretion does not occur. Certainly $[Ca^{2+}]_i$ does not appear to rise into the uM range [29,30]. We hypothesise that the importance of thromboxane A_2 under these conditions is not to provide a sufficient elevation of $[Ca^{2+}]_i$, but rather to promote the hydrolysis of inositol lipids and generate the requisite amounts of diacylglycerol to synergise with the already elevated $[Ca^{2+}]_i$.

CYCLIC NUCLEOTIDES

Agents such as prostaglandin I_2 and forskolin that stimulate adenylate cyclase are potent inhibitors of platlet activation [see eg. ref. 52]. cAMP, assumed as always to act by disinhibiting cAMP-dependent protein kinase, seems to exert its main effect by shutting off the supply of second messengers. PGI_2 and forskolin are potent inhibitors of $[Ca^{2+}]_i$ rises evoked by all agonists so far tested including thrombin [53,54], arachadonic acid, vasopressin, ADP, and platelet-activating factor [55]. Elevated cAMP levels also inhibit agonist-induced turnover of inositol lipids [46,51] and furthermore inhibit the formation of thromboxane. cAMP seems much less effective at suppressing the effects of putative second messengers. For instance cAMP does not decrease the Ca^{2+} sensitivity of secretion in permeabalised platelets in control conditions [39]; cAMP does inhibit the ability of thrombin to enhance the sensitivity, an effect attributed to inhibition by cAMP of the thrombin-evoked formation of diacylglycerol [39]. In intact platelets we find that PGI_2 and forskolin, at concentrations that suppress $[Ca^{2+}]_i$ rises and secretion evoked by thrombin, only slightly reduce the secretory responses to TPA and OAG [56].

The role of cGMP remains unclear. It appears slightly to enhance the effect of thrombin in permeabalised cells but the evidence from intact cells does not suggest an excitatory role for this cyclic nucleotide [39].

DIFFERENTIAL RELEASE FROM AMINE-STORAGE GRANULES AND ALPHA GRANULES

It is known that secretion of proteins from   -granules occurs at lower concentrations of thrombin than are required to stimulate discharge from amine-storage granules [2]. Since low doses of thrombin stimulated secretion of β-thromboglobulin without any increase in quin2 signal (see above), we wondered if this might reflect preferential triggering of α-granule secretion by the C-kinase pathway. This notion was supported by finding that the EC_{50} for TPA to stimulate secretion of β-thromboglobulin was lower than the EC_{50} for TPA to stimulate secretion of ATP from amine-storage granules. We have also mentioned the finding that secretion from amine-storage granules is stimulated at concentrations of thrombin that do not stimulate release from lysosomes. The results of Knight et al. [25] suggest that this is not due to differential sensitivity to calcium ions; it may be due to a differential sensitivity to the C-kinase pathway, but we have not yet investigated this point experimentally.

BEYOND THE SECOND MESSAGE

We have discussed stimulus-secretion coupling only as far as the generation of the second message. The next questions are "what are the targets for these second messengers and what are the molecular mechanisms underlying exocytosis?" We will comment only briefly and only on the first of these points. The mechanisms of secretory exocytosis are still mysterious and (fortunately) beyond the scope of this account. Some current hypotheses are outline in a recent article discussing exocytosis in adrenal chromaffin cells [57].

A great deal of intracellular regulation is done by increasing and reducing the degree of phosphorylation of key proteins. The only established target for diacylglycerol in platelets is protein kinase-C and the major substrates for this kinase are the 47 kilodalton proteins. These proteins have now been isolated and purified but nothing is yet known of their function [45]. Their likely importance in secretion is shown by the fact that so far no conditions have been found in which secretion occurs without phosphorylation of these proteins. Whether the phosphorylation of these proteins is a sufficient stimulus to secretion is more debatable, since one can find conditions in which there is substantial, if not complete, phosphorylation without any secretory exocytosis [45]. However, since there are several related proteins, each with several sites for phosphorylation, it may be that a critical phosphorylation is required for these proteins to be an effective stimulus to secretory exocytosis. Other possibilities are that there are other important substrates for protein kinase-C which are difficult to pick up with current methods, or that OAG and TPA have other cellular targets. If so, the evidence in platelets is that those targets are not phospholipase A_2, or phospholipase-C [58].

The targets for calcium ions are if anything less clear than those for diacylglycerol. It is not clear whether calcium

320

stimulation of exocytosis in platelets works via protein kinase-C. Evidence from permeabalised cells indicates that this may not be so. There was less phosphorylation with substrate of protein kinase-C when uM levels of calcium evoked substantial secretion, than when addition of OAG at low [Ca^{2+}] evoked only a small secretory response [40]. It is of course possible that Ca picks out for phosphorylation just the critical sites. Calcium ions also activate calmodulin-dependent protein kinases; in platelets the main such enzyme appears to be myosin light chain kinase [59]. Phosphorylation of myosin light chains is a feature of platelet activation but it is not clear that this phosphorylation plays a critical role in secretion. Certainly phosphorylation of myosin light chains can occur under conditions in which there is no secretion at all [59,60]. One can also see secretion in the conditions in which the substrate protein kinase-C is substantially phosphorylated but there is little phosphorylation of myosin light chains, for instance when secretion is stimulated by TPA or by collagen in aspirin-treated platelets [48, 59, 60]. However, it is worth re-emphasising that the secretion evoked under these kind of conditions is sluggish and delayed, and the secretion that occurs under normal conditions may very well require activation of protein kinase-C and other cellular processes such as myosin phosphorylation and internal contraction [17].

Despite recent advances, it is clear that we have moved only a short way down the path of the stimulus-secretion coupling mechanism and a great deal more remains to be discovered.

ACKOWLEDGEMENTS

Out work has been supported by grants from the SERC and from Ciba-Geigy. We thank Nuala O'Connor for skilled assistance, R. Y. Tsien for OAG and The Upjohn Company for TMB-8.

REFERENCES

1. Da Prada M, Richards, JG, Kettler R (1981) In: Gordon JL (ed)
    Platelets in Biology and Pathology - 2. Elsevier, Amsterdam,
    pp 107-145

2. Kaplan KL (1981) In: Gordon JL (ed) Platelets in Biology and
    Pathology - 2. Elsevier, Amsterdam, pp 77-88

3. Burnstock G (1981) J. Physiol 313:1-35

4. Pearson JD, Slakey LL, Gordon JL (1983) Biochem J 214:273-276

5. Hays RM, Levine SD (1981) In: Brenner BM, Rector FC,(eds) The
    Kidney. Saunders, Philadelphia, pp 777-840

6. Jiron C, Romano M, Michelangeli F (1984) J Membrane Biol 80:119-
    134.

7. Katz B, (1969) The Release of Neural Transmitter Substances.
    Liverpool University Press

8. Douglas WW, (1968) Brit J Pharmac 34:451-474

9. Rink TJ, Smith SW, Tsien RY (1982) FEBS Lett 148:21-26

10. Rink TJ, Sanchez A, Hallam TJ (1983) Nature 305:317-319

11. Pozzan T, Lew DP, Wollheim CB, Tsien RY (1983) Science 221:1413-
    1415

12. De Virgilio F, Lew DP, Pozzan T (1984) Nature 310:691-693

13. White JR, Ishizaka T, Ishizaka K, Sha'afi RI (1984) Proc Nat
    Acad Sci USA 81:3978-3982

14. Shoback DM, Thatcher J, Leombruno R, Brown EM (1984) Proc Nat
    Acad Sci (USA) 81:3113

15. Meldolesi J, Huttner WB, Tsien RY, Pozzan T (1984) Proc Nat Acad
    Sci USA 81:620-624

16. Massini P, Luscher E (1974) Biochem Biophys Acta 372:109-121

17. White JG, Rao GHR, Gerrard JM (1974) Am J Path 77:135-149

18. Massini P Luscher EF (1976) Biochim Biophys Acta 436:652-663

19. Feinstein MB, Hadjan RR (1981) Mol Pharmacol 21:422-431

20. Sanchez A, Hallam TJ, Rink TJ (1983) FEBS Lett 164:43-46

21. Chiou CY, Malagodi MH (1975) Br J Pharmac 53:279-285

22. Simpson AWM, Hallam TJ, Rink TJ submitted to FEBS Lett

322

23. Clapham DE, Neher E (1984) J Physiol 353:541-564

24. Knight DE, Scrutton MC (1980) Thrombosis Res 20:437-446

25. Knight DE, Hallam TJ, Scrutton MC (1982) Nature 296:256-257

26. Tsien RY (1980) Biochem 19:2396-2404

27. Tsien RY (1981) Nature 290:527-528

28. Tsien RY, Pozzan T, Rink T (1982) J Cell Biol 94:325-334

29. Rink TJ, Hallam TJ (1984) Trends in Biochem Sci 9:215-219

30. Rink TJ, Tsien RY, Sanchez A, Hallam TJ (1984) In: Rubin RP, Weiss G, Putney JW Jr (eds) Calcium in Biological Systems. Plenum, New York

31. Hallam TJ, Sanchez A, Rink TJ (1984) Biochem J 218:819-827

32. Hallam TJ, Thompson NT, Scrutton MC, Rink T (1984) Biochem J 221:897-901

33. Hallam TJ, Pipili E, Rink TJ Unpublished observations

34. Gomperts BD (1983) Nature 306:64-66

35. Streb H, Irvine RF, Berridge MJ, Schultz I (1983) Nature 306:67-69

36. Berridge MJ (1984) Biochem J 220:345-360

37. Tsien RY, Pozzan T, Rink TJ (1984) Trends in Biochem Sci 9:263-266

38. Simpson AWM, O'Connor N, Hallam TJ, Rink TJ, unpublished observations

39. Knight DE, Scrutton MC (1984) Nature 309:66-68

40. Knight DE, Nigli V, Scrutton MC (1984) Eur J Biochem In press

41. Haslam R, Davidson MML (1984) Biochem J 222:351-361

42. Michell R (1982) Cell Calcium 3:285-294

43. Nishizuka Y (1984) Nature 308:693-698

44. Haslam RJ, Lynham JA (1977) Biochem Biophy Res Commun 77:714-722

45. Imaka T, Lynham JA, Haslam RJ (1983) J Biol Chem 258:11401-11414

46. Kaibuchi K, Sano K, Hoshijima M, Takai Y, Nishizuka Y (1982) Cell Calcium 3:323-335

47. Kaibuchi K, Takai Y, Sawamura M, Hoshijima M, Fiyikura T, Nishizuka Y (1983) J. Biol Chem 258:2010-2013

48. Castagna M, Takai T, Kaibuchi K, Sano K, Kikkawa N, Nishizuka Y (1981) J Biol Chem 257:7847-7851

49. Gardner JD, Jensen RT (1981) Phil. Trans. Roy. Lond. B 296: 17-26

50. Keeton, TK, Campbell WB (1981) Pharmac Rev 32:81-227

51. Haslam, RJ, Davidson MML, Davies T, Lynham JA, McClenaghan MD (1978) Adv Cyclic Nuceotide Res 9:533-552

52. Rittenhouse SE (1982) Cell Calcium 3:311-322

53. Rink T, Smith SW (1983) J Physiol 338:66-67

54. Zavoico GB, Feinstein MB (1984) Biochem Biophys Res Commun 120:579-585

55. Sage SO, Hallam TJ, Simpson AMW, RinkTJ Unpublished observations

56. Rink TJ, Sanchez A (1984) Biochem J 222:833-836

57. Baker PF, Knight DE (1984) Trends in Neurosci 7:120-126

58. Watson SP, Ganong BR, Bell RM, Lapetina EG (1984) Biochem Biophys Res Commun 121:386-391

59. Daniel JD, Molish IR, Rigmaiden M, Steward G (1984) J Biol Chem (in press)

60. Hallam TJ, Daniel JD Unpublished observations

RESUME

Les résultats de 2 types d'expériences montrent que si on élève la $[Ca^{2+}]_i$ dans les plaquettes humaines au niveau micromolaire on stimule l'exocytose sécrétoire. Le EC_{50} pour l'activation par le Ca^{2+} de la libération de 5-HT à partir de plaquettes rendues perméables aux tampons Ca^{++} par une décharge de haut voltage est environ 2-4 µM et on trouve une relation similaire Ca^{2+}-sécrétion dans des plaquettes intactes chargées de quin-2 et stimulées par le Ca^{++}-ionophore. Néanmoins, les ions Ca^{2+} ne sont apparamment pas le seul messager intracellulaire excitatoire dans le couplage stimulus-sécrétion.

Dans différentes conditions expérimentales nous voyons une sécrétion provoquée par un agoniste quand la $[Ca^{2+}]_i$ dans des plaquettes chargées de quin 2 reste au, ou proche du, niveau basal (100 nM). Le diacylglycerol formé par l'hydrolyse des lipides contenant de l'inositol pourrait jouer un rôle clé dans cette réponse

324

sécrétoire. Le diacylglycérol exogène stimule les plaquettes chargées de quin 2 à sécréter sans augmenter la $[Ca^{2+}]_i$ comme le fait aussi un autre activateur de la protéine kinase C, l'ester de phorbol. Des sécrétions similaires, "Ca-indépendantes" ont lieu dans d'autres types cellulaires, y compris les polymorphes. Une élévation modeste du $[Ca^{2+}]_i$ par l'ionomycine accélère la sécrétion par des plaquettes chargées de quin 2, provoquée par le diacylglycérol ou l'ester de phorbol. De plus, le diacylglycérol, l'ester de phorbol ou la thrombine peuvent fortement augmenter la sensibilité au Ca^{2+} de la sécrétion par des plaquettes perméabilisées. La sécrétion provoquée par des agonistes naturels pourrait réfléter l'action combinée du diacylglycérol et des ions Ca^{2+}.

Hormones and Cell Regulation, Volume 9
INSERM European Symposium
J.E. Dumont, B. Hamprecht and J. Nunez editors
© 1985 Elsevier Science Publishers B.V. (Biomedical Division)

HORMONAL AND NEUROTRANSMITTER REGULATION OF Ca^{2+} MOVEMENTS IN PANCREATIC ACINAR CELLS

I. Schulz, H. Streb, E. Bayerdörffer, K. Imamura
Max-Planck-Institut für Biophysik, Kennedyallee 70,
6000 Frankfurt (Main) 70, FRG

INTRODUCTION

Enzyme, electrolyte and fluid secretion from exocrine pancreatic cells is stimulated by neurotransmitters as well as by a variety of peptide hormones (10,17,18,22,33). One group of secretagogues such as secretin, vasoactive intestinal polypeptide (VIP) and epinephrine (ß agonist) stimulate enzyme secretion by increasing intracellular cAMP. Another group including acetylcholine, cholecystokinin-pancreozymin and bombesin stimulates enzyme secretion via an increase in the cytosolic free calcium concentration (10,16,22). In addition these receptors might also influence enzyme secretion by production of diacylglycerol and activation of protein kinase C (7,30).

For initiating enzyme secretion release of Ca^{2+} from intracellular stores seems to be important (10,22), whereas for sustained secretion extracellular Ca^{2+} is required (19,23). On the basis of experiments with A23187 to raise intracellular free Ca^{2+} concentration and phorbolesters to stimulate protein kinase C the hypothesis has been put forward that Ca^{2+}-calmodulin processes are responsible only for the initial phase of cell response, whereas protein kinase C-dependent processes are responsible for the sustained phase of response (20).

In this report we describe the role of Ca^{2+} and cellular events which regulate the cytosolic free Ca^{2+} concentration at rest and at stimulation in rat pancreatic acinar cells. This includes the characterization of the site and mechanism of secretagogue-induced Ca^{2+} release and Ca^{2+} uptake into intracellular structures.

The data suggest that at rest both the rough endoplasmic reticulum and the plasma membrane regulate the cytosolic free Ca^{2+} to a value of $\sim 2 \times 10^{-7}$ mol/l (Fig. 3, and ref. 16). During stimulation Ca^{2+} release occurs from the endoplasmic reticulum by which a rise of cytosolic free Ca^{2+} concentration to $\sim 8 \times 10^{-7}$ mol/l occurs. The intracellular messenger for secretagogue-stimulated Ca^{2+} release is most likely inositol- 1,4,5-trisphosphate (IP_3) (27) which is produced in pancreatic acini during receptor-activated hydrolysis of phosphatidylinositol-4,5-bisphosphate (21).

METHODS

Isolation of acinar cells

Pancreatic acinar cells were prepared by controlled digestion of pancreatic tissue from rats (3,12) using collagenase (Worthington, Freehold, NJ, USA).

$^{45}Ca^{2+}$-flux measurements in isolated "intact" acinar cells

Cells were preloaded with $^{45}Ca^{2+}$ (1 μCi/ml) in a Krebs-Ringer bicarbonate buffer, containing 1.25 mmol/l Ca^{2+}, until a quasi steady state for $^{45}Ca^{2+}$ was reached. To separate cells from the incubation medium at different time intervals, aliquots of 400 μl were removed from the incubation medium, layered over 100 μl of silicone oil (density 1.034) and centrifuged in a Beckman microfuge at 15 000 g for 30 s. The cell pellet was deep-frozen in liquid nitrogen, the tip of the tube was cut through the oil layer and radioactivity was counted, similar to the method described previously (25).

Estimation of cytosolic free Ca^{2+} concentration in "intact" cells

Cytosolic free Ca^{2+} concentration was estimated using the "quin 2 method" (31). Cells were preincubated in Krebs-Ringer Hepes buffer containing 100 μmol/l of quin 2 tetracetoxymethylester (quin 2 AM) for 20 min and additional 40 min with 25 μmol/l of quin 2 AM. Fluorescence measurements were performed in a Perkin-Elmer LS-5 fluorometer at 339 nm and 488 nm wave lengths for excitation and emission respectively.

Determination of free Ca^{2+} concentrations in the surrounding medium of "leaky" cells

Following isolation, cells were permeabilized by washing cells twice in a nominally "Ca^{2+}-free solution" (29). This method rendered cells leaky as judged by trypan blue uptake into 90% of cells and release of the cytosolic enzyme lactate dehydrogenase (LDH) by 80% of total LDH present in cells before washing in "Ca^{2+}-free solution" (29). The free Ca^{2+} concentration of the incubation medium was recorded continuously with a Ca^{2+}-specific electrode (1), using Ca^{2+}-selective membranes which contained the neutral carrier N,N'-di((11-ethoxycarbonyl)undecyl)-N,N'-4,5-tetramethyl-3,6-dioxaoctane amide (Glasblaeserei W. Moeller, Zuerich, Switzerland). Electrodes were calibrated at free Ca^{2+} concentrations between $6x10^{-8}$ and 10^{-2} mol/l using the chelators nitrilotriacetic acid (NTA) or ethylene-diamine-tetra-acetic acid (EDTA) as described previously (29).

Preparation of subcellular fractions

For preparation of rough endoplasmic reticulum isolated acinar cells were homogenized in an isotonic mannitol buffer, the homogenate was centrifuged at 1 000 g and 11 000 g for 12 and 15 min, respectively, and the resulting supernatant for 15 min at 27 000 g. The pellet of the last centrifugation step was mixed with 11% Percoll in mannitol buffer and a density gradient was formed by spinning the tubes at 41 000 g for 40 min as described previously (6). Purified rough endoplasmic reticulum was localized in the most dense fractions with an average density of 1.055 g/cm^3.

For preparation of a plasma membrane enriched fraction, a fluffy layer of the 11 000 g pellet (see above) was brought to 1.25 mol/l of sucrose and the sample was layered over a 2 mol/l sucrose cushion and overlayered with 0.3 mol/l sucrose. The step gradient was centrifuged at 150 000 g for 90 min and the material banding at the interfaces between 0.3 mol/l and 1.25 mol/l sucrose ("S 1" enriched in plasma membranes) and 1.25 mol/l and 2 mol/l sucrose ("S 2" enriched in endoplasmic reticulum) were collected. Alternatively total homogenate was suspended in a 280 mmol/l mannitol-Hepes buffer and MgCl$_2$ (11 mmol/l) was added. It was then spun at 400 g for 10 min and after resuspension the resulting supernatant and precipitate were spun down at 25 000 g for 15 min.

Measurement of $^{45}Ca^{2+}$ uptake into isolated membrane vesicles

Isolated membrane vesicles were incubated in a 130 mmol/l KCl-Hepes buffer at different free $^{45}Ca^{2+}$ concentrations in the presence of MgATP or without ATP in the presence of Na$^+$ gradients as indicated in the legends to the figures. At given time points, triplicate samples were filtered through polycarbonate or cellulose nitrate filters, the filters were washed and the radioactivity was quantitated by a standard liquid scintillation procedure (13).

Preparation and labelling of inositol-1,4,5-trisphosphate

Inositol-1,4,5-trisphosphate (IP$_3$) was prepared by Dr. Robin Irvine, Cambridge, UK, by incubating human red blood cell ghosts with CaCl$_2$ followed by a Dowex-formate column separation, and desalted by elution from a Dowex-Cl column with 1 mol/l LiCl, followed by removal of the LiCl with ethanol (9). ^{32}P-labelled IP$_3$ was prepared from human erythrocytes by the method of Downes et al. (9) from ^{32}P-prelabelled erythrocyte ghosts (8).

Protein phosphorylation and -separation by polyacrylamide gel electrophoresis

The phosphorylation procedure was similar to that described by Amory et al. (2). The phosphorylation reaction was started by adding purified rough endoplasmic reticulum membranes to a Hepes-Tris buffer containing γ-^{32}P-ATP (5x10^{-6} mol/l) and ions as indicated. Free Ca^{2+} and Mg^{2+} concentrations were adjusted with 3 mmol/l EGTA or EDTA as described previously (29). At different incubation times the reaction was stopped by addition of 10% trichloracetic acid (TCA) containing 10 mmol/l KH$_2$PO$_4$ and 1 mmol/l ATP and washed subsequently with the same solution, followed by twice washing with a solution of 50 mmol/l KH$_2$PO$_4$/H$_3$PO$_4$ (pH 2.0). The TCA precipitated pellet was solubilized in 250 mmol/l sucrose with 50 mmol/l tetradecyltrimethylammonium bromide (TDAB) and submitted to polyacrylamide gel electrophoresis at acidic pH as described previously (2). After electrophoresis, the gel was incubated in 1% (v/v) glycerol, dried and exposed to Kodak X-Omat AR 5 film for 15 to 18 hours at -70°C. ^{32}P-labelled proteins were excised from gels and their radioactivity was counted by a standard liquid scintillation procedure (13).

Measurement of Ca^{2+}-(Mg^{2+})-ATPase activity

Adenosine triphosphatase activity was measured subsequently to the phosphorylation reaction according to the method of Bais (4). Briefly, after stopping the phosphorylation reaction by addition of 10% TCA, samples were centrifuged for 5 min at 3500 rpm at 4°C. Ten µl aliquots of the resulting supernatant were mixed with 500 µl of a charcoal suspension (125 mg of activated charcoal/ml 1N HCl) and centrifuged at 2500 g for 10 min at 4°C. Aliquots of the supernatant were counted for radioactivity as described (13). Radioactivity obtained in the absence of membranes were subtracted from each sample. Ca^{2+}-dependent $^{32}P_i$ liberated from γ-^{32}P-ATP was obtained after subtraction of the value obtained in the absence of Ca^{2+}. Liberated $^{32}P_i$ is expressed as pmol/mg membrane protein x time of incubation.

RESULTS

Ca^{2+} movements in isolated "intact" acinar cells

When intact acinar cells are loaded with $^{45}Ca^{2+}$ until tracer equilibrium is obtained, stimulation with a Ca^{2+}-mobilizing secretagogue, such as carbamyl-choline, causes rapid release of $^{45}Ca^{2+}$ from the cells (1st phase) and this decrease is followed by a slower $^{45}Ca^{2+}$ reuptake into cells (2nd phase) (Fig. 1). Subsequent restimulation by a different secretagogue such as cholecystokinin-pancreozymin (CCK-Pz) does not cause a second release of $^{45}Ca^{2+}$. When the antagonist atropine is given subsequently to the agonist carbamylcholine, a rapid increase in $^{45}Ca^{2+}$ content occurs followed by a slower decrease back to the prestimulation level (Fig. 1). At that time a second release of $^{45}Ca^{2+}$ by CCK-Pz can be induced. We have interpreted this experiment to mean that during stimulation Ca^{2+} is released from an intracellular store and extruded from the cell, which results in a decrease in cellular calcium content. Due to increased Ca^{2+} permeability of the plasma membrane (12) increased Ca^{2+} influx occurs into cells, which leads to Ca^{2+} uptake and increase in cellular Ca^{2+} during the second slower phase of stimulation. Ca^{2+} is taken up into a Ca^{2+} pool, different from the one, which is emptied by secretagogues during the first phase of stimulation. A second Ca^{2+} release cannot be induced by a different secretagogue (Fig. 1). However, when the antagonist to the first agonist is added, refilling of the first pool ("trigger Ca^{2+} pool") occurs and the second pool releases its Ca^{2+} until the prestimulation level is reached. A second Ca^{2+} release by another secretagogue is then possible (25). In the light of the present knowledge

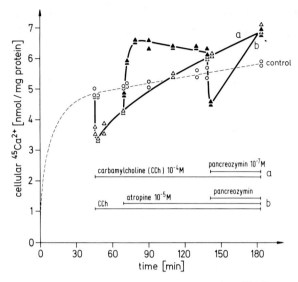

Fig. 1. Effect of stimulation and restimulation on $^{45}Ca^{2+}$ content. Curve a: subsequent addition of carbamylcholine (CCh) and pancreozymin. Curve b: subsequent addition of CCh and pancreozymin with interposition of atropine. For each time point duplicates are taken and data from 1 of 6 similar experiments are shown. (From 25).

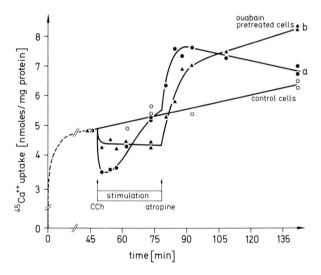

Fig. 2. Effect of ouabain (2 mmol/l) on carbamylcholine (CCh) and atropine-induced changes in cellular $^{45}Ca^{2+}$. Curve a: effect of CCh (10^{-5} mol/l) and atropine (10^{-5} mol/l) on $^{45}Ca^{2+}$ content in cells that had been incubated for 45 min at 1.25 mmol/l $^{45}Ca^{2+}$ without ouabain. Curve b: Cells preincubated for 30 min in presence of ouabain (2 mmol/l). They were further incubated for 45 min in presence of $^{45}Ca^{2+}$ (1.25 mmol/l). Subsequently CCh (10^{-5} mol/l) and atropine (10^{-5} mol/l) were added at indicated times. (From 25).

(27) these observations can also be interpreted to mean that a second stimulation does not cause a second Ca^{2+} release since the second messenger for secretagogue action is not resynthetized in sufficient amounts.

When cells are pretreated with ouabain, $^{45}Ca^{2+}$ release is smaller during the 1st phase and $^{45}Ca^{2+}$ reuptake is inhibited during the 2nd phase (Fig. 2). Addition of atropine causes rapid $^{45}Ca^{2+}$ uptake over the control, however, return of cellular $^{45}Ca^{2+}$ content back to the control does not occur and $^{45}Ca^{2+}$ content remains at a higher level than before stimulation. Similar results were obtained in the presence of high K^+ and reduced Na^+ concentrations in the medium and were interpreted to mean that Ca^{2+} extrusion may be partially achieved by a plasma membrane located Na^+/Ca^{2+} countertransport. The missing reuptake in the second phase after carbachol addition under these conditions might be explained by reduced Ca^{2+} uptake into mitochondria in the presence of increased cytoplasmic Na^+ concentrations. Because Ca^{2+} uptake as induced by atropine (i.e. refilling of the "trigger Ca^{2+} pool") is not impaired (Fig. 2) one can assume that the trigger pool had been depleted by carbamylcholine. However, Ca^{2+} released into the cytosol could not completely be extruded by the cell because of reduced Na^+/Ca^{2+} exchange under ouabain.

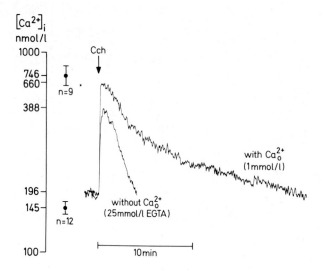

Fig. 3. Effect of carbachol (CCh, 10^{-6} mol/l) on the fluorescence of intracellular quin 2 in pancreatic acini in the presence and absence of extracellular calcium. Cells were preincubated for 20 min in Krebs-Ringer Hepes buffer with 100 µmol/l quin 2-tetracetoxymethylester (quin 2 AM) and for additional 40 min with 25 µmol/l quin 2 AM at 37°C under oxygenation. Fluorescence measurement was made in a Perkin-Elmer LS-5 fluorometer at wave lengths of 339 nm for excitation and 488 nm for emission.

Cytosolic free Ca^{2+} concentration at rest and stimulation

As shown in Fig. 3 the intracellular free Ca^{2+} concentration as measured with quin 2 is at rest ∿145 nmol/l and it increases to ∿780 nmol/l, when the cells are stimulated with carbachol. In the absence of extracellular Ca^{2+} in the medium, the increase in free Ca^{2+} concentration is smaller and the decrease to the prestimulation level is more rapid than compared to stimulation in the presence of 1 mmol/l of calcium in the surrounding medium. This observation confirms previously made conclusions that during stimulation Ca^{2+} is released from intracellular stores (10,22) by which event secretion is initiated. Due to increased Ca^{2+} permeability of the plasma membrane Ca^{2+} influx into the cell is increased (12) and cytosolic free Ca^{2+} concentration is maintained at a higher level as compared to the unstimulated state. This increased free Ca^{2+} concentration makes sustained secretion possible, whereas after an initial burst of secretion, prolonged secretion is not maintained by secretagogues in the absence of extracellular Ca^{2+} (19,23).

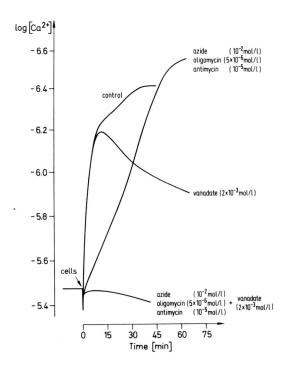

Fig. 4. Effect of vanadate and mitochondrial inhibitors on Ca^{2+} uptake into leaky acinar cells. Cells were rendered leaky by washing them twice in a nominally "Ca^{2+}-free solution" and were then incubated in a standard incubation buffer with 120 mmol/l KCl, 10 mmol/l MgATP and an ATP regenerating system in the presence of indicated substances. (From 29).

Intracellular Ca^{2+} pools involved in the regulation of free Ca^{2+} concentration

The nature of intracellular Ca^{2+} pools involved in the regulation of cytosolic free Ca^{2+} concentration has been investigated using isolated permeabilized acinar cells. As shown in Fig. 4 permeabilized cells take up Ca^{2+} from a medium that contains ATP and a regenerating system until a steady state of $\sim 4 \times 10^{-7}$ mol/l free $[Ca^{2+}]$ is reached. In the presence of mitochondrial inhibitors the same steady state free Ca^{2+} concentration is reached as in the control. In the presence of vanadate which inhibits Ca^{2+}-(Mg^{2+})-ATPases but does not effect mitochondria (29) the initial rate of Ca^{2+} uptake is the same as in the control but the steady state free Ca^{2+} concentration is not reached. This suggests that a non-mitochondrial Ca^{2+} pool regulates the free Ca^{2+} concentration at 4×10^{-7} mol/l.

The nature of this nonmitochondrial Ca^{2+} pool has been identified by measuring $^{45}Ca^{2+}$ uptake into isolated rough endoplasmic reticulum of pancreatic acinar cells. As shown in Fig. 5, MgATP-dependent Ca^{2+} uptake shows similar dependencies on free Ca^{2+} concentration in both leaky cells and isolated rough endoplasmic reticulum. Similarly other parameters tested such as cation-, anion-, $[Mg^{2+}]$-, ATP-concentration, and pH-dependencies are the same for both

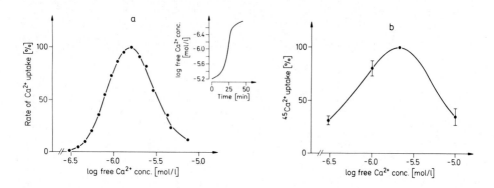

Fig. 5. Dependence of calcium uptake into leaky acinar cells and isolated membrane vesicles from rough endoplasmic reticulum on the free Ca^{2+} concentration in the medium. The experiments were performed as described in "Results". For permeabilized cells one typical curve (from n=6) of the calcium uptake rate is shown (a). The inset shows the original trace of calcium uptake against time as measured with the calcium-specific electrode, on which the calculation of the calcium uptake rate was based. 100% were equivalent to 8 ± 1.2 nmol/mg protein per min. The values in (b) represent the means of $^{45}Ca^{2+}$ uptake after 20 min of 4 experiments $\pm$ S.E. performed with rough endoplasmic reticulum vesicles. When the $^{45}Ca^{2+}$ uptake rate of the initial 2 min were considered a similar curve was obtained. ATP-dependent calcium uptake is expressed in % of the highest uptake, 100% were equivalent to a specific uptake of 5.1 ± 0.3 nmol/mg protein per 20 min. (From 6).

preparations (6). This suggests that the rough endoplasmic reticulum is the nonmitochondrial Ca^{2+} pool into which Ca^{2+} is taken up. Further support for this assumption was obtained by electron microscopy showing Ca^{2+} oxalate precipitates in the rough endoplasmic reticulum in leaky acinar cells when incubated in the presence of Ca^{2+}, oxalate and MgATP. These precipitates were strongly reduced in the presence of the Ca^{2+} ionophore A23187 or in the absence of MgATP (32).

Underlying mechanism for Ca^{2+} uptake into rough endoplasmic reticulum

The MgATP dependence of Ca^{2+} uptake into the rough endoplasmic reticulum (RER) suggests that a $Ca^{2+}-(Mg^{2+})$-ATPase is involved. Indeed we could demonstrate $Ca^{2+}-(Mg^{2+})$-ATPase activity and a phosphorylated intermediate of this enzyme in a purified fraction of RER (Fig. 6). Both the $Ca^{2+}-(Mg^{2+})$-ATPase and P_i incorporation are Ca^{2+} dependent in the same range of free Ca^{2+} concentration as for Ca^{2+} uptake into leaky cells and isolated RER with an optimum at

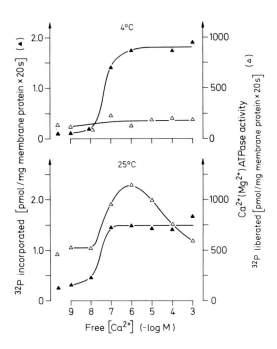

Fig. 6. Dependence of ^{32}P-incorporation into the 130 kDa protein of rough endoplasmic reticulum (▲) and of $Ca^{2+}-(Mg^{2+})$-ATPase activity (△) on the free Ca^{2+} concentration. Incubation medium was Hepes-Tris buffer (pH 7.0) containing in mol/l: KCl 126×10^{-3}, free $[Mg^{2+}]$ 10^{-5}, $[\gamma^{32}P]ATP$ 5×10^{-6}, ouabain 2×10^{-3}, NaN_3 5×10^{-3}, free $[Ca^{2+}]$ was adjusted with 3×10^{-3} mol/l EGTA (pH 7.0), 20 s incubation at 4°C and 25°C.

334

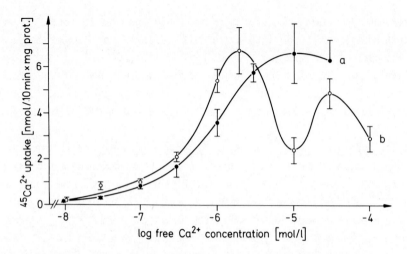

Fig. 7. Dependence of calcium uptake into plasma membranes (a) and rough endoplasic reticulum (b) on the free Ca^{2+} concentration. The values represent means ± S.E. of 5 preparations for (a, ●) and 8 preparations for (b, o). Highest specific $^{45}Ca^{2+}$ uptake after 10 min was 6.6 ±1.3 nmol/mg protein in (a) and 6.7 ±1.0 nmol/mg protein in (b). The medium free Ca^{2+} concentrations were buffered with EGTA/ATP, or EDTA/ATP, or ATP using association constants from reference (24). The free Mg^{2+} concentration was kept constant at 0.3 mmol/l. To minimize the source of error the calcium concentrations were adjusted exclusively with $^{45}CaCl_2$ and checked with a Ca^{2+}-specific electrode for each tracer batch as described (29). All experiments were performed with the same solutions which were stored at -25°C between the experiments. (From 5).

10^{-6} mol/l. Similar to Ca^{2+} uptake Ca^{2+}-(Mg^{2+})-ATPase activity decreases with higher free Ca^{2+} concentrations (Fig. 6). Taken together these findings strongly indicate that the RER is the Ca^{2+} pool into which Ca^{2+} is transported by means of a Ca^{2+}-(Mg^{2+})-ATPase which is involved in the regulation of the cytosolic free Ca^{2+} concentration.

Properties of Ca^{2+} extrusion from the cell

The low cytosolic free Ca^{2+} should finally be determined by Ca^{2+} transport systems located in the plasma membrane with K_m values for Ca^{2+} transport similar to those in the rough endoplasmic reticulum which determines the steady state free Ca^{2+} concentration in leaky cells. As shown in Fig. 7, MgATP-dependent Ca^{2+} uptake into both isolated plasma membrane vesicles and RER vesicles are similar concerning the Ca^{2+} concentrations at which Ca^{2+} uptake is maximal, as well as for the K_m of free Ca^{2+} concentration. The Ca^{2+} dependency, however, is quite different in both preparations. In the RER Ca^{2+} uptake shows a maximum at $2x10^{-6}$ mol/l, decreases with higher free Ca^{2+} concentrations and

TABLE 1

Transport characteristics of the Ca^{2+}-ATPases in plasma membrane and rough endoplasmic reticulum vesicles prepared from pancreatic acinar cells. The table summarizes the main findings from ref. 5.

			plasma membrane	rough endoplasmic reticulum
for Ca^{2+} uptake				
$[Ca^{2+}]$	_ / μmol _ _ o E μ	max. K_m	10 0.88	2 0.54 Ca^{2+} uptake decreases above 2 μmol/l Ca^{2+}
$[Mg^{2+}]$		max. K_m	3000 30	200 8
$[ATP]$		max. K_m	5000 2000	1000 10
pH optimum			6.5 - 7.0	6.5 - 7.0
substrate specificity			ATP	ATP >> GTP > UTP > ITP > CTP
specificity for Mg^{2+}			$Mg^{2+} > Mn^{2+} >> Zn^{2+}$	$Mg^{2+} >> Mn^{2+}$
cation dependence			$K^+ > Rb^+ > Na^+ > Li^+ > choline^+$	$Rb^+ \gtreqqless K^+ \gtreqqless Na^+ > Li^+ > choline^+$
anion dependence			$Cl^- \gtreqqless Br^- \gtreqqless I^- > SCN^- > NO_3^- > ise$-thionate$^- >$ gluconate$^- >$ cyclamate$^- > SO_4^{2-} \gtreqqless$ glutarate^{2-}	$Cl^- > Br^- >$ gluconate$^- > SO_4^{2-} \gtreqqless NO_3^- > I^- >$ cyclamate$^- \gtreqqless SCN^-$
oxalate dependence			no	yes

increases again. Ca^{2+} uptake into plasma membrane vesicles, however, shows a maximum at 10^{-5} mol/l free Ca^{2+} concentration and remains high. Other differences in Ca^{2+} uptake properties are listed in Table 1.

Most striking are the anion dependencies in both systems. Plasma membranes are tight for oxalate, and anion dependence for Ca^{2+} uptake follows a lyotropic series. In RER, however, oxalate, gluconate, sulfate and cyclamate, usually considered as weakly permeant anions, stimulate Ca^{2+} uptake more than SCN^-, a highly permeant anion.

The anion sequence to stimulate Ca^{2+} uptake into plasma membrane vesicles suggests that Ca^{2+} uptake is electrogenic. Indeed a detailed study under ion

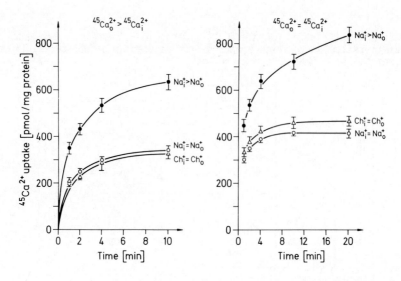

Fig. 8. Dependence of $^{45}Ca^{2+}$ uptake into isolated plasma membrane vesicles in the absence of ATP and in the presence of a Na^+ gradient. Plasma membrane vesicles were preincubated for 60 min at 25°C in the presence of Na_2SO_4 or cholinesulfate (100 mmol/l each), 33 mmol/l K_2SO_4 and 30 mmol/l Hepes-Tris (pH 7.0) with or without $^{45}CaCl_2$ (10^{-5} mol/l). They were then diluted by 40-fold into a medium containing 100 mmol/l Na_2SO_4 or 100 mmol/l choline sulfate, 33 mmol/l K_2SO_4, 30 mmol/l Hepes-Tris (pH 7.0), 10^{-5} mol/l $^{45}CaCl_2$ and 2×10^{-6} mol/l valinomycin, thus creating a Na^+-gradient directed vesicle inside to outside ($Na^+_i > Na^+_o$) or equal Na^+ or choline$^+$ concentrations on both sides of the vesicle membrane. Left panel: Vesicles were preincubated without calcium and transferred to indicated media containing $^{45}Ca^{2+}$. Right panel: Vesicles were preincubated in the presence of $^{45}Ca^{2+}$ and transferred to indicated medium of the same $^{45}Ca^{2+}$ concentration.

gradient conditions to impose diffusion potentials showed that MgATP-dependent Ca^{2+} uptake into plasma membrane vesicles is electrogenic (5). Furthermore, a Na^+/Ca^{2+} countertransport is present in the plasma membrane of pancreatic acinar cells. As shown in Fig. 8, in the absence of MgATP, a vesicle outward directed Na^+ gradient increases Ca^{2+} uptake as compared to conditions in which Na^+ concentrations are equal on both sides of the vesicle membrane. Similarly $^{45}Ca^{2+}$ efflux from preloaded plasma membrane vesicles is stimulated in the presence of an inward directed Na^+ gradient (data not shown). Li^+ can partially replace Na^+, whereas in the presence of choline$^+$ Ca^{2+} transport is minimal (Bayerdörffer et al., manuscript submitted to J Membr Biol).

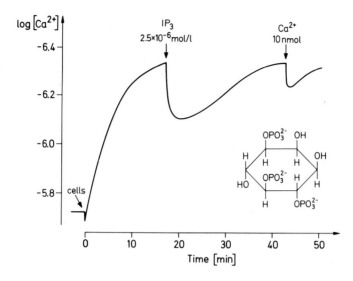

Fig. 9. Effect of inositol 1,4,5-trisphosphate (IP$_3$, 2.5 μmol/l) on intracellular Ca^{2+} stores of leaky acinar cells. CaCl$_2$ (10 nmol) was added as calibration pulse where indicated. (From 27).

Intracellular messenger for secretagogue-induced Ca^{2+} mobilization in acinar cells

Recently we have obtained evidence (27) that the intracellular messenger for secretagogue-induced Ca^{2+} release is inositol-1,4,5-trisphosphate (IP$_3$) which is produced during receptor-activated hydrolysis of phosphatidylinositol-4,5-bisphosphate in a variety of tissues (21).

As shown in Fig. 9 addition of IP$_3$ to leaky acinar cells at steady state free Ca^{2+} concentration results in cellular Ca^{2+} release and transient increase in medium free Ca^{2+} concentrations followed by reuptake to the prestimulation value similar to the observation made with addition of secretagogues (29). The dose response relationship of IP$_3$-induced Ca^{2+} release showed maximal release at 5x10^{-6} mol/l IP$_3$ and an apparent K$_m$ of 1.1x10^{-6} mol/l. The same result was obtained in the presence of mitochondrial inhibitors (27) suggesting that Ca^{2+} is released from a non-mitochondrial Ca^{2+} pool. If the acetylcholine analog, carbachol, was added after IP$_3$, the carbachol-induced Ca^{2+} release was completely abolished. If, on the other hand, IP$_3$ was added after carbachol, the IP$_3$-induced release of calcium was descreased by about one third, the sum of both effects, however, was constant (27). This suggests that both agents act on the same pool of releasable calcium and that IP$_3$ is a second messenger for secretagogue-stimulated Ca^{2+} release. This assumption became more substantial

338

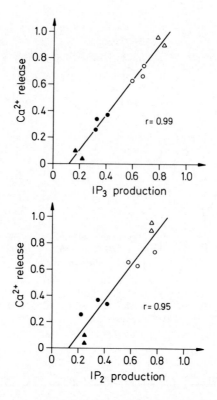

Fig. 10. Correlation of cholecystokinin-octapeptide induced Ca^{2+} release and formation of inositol phosphates at different free Mg^{2+} concentrations. Cholecystokinin-octapeptide (CCK-OP) induced Ca^{2+} release and production of inositol phosphates was determind at 0.3 mmol/l and 1.3 mmol/l free Mg^{2+} concentration. Results were standardized by setting the sum of CCK-OP induced Ca^{2+} release and increase in IP$_3$ and IP$_2$ concentration measured for each pair of incubation at 1.3 mmol/l and 0.3 mmol/l free Mg^{2+} concentration to 1. Open symbols = 0.3 mmol/l free Mg^{2+} concentration. Closed symbols = 1.3 mmol/l free Mg^{2+} concentration. Different symbols refer to different cell preparations. r = correlation coefficient. (From 28).

by showing that IP$_3$ was also produced in leaky acinar cells during secretagogue stimulation (28). The amount of Ca^{2+} release is well correlated to IP$_3$ production at different conditions (Fig. 10). Furthermore, IP$_3$ releases Ca^{2+} from isolated endoplasmic reticulum in a way similar to isolated leaky cells (26). There is no IP$_3$-induced Ca^{2+} release from other cell organelles such as mitochondria or plasma membranes.

Taken together these experiments suggest that IP$_3$ is the intracellular messenger for secretagogues to release Ca^{2+} from the endoplasmic reticulum of pancreatic acinar cells.

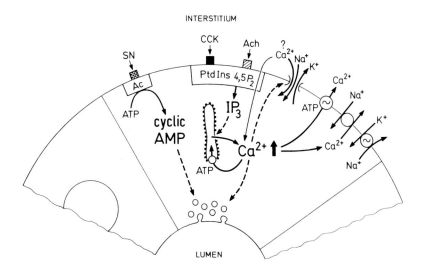

Fig. 11. Model for the pathways involed in "stimulus-secretion coupling". For explanation see text.

DISCUSSION

Regulation of cytosolic free Ca^{2+} concentration at rest and at stimulation is of crucial importance in receptor-activated processes which lead to enzyme secretion from the exocrine pancreas. Our data show that both the plasma membrane and the rough endoplasmic reticulum (RER) play an important role in the regulation of free Ca^{2+} concentration.

The steady state level of $\sim 4 \times 10^{-7}$ mol/l for free Ca^{2+} concentration as adjusted by the RER in leaky cells is close to the value for the cytosolic free Ca^{2+} in intact cells of $1-2 \times 10^{-7}$ mol/l as measured by the quin 2 method (Fig. 3).

It is therefore evident that plasma-membrane bound extrusion mechanisms regulate the final cytosolic free Ca^{2+} concentration to the same or somewhat lower level than the steady state adjusted by the RER.

Ca^{2+} uptake into RER is promoted by a $Ca^{2+}-(Mg^{2+})$-ATPase (6) with features similar to that in the better investigated sarcoplasmic reticulum (11,14,15), whereas the plasma membrane $Ca^{2+}-(Mg^{2+})$-ATPase shows different Ca^{2+} uptake characteristics (Table 1). The main differences between both Ca^{2+} uptake mechanisms are their dependencies on free Ca^{2+} concentration (Fig. 7) and on anions (Table 1). The finding that in RER calcium uptake (Fig. 5) and $Ca^{2+}-(Mg^{2+})$-ATPase activity (Fig. 6) decrease at free Ca^{2+} concentration higher than 2 μmol/l seems to be an individual characteristic of the RER in pancreatic acinar cells. Regarding the increase in the cytosolic free Ca^{2+} concentration level to trigger enzyme secretion as the physiological function of the RER such a characteristic appears to be reasonable. During stimulation of enzyme secretion, increased cytosolic calcium concentration would suppress the calcium sequestering activity of the RER. The steady state level at sustained stimulation would therefore be determined by Ca^{2+} influx and Ca^{2+} extrusion over the plasma membrane, the latter being achieved by both an electrogenic $Ca^{2+}-(Mg^{2+})$-ATPase as well as a Na^+/Ca^{2+} countertransport (Fig. 8).

Our data provide strong evidence that inositol-1,4,5-P_3 functions as a second messenger for secretagogues to mobilize intracellular calcium from the endoplasmic reticulum, to initiate enzyme secretion most likely from the same site into which Ca^{2+} is taken up at rest. The model shown in Figure 11 illustrates intracellular events involved in Ca^{2+} metabolism. During stimulation the following steps are involved: secretagogues interact with their specific receptors on the plasma membrane. This induces hydrolysis of phosphatidyl-4,5-bisphosphate and an increase in the hydrolysis product inositol-1,4,5-trisphosphate (IP_3). This increase causes release of Ca^{2+} from the endoplasmic reticulum. The inreased cytosolic free Ca^{2+} concentration triggers enzyme secretion.

In intact cells continuous turnover of phosphatidylinositides during stimulation results in continuous production of IP_3. Thus, following an initial peak of IP_3 concentration in the cell, during prolonged stimulation IP_3 might be maintained at a higher level as compared to unstimulated cells. Consequently refilling of the "trigger Ca^{2+} pool", i.e. the endoplasmic reticulum would be inhibited during stimulation. Increased Ca^{2+} influx into the cells, which probably does not involve IP_3 and whose mechanism is not yet known, would be buffered by a second pool, probably mitochondria (Figs. 1,2). When stimulation is stopped, by addition of an antagonist, refilling of the "trigger-IP_3-Ca^{2+} pool" would occur. This would transiently lead to increased Ca^{2+} content of the cell until the second pool releases its Ca^{2+} taken up during stimulation and the resting state is reached again (Fig. 1).

Ca^{2+} extrusion from the cell can be achieved by a $Ca^{2+}-(Mg^{2+})$-ATPase (5) and a Ca^{2+}/Na^+ exchange mechanism located in the plasma membrane (Fig. 8).

In "leaky" cells the situation is different. Since measurements of Ca^{2+} fluxes are made at very low Ca^{2+} concentrations near to the steady state and the plasma membrane barrier is gone, "Ca^{2+} influx" as in "intact cells" does not occur during stimulation. Furthermore leaky cells do not produce continuously IP_3. It is rather a "one shot" phenomenon, in which phosphatidylinositides, present in the plasma membrane, are hydrolyzed but not resynthetized during stimulation. Consequently higher IP_3 levels as in stimulated cells are not maintained. IP_3 is degraded due to IP_3 hydrolyzing phosphatases and despite continuous presence of secretagogues in the incubation medium, IP_3 level rapidly returns to that of the resting state (28). Ca^{2+} reuptake into the same pool, from which it is released, can therefore occur in leaky cells (Fig. 9).

Detailed mechanisms how increased cytosolic free Ca^{2+} concentrations trigger enzyme secretion are not yet known. It appears that Ca^{2+} is not the only trigger and that diacylglycerols, which are produced by hydrolysis of phosphatidylinositides in the same step as IP_3, as well as activation of protein kinase C by diacylglycerols play an equally important role.

It is also not clear in what step cAMP is involved in enzyme secretion. The interplay between Ca^{2+}, diacylglycerol and cAMP-mediated processes in secretory events is one of the main interests for future work.

REFERENCES
1. Affolter H, Sigel E (1979) Anal Biochem 97:315-319
2. Amory A, Foury F, Goffeau A (1980) J Biol Chem 255:9353-9357
3. Amsterdam A, Jamieson JD (1972) Proc Nat Acad Sci USA 69:3028-3032
4. Bais R (1975) Analytical Biochem 63:271-273
5. Bayerdörffer E, Eckhardt L, Haase W, Schulz I, J Membr Biol, in press
6. Bayerdörffer E, Streb H, Eckhardt L, Haase W, Schulz I (1984) J Membr Biol 81:69-82
7. De Pont JJHHM, Fleuren-Jakobs AMM (1984) FEBS Letters 170:64-68
8. Downes CP, Michell RH (1981) Biochem J 198:133-140
9. Downes CP, Mussat MC, Michell RH (1982) Biochem J 203:169-177
10. Gardner JD (1979) Ann Rev Physiol 41:55-66
11. Jones LR, Besch HR, Watanabe AM (1977) J Biol Chem 252:3315-3323
12. Kondo S, Schulz I (1976) Biochim Biophys Acta 419:76-92
13. Kribben A, Tyrakowski T, Schulz I (1983) Am J Physiol 244:G480-G490
14. Meissner G (1981) J Biol Chem 256:636-643
15. Meissner G, McKinley D (1982) J Biol Chem 257:7704-7711
16. Ochs DL, Korenbrot JI, Williams JA (1983) Biochem Biophys Res Comm 117:122-128
17. Pearson GT, Singh J, Petersen OH (1984) Am J Physiol 246:G563-G573
18. Petersen OH (1982) Biochim Biophys Acta 694:163-184
19. Petersen OH, Ueda N (1976) J Physiol 254:583-606
20. Rasmussen H (1984) Cell Calcium 5:270
21. Rubin RP, Godfrey PP, Chapman DA, Putney JW Jr (1984) Biochem J 219:655-659
22. Schulz I (1980) Am J Physiol 239:G335-G347
23. Schulz I, Heil K, Kribben A, Sachs G, Haase W (1980) In: Ribet A, Pradayrol L, Susini C (eds) Biology of Normal and Cancerous Exocrine Pancreatic Cells, INSERM Symp No. 15. Elsevier, Amsterdam, p 3
24. Sillen LG (1971) Special publication No. 25, Suppl. No. 1 to special publication No. 17, 2nd ed, Chem Soc London
25. Stolze H, Schulz I (1980) Am J Physiol 238:G338-G348
26. Streb H, Bayerdörffer E, Haase W, Irvine RF, Schulz I (1984) J Membr Biol 81:241-253
27. Streb H, Irvine RF, Berridge MJ, Schulz I (1983) Nature 306:67-69
28. Streb H, Heslop JP, Irvine RF, Schulz I, Berridge MJ, J Biol Chem, submitted

29. Streb H, Schulz I (1983) Am J Physiol 245:G347-G357
30. Takai Y, Kishimoto A, Iwasa Y, Kawahara Y, Mori T, Nishizuka Y (1979) J Biol Chem 254:3692-3695
31. Tsien RY, Pozzan T, Rink TJ (1982) J Cell Biol 94:325-334
32. Wakasugi H, Stolze H, Haase W, Schulz I (1981) Am J Physiol 240:G281-G289
33. Williams JA (1980) Am J Physiol 238:G269-G279

RESUME

La régulation de la concentration stationnaire en Ca^{2+} libre au repos et à l'état stimulé a été étudiée dans les cellules acinaires du pancréas, isolées et perméabilisées, en mesurant la concentration en Ca^{2+} libre du milieu d'incubation à l'aide d'une électrode spécifique aux ions Ca^{2+}. Les mécanismes de transport du Ca^{2+} ont été caractérisés plus amplement dans des fractions membranaires subcellulaires en mesurant l'accumulation de $^{45}Ca^{2+}$ dans des vésicules membranaires ainsi qu'en mesurant la phosphorylation de protéines à l'aide de l'électrophorèse en gel de polyacrylamide.

Nos résultats montrent que dans des cellules perméabilisées, la concentration stationnaire en Ca^{2+} libre est réglée à $410^{-7}M$ par le réticulum endoplasmique rugueux (RER). La captation de Ca^{2+} par ce pool est maximale à $210^{-6}M$ en Ca^{2+} libre et est réalisée par une Ca^{2+}-(Mg^{2+})-ATPase dont l'activité maximale se situe à $10^{-6}M$ en Ca^{++}. A des concentrations plus élevées en Ca^{2+}, à la fois la captation de Ca^{2+} et l'activité ATPasique sont inhibées. L'incorporation de ^{32}P à partir de $[\gamma^{32}P]$ ATP dans un intermédiaire protéique de cette Ca^{2+}-(Mg^{2+})--ATPase est maximale à $10^{-7}M$ en Ca^{2+} libre.

La captation de Ca^{2+} par des vésicules de membrane plasmique est différente de celle propre au RER. Elle montre une captation maximale de Ca^{2+} à $10^{-5}M$ en Ca^{2+} libre. Contrairement à la captation de Ca^{2+} dans le RER il n'y a pas d'inhibition pour des concentrations supérieures en Ca^{2+}. La membrane plasmique contient en plus de ce système de transport du Ca^{2+} dépendant du Mg ATP, 1 échangeur Ca^{2+}/Na^+. Nous pensons que le Ca^{2+} est expulsé de la cellule par ces 2 mécanismes.

L'inositol-1,4,5-triphosphate (IP_3), qui est produit lors de l'hydrolyse du phosphatidylinositol-biphosphate par les hormones mobilisant le Ca^{2+}, libère du Ca^{2+} à partir d'un pool de Ca^{2+} identique à celui qui est l'objectif des sécrétagogues dans les cellules perméabilisées.

Nos résultats indiquent que l'IP_3 est le messager intracellulaire des sécrétagogues qui agissent en libérant le Ca^{2+} du réticulum endoplasmique.

NON-CLASSICAL FACTORS IN NEURAL DEVELOPMENT

LES FACTEURS NON-CLASSIQUES DANS LE DEVELOPPEMENT NERVEUX

Hormones and Cell Regulation, Volume 9
INSERM European Symposium
J.E. Dumont, B. Hamprecht and J. Nunez editors
© 1985 Elsevier Science Publishers B.V. (Biomedical Division)   345

PROPERTIES OF A BRAIN GROWTH FACTOR PROMOTING PROLIFERATION AND MATURATION OF
RAT ASTROGLIAL CELLS IN CULTURE

MONIQUE SENSENBRENNER, BRIGITTE PETTMANN, GERARD LABOURDETTE and MARC WEIBEL
Centre de Neurochimie du CNRS and INSERM U44 - 5, rue Blaise Pascal - 67084
Strasbourg Cedex - France.

INTRODUCTION

During brain ontogenesis, the neural cells undergo proliferation, migration,
selective aggregation and subsequent maturation. The regulation of these developmental processes must be influenced by different environmental factors, such
as cellular contacts, hormones and tumoral and trophic factors. Moreover, neural
cells must be also susceptible to extrinsic agents during adulthood, in normal
and pathological conditions.

Since the discovery of nerve growth factor (NGF) thirty years ago (for review,
see refs. 1, 2 and 3), other neurotrophic factors affecting both peripheral and
central neurons have been extensively described and partially characterized. More
detailed information can be obtained from recent reviews (4, 5).

Until recently, there were few indications about factors controlling the development of glial cells. The availability of cell culture systems consisting of a
nearly homogeneous population of either astroglial or Schwann cells has led to a
better experimental approach to understand the dependance of glial cell proliferation and maturation on the presence of growth factors (6, 7, 8). In the past
ten years studies have been initiated to detect and identify factors in nervous
tissue extracts affecting the growth and development of these astroglial and/or
Schwann cells in culture (9, 10, 11).

In 1972, our first study of the effects of chick brain extract on the development of chick astroglial cells in primary culture demonstrated a stimulation of
the morphological maturation (12). Subsequently, we described morphological changes in chick and rat astroglial cultures under the influence of rat brain extract
(13, 14). At the same time other authors reported similar observations (15, 16).
The extension of these investigations showed a stimulatory effect of different
brain extracts not only on the morphology, but also on the biochemical maturation (17, 18, 19) and the proliferation of the astroglial cells (18, 19, 20, 21).

Studies were then undertaken by us (22, 23) and Lim's research group (24, 25,
26) in an attempt to purify and characterize the active factor(s) affecting astroglial cells present in brain extracts. An astroglial growth factor (AGF) and
a glia maturation factor (GMF) have been partially purified and identified in
our and Lim's laboratory, respectively. Experiments performed by other investigators (27, 28) have shown that a partially characterized factor, named glial

growth factor (GGF), extractable from pituitary gland and brain, which stimulates the proliferation of Schwann cells also enhances the multiplication of astroglial cells.

The main object of this chapter is to review the work from our own laboratory on the influence of brain extracts and of the AGF on the development of astroglial cells in culture. The data accumulated on the two other gliotrophic factors (GMF and GGF) will be emphasized. Finally, findings will be discussed on other known growth factors, derived either from neural or nonneural tissues, affecting astroglial cells in primary culture.

MATERIAL AND METHODS

Astroblast cultures

Cultures in serum-containing medium. Astroblast cultures were prepared by a modification of the method of Booher and Sensenbrenner (29). Cerebral hemispheres from newborn Wistar rat, free of meningeal membranes, were dissociated mechanically and the cells were collected in a serum-supplemented medium. This medium consisted of Waymouth's MD 705/1 medium (Flow Laboratories) containing 10 per cent heat-inactivated fetal calf serum (Flow Laboratories), 50 units/ml penicillin and 50 μg/ml streptomycin.

Five or 10 ml of the cell suspension (3.5 or 5.5 x 10^4 cells/ml) were dispensed per 60 mm or 100 mm Falcon Petri dish, respectively. Cultures were incubated at 37°C in a 5 per cent CO_2 humidified atmosphere. The medium was changed after 5 days and subsequently twice a week.

Cultures in chemically defined medium. Cultures were prepared and grown for 5 days in the presence of serum, as described above. They were then rinsed three times with Waymouth's medium and maintained in a serum-free defined medium thereafter. The chemically defined medium (30) consisted of Waymouth's medium supplemented with 5 μg/ml bovine insulin (Sigma) and 0.5 mg/ml fatty acid free bovine serum albumin (Sigma). Antibiotics (50 units/ml penicillin and 50 μg/ml streptomycin) were added.

Preparation of soluble brain extract

Soluble brain extracts were prepared from whole brains of embryonic or adult chick, rat or ox (31). Meninges were removed and brains were homogenized with either a glass homogenizer or a blender in Tyrode's solution (200 mg fresh weight/ 1 ml of Tyrode solution). The homogenate was centrifuged at 105,000 g for 1 hour at + 4°C, small portions of the supernatant (soluble brain extract) were sterilized by filtration through 0.22 μm pore-size Millipore membranes (Millex GV) and were frozen at - 70°C until use. One-half ml or 1.0 ml (8 mg protein/ml) of the thawed extract was added to 10 ml of nutrient medium.

Immunocytochemistry

Immunocytochemical staining with rabbit antibodies against cell-specific markers was used to identify the different cell types (30). Glial fibrillary acidic
(GFA) protein, S100 protein and glutamine synthetase (GS) were used to characterize astroglial cells. For the identification of fibroblasts, endothelial cells
and oligodendrocytes, antibodies to fibronectin, to factor VIII and to the iso-
enzyme II of carbonic anhydrase (CAII) were applied, respectively.

Biochemical analysis

GFA protein was quantified by rocket immunoelectrophoresis (18). S100 protein
was measured by radioimmunoassay (30). The glutamine synthetase assay was performed as described by Miller et al. with slight modifications (30). Tritiated thymidine assay for mitogenic activity was performed as previously described (32).

Purification procedure of the AGF

Preparation of crude brain extract. Bovine brains were obtained from the slau-
ghterhouse and transported on ice. Blood clots and outer membranes of the meninges
were removed. The whole brains were homogenized in 1 volume of 60 mM citric acid
containing 0.5 mM phenylmethylsulfonylfluoride (PMSF) using a blender. After centrifugation (3500 g, 15 min), the pellet was rehomogenized in 1 volume of 25 mM
citric acid, 0.5 mM PMSF and centrifuged as above. The supernatants were pooled
and Tris and EDTA solutions were added to obtain final concentrations of 20 mM
and 5 mM, respectively. The mixture was adjusted to pH 7.8 with 10N NaOH. After
filtration on glass microfibre filters GF/F, the extract (crude brain extract)
was submitted to chromatographic separation (33).

Fractionation procedure. The successive chromatographic steps for the purification of the active factor (astroglial growth factor : AGF) were blue-Trisacryl M
(IBF) affinity chromatography, adsorption chromatography on hydroxyapatite-Ultro-
gel (IBF), FPLC (Fast Protein Liquid Chromatography) (Pharmacia) cation-exchange
chromatography on a Mono S HR5/5 column followed by a FPLC reversed-phase chromatography on a ProRPC HR5/10 column. Final purification was achieved by polyacrylamide gel electrophoresis (PAGE) in sodium dodecyl sulfate (SDS). Parallel
gels were silver stained to detect proteins.

After the different purification steps, the fractions were sterilized by filtration through 0.22 μm pore-size Millipore membranes (Millex GV) and assayed for
their ability to stimulate proliferation and to produce morphological modification
in astroblast cultures.

RESULTS AND DISCUSSION

The experimental in vitro system : rat astroblast cultures

When dissociated cells from the brain of newborn rat were seeded at low density
directly on the plastic surface of the Petri dish and grown in a serum-containing

medium flat polygonal-shaped cells predominantly developed in these cultures (19) (Fig. 1). In a serum-free chemically defined medium most cells were also flat and polygonal, while others were smaller, more elongated with processes (30). These cells have been referred to as astroblasts. Between 3 and 4 weeks of culture the cells formed a continuous monolayer. In the serum-supplemented medium the cells survived for several months and in the defined medium for 5 weeks.

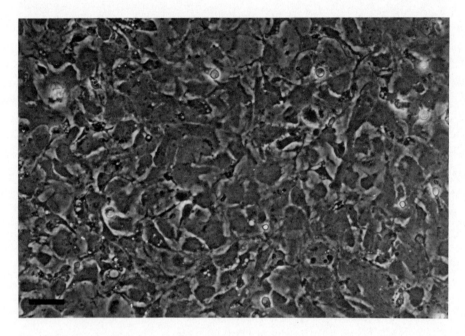

Fig. 1. Rat astroglial cells cultured during 14 days in Waymouth's medium plus 10 per cent fetal calf serum - Bar = 50 μm.

Electron microscopic observations have revealed, in most of the cells, the presence of intermediate filaments, which accumulate progressively with time in culture. These filaments are identical to the gliofilaments present in mature fibrous astrocyte in vivo. Numerous glycogen granules were seen throughout the cytoplasm. This ultrastructural aspect strongly indicates the astroglial nature of the cells (18).

These cells have been further characterized as astroglial cells immunocytochemically by using specific markers (30, 34). Almost all the cells stained for GFA and S100 proteins, and many were positive for GS (Figs. 2a and 2b). Under our culture conditions fibroblasts, endothelial cells and oligodendroglial cells were scarce. Ependymal cells, recognized by their beating cilia, were present in only small numbers.

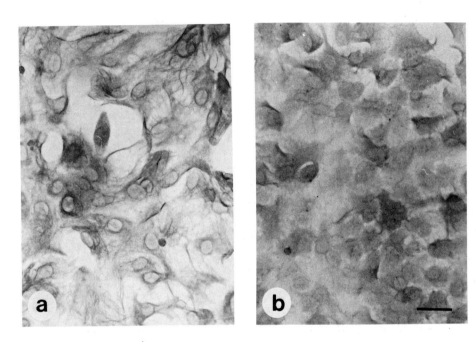

Fig. 2. Immunoperoxidase staining of rat astroglial cells for GFA protein (a) and GS (b). Three-week-old culture - Bar = 50 μm.

Such primary cultures consisting of a nearly homogeneous population of astro-glial cells are adequate systems to investigate the effects of gliotrophic fac-tors and have been used throughout our studies. Several research groups have employed similar astroglial cultures for investigations of growth factors, while others have used secondary cultures. In such a model system the cultures have been derived from rat embryonic brain and when the cells attain confluency they have been subcultured. After about one week a nearly pure population of astroglial cells had then developed (15).

Influence of brain extracts on rat astroblasts

Morphological changes. The addition of embryonic or adult chick soluble brain extracts to rat astroblast cultures after 5 days and at each medium change indu-ced within a short time period (2-3 days) morphological alterations of the cells (6, 19). The flat polygonal shaped cells became more elongated, developed cyto-plasmic processes and acquired a star-like appearance, resembling mature astro-cytes. Under the influence of brain extract from rat similar morphological

changes occured, and have been described by us (6, 7, 35) (Fig. 3) and by other
investigators (15, 16). It has been also shown that pig and bovine brain extracts
affected the behaviour of astroglial cells in culture (35, 36). When brain extract was not added continuously, but was present only during the initial ten
days the morphology of the cells on the whole returned to the original flat polygonal aspect. Thus, the morphological change was dependent on the continued presence of brain extract.

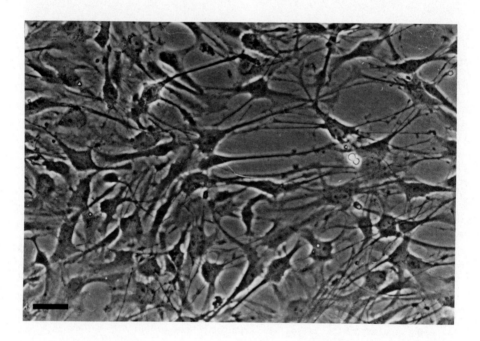

Fig. 3. Rat astroglial cells cultured in the presence of brain extract from
adult rat. Two-week-old culture - Bar = 50 μm.

At the ultrastructural level we observed that in the treated cultures the intermediate filaments became very abundant in the cells between 3 and 4 weeks, and
formed large bundles in the cell bodies as well as in the processes (7, 18, 19).
However, direct evidence that the intermediate filaments correspond to GFA containing gliofilaments has not yet been obtained. In the treated cells numerous
free ribosomes and well developed rough endoplasmic reticulum were seen. Similar
effects of brain extract on the ultrastructural aspect of astroglial cells have
been described by Lim et al. (37) and by Haugen and Laerum (16). However, their
studies concerned only one time in the development of the culture and within the
first seven days.

All these studies indicate that under the influence of brain extract the astroglial cells undergo considerable morphological changes, consistent with maturation processes.

Biochemical changes. Biochemical investigations have shown a constant rise of the protein content in both control cultures and cultures treated with brain extract until about 5 weeks (18, 19). This increase occured in a parallel fashion to an enrichment of free and bound ribosomes in the cytoplasm of the cells. In treated cultures the protein level was always higher compared to controls (18, 19), indicating enhanced protein synthesis.

Several reports have dealt with changes in specific proteins for astroglial cells. It has been shown that the amount of extractable GFA present in the cells increased with time of culture (11, 18). This rise of GFA paralleled the morphological accumulation of the intermediate filaments. The values in the cultures treated with brain extract were always higher than in controls, but not significantly different.

Another glial-specific protein, S100 protein, has been investigated. The content of S100 did not increase much and remained low in control cultures during the whole culture period studied (50 days) (11, 19). In the continuous presence of brain extract the S100 level increased rapidly until about six weeks. Thereafter, it remained constant at a value about 3.5 times higher than in untreated cultures.

Embryonic and adult chick brain extract, newborn rat and adult rat, pig and bovine extracts all induced a strong stimulation of S100 protein level. Whereas, brain extracts from rat embryos were shown to be less active (unpublished data).

A stimulation was still obtained when cultures were treated for short periods of time ; a 24-hour treatment was sufficient to induce a long-term increase of S100 protein. However, when treatment was started only at day 25, S100 protein synthesis was still enhanced. These results suggest that a short treatment of the cells allows an irreversible process of maturation in proliferating cells, and that the cells in stationary phase are still receptive to the brain factors at least up to 25 days after seeding. Stimulatory effects on the S100 protein content in astroglial cultures under the influence of a pig brain extract have been reported by Lim et al. (17).

These results indicate that in untreated cultures the astroglial cells undergo biochemical maturation only to a limited extent. But, under the influence of different brain extracts some specific markers of astroblasts were stimulated, especially the S100 protein, indicating enhanced biochemical maturation of the cells.

Mitogenic activity. In control rat astroglial cultures the DNA content increased regularly for about 3 weeks (18, 19). Under the influence of brain extract, prepared either from embryonic or adult chick and rat, the proliferation of the cells was stimulated already 24 hours after the addition of the extract to the medium. The DNA amount increased sharply during the first two weeks of culture and after 3 weeks reached a level 50 per cent higher than in controls. Thereafter, the DNA content of the untreated and treated cultures remained rela tively constant and most of the cells ceased dividing. Such a mitogenic activity for astroglial cells was also elicited by brain extract derived from adult pig as reported by Lim et al. (38). Morrison et al. (39) found mitogenic activity in brain extracts prepared from prenatal, neonatal and adult rat brains. These au- thors showed that the postnatal and the adult extracts possessed the highest and lowest specific activity, respectively.

In conclusion, one or several factors present in both embryonic and adult brain extracts from different species can regulate the proliferation and the maturation of astroglial cells in primary culture.

All the above mentioned studies on brain extracts have been carried out on astroglial cells grown in serum-containing medium. The recent availability of serum-free chemically defined media for astrocytes will allow more precise stu- dies on the different effects of brain extracts and will facilitate the identi- fication of the brain factor(s) affecting cell division and maturation.

Neural tissue-derived glial cell growth factors

The results described in the preceding section indicate that central nervous tissue contain factors affecting the growth and development of cultured astro- glial cells. Studies were performed to isolate and characterize these trophic substances.

The present section will be devoted to the three recently partially purified and characterized glial factors isolated from brain and pituitary : the astroglial growth factor (AGF), the glia maturation factor (GMF) and the glial growth factor (GGF).

Isolation and characterization. By using classical techniques for protein separation a first procedure was carried out by us to purify an active factor for astroglial cells from bovine whole brain, published in 1982 (22). However, the growth factor was partially purified only about 700-fold and the yield of this purification procedure was poor. We had, therefore, to modify the isolation method.

We have since reported an improved method (23) and a growth factor, tentatively named astroglial growth factor (AGF) was partially purified from bovine brain on the basis of its ability to induce morphological changes of the cultured rat astroblasts. The isolation procedure consisted in four successive chromatographic steps and a final fractionation by isoelectric focusing on polyacrylamide gel. The active fraction was purified approximately 16,000-fold.

Recently, we have purified the AGF approximately 100,000-fold by a combination of column chromatography steps and SDS gel electrophoresis as described in Material and Methods (33). The importance of the two FPLC steps in this isolation procedure should be pointed out. Indeed, this method for the purification of the AGF provides improved purification and a greatly improved yield.

The AGF is a heat-labile protein, sensitive to pronase and trypsin. It is an acidic protein with an isoelectric point between 4.8 and 5.2 and an apparent molecular weight (M_r) between 45,000 to 50,000 daltons.

In 1975, Lim and Mitsunobu (40) already reported a partial purification of a morphological transforming factor from pig brain. A 400-fold purification was indicated and the partially purified factor was described to be a protein of a high molecular weight (estimated M_r of 350,000 daltons). Later, this factor was designated as glia maturation factor (17).

Multiple molecular forms, a large molecular weight form (200,000 daltons) and a smaller form (40,000 daltons), of the GMF from the pig brain have then been reported (24). The same investigators have also purified a GMF from bovine brain (25) and this partially purified protein has an apparent M_r of 23,000 daltons. An improved method for the purification of the GMF from bovine brain was recently published by Lim and Miller (26). This procedure gave a 10,000-fold purification. The isolated GMF had finally a M_r of 13,000 daltons and an isoelectric point of about 5.4. This acidic protein is heat-labile and relatively resistant to trypsin. This procedure was claimed to be faster and more economical than the previous methods, and to give a better yield and a higher specific activity of the substance.

Another protein growth factor, termed glial growth factor (GGF), present in the bovine brain and pituitary, has been first purified from the pituitary gland by Brocke's research group (27, 28). The GGF was purified approximately 100,000-fold to apparent homogeneity. It is a basic protein, relatively heat-stable and its M_r is 31,000 daltons. The same GGF was then partially purified from the brain (28).

Some of the characteristics of these three purified factors are summarized in Table I. The molecular weights reported for the three factors differ widely. However, one should be aware of the difficulties in making such determinations for growth factors. For instance, molecular weights determined by gel filtration of the AGF range from 10,000 to 150,000 daltons, depending on the composition of the elution buffer, and particularly on its ionic strength. Determinations by SDS-PAGE give more meaningful results. Molecular weights of AGF and GGF determined by this method appear somewhat different while the isoelectric points of AGF and GMF, determined directly, are almost identical. The basicity of GGF was estimated by its ion-exchange chromatographic behaviour, but this behaviour may be due to interactions with the gel matrix and not to the electric charge. To demonstrate firmly the similarity or distinction of these three factors a complete purification of the factors and subsequent information about their amino acid sequence will be necessary.

TABLE I

PHYSICOCHEMICAL PROPERTIES OF PURIFIED GROWTH FACTORS FOR ASTROGLIAL CELLS

Factor	Tissue extract	Apparent molecular weight	Isoelectric point	Stability
AGF	Bovine brain	45,000 - 50,000	4.8 - 5.2	Heat-labile ; trypsin-resistant (3 hours treatment) ; trypsin-sensitive (18 hours treatment)
GMF	Bovine brain	13,000	5.4	Heat-labile ; trypsin-resistant (3 hours treatment)
GGF	Bovine pituitary and brain	31,000	basic	Heat-stable ; sensitive to proteolytic enzymes

Trophic influences on cultured rat astroblasts. Early experiments on effects of the partially purified AGF have been made on rat astroblasts cultured in a serum-containing medium (11, 35). It was shown that AGF is a potent mitogen as well as a maturation factor for these cells. Subsequently, we performed detailed studies with the most purified factor using astroglial cultures grown in defined media. The rat astroblasts were cultured either in a serum-free medium, containing only insulin and/or transferrin (32), or in a simple defined medium developed by us for the growth of astroglial cells (30, 41).

The addition of AGF to the culture medium at day 5, then at each medium change stimulated the proliferation of the cells as demonstrated by an increase of the cell number, of the DNA content and of tritiated thymidine incorporation. A treatment as short as 24 hours was able to induce a rise of the DNA level. A concentration of about 0.5 ng of AGF per ml culture medium was required to obtain maximal mitogenic activity.

The AGF was also shown to induce a morphological change of the astroblasts 24 to 48 hours after exposure to the factor (32). Many cell bodies retracted and became smaller, long processes developed and most cells had a fibrous appearance (Fig. 4). At the ultrastructural level intermediate filaments accumulated in the somata and processes of the treated cells, forming large bundles. Numerous rough endoplasmic reticular membranes and glycogen granules were observed. The morphological alteration was maximally stimulated by 5 ng/ml AGF.

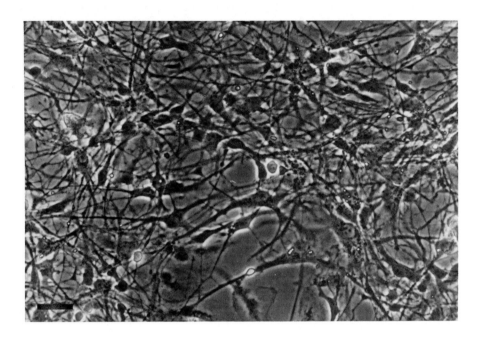

Fig. 4. Rat astroglial cells grown for 5 days in serum-supplemented Waymouth's medium, then for 10 days in Waymouth's medium supplemented with insulin (5 µg/ml), bovine serum albumin (0.5 mg/ml) and AGF (50 ng/ml) - Bar = 50 µm.

By studying specific astroglial markers we demonstrated that the biochemical maturation of the astroblasts was also enhanced under the effect of AGF (Table II) (32, 41). While the amount of GFA protein was only slightly increased, the level of S100 protein and the GS activity were strongly stimulated from 2 to 9 fold.

Stimulatory effects on both proliferation and maturation of astroblasts have also been observed in the presence of GMF by Lim and his coworkers. The rat astroglial cells underwent morphological changes within the first week after stimulation with GMF (38, 40, 42). A rise in the glia-specific protein S100 was reported (17). In addition, it was described that GMF treatment promotes DNA synthesis and cell proliferation within the first 12 to 18 hours (42).

While the GGF was found primarily to stimulate rat Schwann cell proliferation, it was shown subsequently that this protein also enhanced division of rat astro-glial cells (27, 28) ; but no morphological effect was reported.

All these results clearly demonstrate that AGF, GMF and GGF, like the soluble brain extract, promote proliferation of cultured rat astroblasts. Moreover, the

TABLE II

GFA AND S100 PROTEIN LEVELS, AND GS ACTIVITIES IN CULTURES OF ASTROGLIAL CELLS.
Average ± S.E.M. Cells were grown in serum-supplemented medium for 5 days, then
in the defined medium with or without (control) astroglial growth factor (AGF).

Days in culture	Culture condition	GFA (a.u./mg protein)	S100 (μg/mg soluble protein)	GS (nmol/ min/mg protein)
14	Control	13.00 ± 3.24	0.11 ± 0.01	21.9 ± 1.6
	+ AGF	14.75 ± 0.75	0.29 ± 0.05	106.8 ± 9.7
21	Control	16.50 ± 1.04	0.32 ± 0.09	21.2 ± 2.2
	+ AGF	18.25 ± 1.70	0.68 ± 0.02	185.8 ± 7.0

AGF and GMF also induce morphological changes compatible with maturation processes
and influence some astroglial specific protein markers, in the direction of biochemical differentiation.

More recently, it was reported that AGF, GMF and GGF are also mitogenic for
other cell types. Thus, the three growth factors stimulate the multiplication of
fibroblast (26, 28 and unpublished data). GMF like GGF was found to enhance
Schwann cell proliferation (43). Microglia and oligodendroglial cells are not
influenced by GGF (27), while the latter cell type respond to the partially purified AGF (11, 44). Further studies are needed to find out if AGF, GMF and GGF
are growth factors for all the various types of glial cells.

The presence of AGF- and GMF-like activity was demonstrated not only in brain
extract, but was also detected in several rat non-neural tissues by us (unpublished data) and by others (45). However, the activity in these other organs is
lower compared to that found in the brain. In contrast, the GGF activity was
detected only in brain and pituitary (10).

Comparison with other known growth factors

A polypeptide growth factor produced by non-neural tissue (Concanavalin A -
stimulated mouse spleen cells) that stimulates both proliferation and maturation
of astrocytes was isolated recently by Fontana et al. (46). This partially purified fraction was named glia cell stimulating factor (GSF). It is a protein of
an apparent M_r between 60,000 and 80,000 daltons. On the basis of chromatographic
separation GSF was found to differ from the bovine brain-derived GMF.

A growth factor extracted from ocular tissues, termed eye-derived growth factor
or EDGF, that stimulates proliferation and induces morphological changes in

cultured bovine epithelial lens cells (47), was shown in our laboratory to promote multiplication and morphological maturation of rat astroblasts in culture
(unpublished data). EDGF has an apparent M_r of 17,500 and an isoelectric point
of 4.5. This factor possesses some physicochemical and biological properties
close to those of bovine brain-derived AGF and GMF.

Several well characterized growth factors, that are mitogenic for various cell
types, have been demonstrated to stimulate also the proliferation of astroglial
cells in primary culture. It has been observed that the epidermal growth factor
(EGF), purified from male mouse salivary gland, stimulates DNA synthesis and cell
division in human and rat astrocyte cultures (48, 49). The fibroblast growth factor (FGF) isolated from bovine whole brain and pituitary is a potent mitogenic
agent for fibroblasts and other cell types, including astroglial cells (50). Finally, the human platelet-derived growth factor (PDGF), which is mitogenic for
fibroblasts also stimulates the proliferation of human normal glial cells (51).
Although all these factors affect the proliferation of astroglial cells the data
concerning their physicochemical properties indicate that these classical well
defined growth factors are different from the recently isolated neural tissue-
derived AGF, GMF and GGF. Only PDGF and GGF have similar physicochemical properties and both act on fibroblasts, but are distinct in that PDGF has no mitogenic
effect on Schwann cells (28).

CONCLUSION

Recently, protein growth factors affecting proliferation and maturation of astroglial cells in primary culture have been detected in neural tissues (brain
and pituitary). These factors have been partially purified or to apparent homogeneity and characterized in three different laboratories, and have been termed
astroglial growth factor (AGF), glia maturation factor (GMF) and glial growth
factor (GGF). All three factors possess similar biological activities in stimulating proliferation of astroglial cells as well as of fibroblasts. Two of them
(GGF and GMF) promote also Schwann cell multiplication ; AGF was not yet tested.
AGF and GMF were shown to be both mitogenic and maturation agents for astroblasts.
The brain-derived factors, AGF and GMF, share some physicochemical properties and
are both acidic proteins. Basicity of pituitary- or brain-derived factor, GGF,
is not established well enough to assert that this factor is different from the
two others. Complete purification and detailed characterization of these factors
are necessary to afford clear conclusions about their similarities or dissimila-
rities. However, these new purified factors seem to be distinct from the classical well-characterized growth factors that are mitogenic for various cell types,
including astroglial cells.

The observations that AGF and GMF have been found in embryonic and adult brain tissues suggest a possible role of these agents in astroblast development during neurogenesis as well as in regeneration and pathological events during adulthood. Similar potential role for GGF have been suggested. However, it is not yet possible to affirm the importance and precise role of these factors in vivo.

Studies on isolation of astroglial cell stimulating factors have been facilitated by the availability of nearly pure astroglial cell cultures. Further progress along this line will be possible by growing these cells in serum-free chemically defined media.

ACKNOWLEDGEMENTS

We thank Dr. E. Bock for the determination of GFA level and for the gift of GFA antiserum. We are grateful to Dr. B. Hamprecht for kindly supplying GS antiserum. We thank Mrs. M.F. Knoetgen for technical assistance. This work was supported by D.G.R.S.T. Grant n° 82.V.0012.

REFERENCES

1. Varon S (1975) Exptl Neurol 48:75-92

2. Thoenen H, Barde YA (1980)Physiol Rev 60:1284-1335

3. Yankner BA, Shooter EM (1982) Ann Rev Biochem 51:845-868

4. Special Issue on Growth and Trophic Factors (1982) J Neurosci Res, Vol 8, Number 2/3

5. Barde YA, Edgar D, Thoenen H (1982) In: Pfeiffer SE (ed) Neuroscience Approached through Cell Culture. CRC Press, Boca Raton, Vol I, pp 69-82

6. Sensenbrenner M (1977) In: Fedoroff S, Hertz L (eds) Cell, Tissue and Organ Cultures in Neurobiology. Academic Press, New York, pp 191-213

7. Sensenbrenner M, Labourdette G, Delaunoy JP, Pettmann B, Devilliers G, Moonen G, Bock E (1980) In: Giacobini E, Vernadakis A, Shahar A (eds) Tissue Culture in Neurobiology. Raven Press, New York, pp 385-395

8. Brockes JP, Fields KL, Raff MC (1979) Brain Res 165:105-118

9. Lim R, Turriff DE, Troy SS, Kato T (1977) IN: Fedoroff S, Hertz L (eds) Cell, Tissue and Organ Cultures in Neurobiology. Academic Press, New York, pp 223-235

10. Raff MC, Abney E, Brokes JP, Hornby-Smith A (1978) Cell 15:813-822

11. Sensenbrenner M, Barakat I, Delaunoy JP, Labourdette G, Pettmann B (1982) In: Pfeiffer SE (ed) Neuroscience Approached through Cell Culture. CRC Press, Boca Raton, Vol I, pp 87-105

12. Sensenbrenner M, Springer N, Booher J, Mandel P (1972) Neurobiol 2:49-60

13. Athias P, Sensenbrenner M, Mandel P (1974) Differentiation 2:99-106

14. Sensenbrenner M, Moonen G, Delaunoy JP, Bock E, Poindron P, Mandel P (1977) Trans Am Soc Neurochem 8:43

15. Lim R, Mitsunobu K, Li WKP (1973) Exptl Cell Res 79:243-246

16. Haugen A, Laerum OD (1978) Brain Res 150:225-238

17. Lim R, Turriff DE, Troy SS, Moore BW, Eng LF (1977) Science 195:195-196

18. Sensenbrenner M, Devilliers G, Bock E, Porte A (1980) Differentiation 17:51-61

19. Pettmann B, Labourdette G, Devilliers G, Sensenbrenner M (1981) Dev Neurosci 4:37-45

20. Kato T, Yamakawa Y, Lim R, Turriff DE, Tanaka R (1981) Neurochem Res 6:401-412

21. Morrison RS, Saneto RP, de Vellis J (1982) J Neurosci Res 8:435-442

22. Pettmann B, Sensenbrenner M, Labourdette G (1980) FEBS Letters 118:195-199

23. Pettmann B, Labourdette G, Weibel M, Daune G, Sensenbrenner M (1983) J Neurochem 41, Suppl : S107D

24. Kato T, Chin TC, Lim R, Troy SS, Turriff DE (1979) Biochim Biophys Acta 579:216-227

25. Kato T, Fukui Y, Turriff DE, Nakagawa S, Lim R, Arnason BGW, Tanaka R (1981) Brain Res 212:393-402

26. Lim R, Miller JF (1984) J Cell Physiol 119:255-259

27. Brokes JP, Lemke GE, Balzer DR (1980) J Biol Chem 255:8374-8377

28. Lemke GE, Brockes JP (1984) J Neurosci 4:75-83

29. Booher J, Sensenbrenner M (1972) Neurobiol 2:97-105

30. Weibel M, Pettmann B, Daune G, Labourdette G, Sensenbrenner M (1984) Intern J Dev Neurosci 2:355-366

31. Pettmann B, Delaunoy JP, Courageot J, Devilliers G, Sensenbrenner M (1980) Dev Biol 75:278-287

32. Pettmann B, Weibel M, Daune G, Sensenbrenner M, Labourdette G (1982) J Neurosci Res 8:463-476

33. Pettmann B, Weibel M, Sensenbrenner M, Labourdette G (in preparation)

34. Bock E, Møller M, Nissen C, Sensenbrenner M (1977) FEBS Letters 83:207-211

35. Sensenbrenner M, Delaunoy JP, Labourdette G, Pettmann B (1982) Biochem Soc Transact 10:424-426

36. Lim R, Mitsunobu K (1974) Science 185:63-66

37. Lim R, Troy SS, Turriff DE (1977) Exptl Cell Res 106:357-372

38. Lim R, Turriff DE, Troy SS (1976) Brain Res 113:165-170

39. Morrison RS, Saneto RP, de Vellis J (1982) J Neurosci Res 8:435-442

40. Lim R, Mitsunobu K (1975) Biochim Biophys Acta 400:200-207

41. Weibel M, Labourdette G, Pettmann B, Daune G, Sensenbrenner M (1984) In: Caciagli F (ed) Physiological and Pharmacological Control of Nervous System Development. Elsevier, Amsterdam (in press)

42. Kato T, Yamakawa Y, Lim R, Turriff DE, Tanaka R (1981) Neurochem Res 6:401-412

43. Bosch EP, Assouline JG, Miller JF, Lim R (1984) Brain Res 304:311-319

44. Delaunoy JP, Hog F, Sensenbrenner M (1984) Int J Dev Neurosci 2:131-141

45. Turriff DE, Lim R (1981) Dev Neurosci 4:110-117

46. Fontana A, Dubs R, Merchant R, Balsiger S, Grob PJ (1981) J Neuroimmunol 2:55-71

47. Barritault D, Plouët J, Comty J, Courtois Y (1982) J Neurosci Res 8:477-490

48. Westermark B (1976) Biochem Biophysic Res Commun 69:304-310

49. Simpson DL, Morrison R, de Vellis J, Herschman HR (1982) J Neurosci Res 8: 453-462

50. Pruss RM, Bartlett PF, Gavrilovic J, Lisak RP, Rattray S (1982) Dev Brain Res 2:19-35

51. Heldin CH, Wasteson A, Westermark B (1977) Exptl Cell Res 429-437

RESUME

Des cultures primaires de cellules astrogliales ont été obtenues à partir d'hémisphères cérébraux de rat nouveau-né. Des études immunocytochimiques et biochimiques ont montré la présence de la protéine acide fibrillaire gliale et de la protéine S100, ainsi que de la glutamine synthétase dans ces astrocytes. En présence d'extraits solubles de cerveau de poulet, de rat et de boeuf la prolifération, ainsi que la maturation morphologique et biochimique des astroblastes sont stimulées. Un facteur actif, dénommé "astroglial growth factor" (AGF) a été partiellement purifié à partir de cerveau entier de boeuf et caractérisé. Ce facteur est une protéine acide de poids moléculaire entre 45.000 et 50.000 daltons. L'AGF stimule à la fois la prolifération et la maturation des astroblastes de rat en culture. Des données concernant d'autres facteurs de croissance, affectant les cellules astrogliales en culture primaire, sont soulignées.

Hormones and Cell Regulation, Volume 9
INSERM European Symposium
J.E. Dumont, B. Hamprecht and J. Nunez editors
© 1985 Elsevier Science Publishers B.V. (Biomedical Division)

NEW APPROACHES TO DETECT NGF-LIKE ACTIVITY IN TISSUES

TED EBENDAL[1], LENA LÄRKFORS[1], CHRISTIANE AYER-LE LIEVRE[2], AKE SEIGER[2] and
LARS OLSON[2]

Department of Zoology[1], Uppsala University, Box 561, S-751 22 UPPSALA, and
Department of Histology[2], Karolinska Institute, Box 60400, S-104 01 STOCKHOLM
(Sweden)

INTRODUCTION

It is generally agreed that a number of chemical factors govern various
processes during development, maintenance and regrowth of the neuronal cell.
Specific classes of factors have been associated with the survival of neurons
after critical stages of development, the stimulation of outgrowth of neurites
from the neuron soma, and the selection of pathway taken by the outgrowing axon
(Fig. 1).

In some instances considerable progress has been made to characterize the
molecular species involved. The substance standing as a model for neuronal
survival factors is NGF (nerve growth factor), a protein so far isolated only
from some rich exocrine sources of unknown relevance for neurobiological function
(for reviews see refs. 1-6). Beside NGF a number of less well characterized
survival factors have been described (reviewed in 7 and 8), two of which have
been purified under denaturing conditions (9, 10).

NGF also stimulates the outgrowth of axons (1, 6, 11). Recently, this effect
has been demonstrated even in a population of neurons where NGF does not support
survival (12).

Concerning the substratum pathways for axons it has been demonstrated that
the basal lamina protein laminin, at least in culture, is highly efficient in
supporting neurite extension (13), and that the cell adhesion molecule N-CAM is
important in the formation of nerve bundles (14, 15).

Considering the advances in knowledge on these "new" factors involved in the
patterning of the nervous system, it may seem surprising that information
regarding the appearance of NGF and NGF-like factors in normal tissues remains
fairly incomplete. The reason is in part due to the small amounts of NGF needed
for a biological effect and to the "stickiness" of the protein (16). The problems
and pitfalls of quantitatively determining low levels of NGF in tissues are
discussed in refs. (4) and (17).

We have been interested to study NGF-like activity in embryonic, fetal and
denervated adult tissues. In such tissues NGF may be hypothesized to fulfill
a role for regulating nerve growth. The studies of embryonic chick homogenates

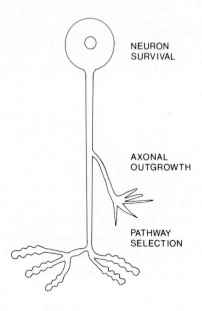

NEURON
SURVIVAL

AXONAL
OUTGROWTH

PATHWAY
SELECTION

Fig. 1. Schematic illustration of some basic processes during development of a neuron: regulation of survival of the cell, initiation of axon elongation, and choice of pathway by the outgrowing nerve fibre.

revealed a growth stimulating activity for sympathetic chick ganglia (18, 19) not sharing the properties of the classical NGF. However, as was reported in brief earlier (20), progressive purification of the embryonic extract unmasked an activity indistinguishable from NGF in biological activity and immunological properties. Another study using the bioassay with explanted chick sympathetic ganglia demonstrated an over tenfold increase in NGF-like activity in the adult rat iris following combined sensory and sympathetic denervation (21). This was later confirmed in our laboratories (22), using a modification of the reliable sandwich radioimmunoassay for βNGF (23).

Further work to determine the localisation, chemical properties, amount, place of synthesis and regulated appearance of NGF-like factors rests heavily on methodological advances. In principle, detection of NGF (and NGF-like substances) can be based firstly on binding to its cell surface receptor (as in bioassay or in radioreceptorassay (24, 25)), secondly on binding to antibodies recognizing NGF and, thirdly on hybridization of the cDNA probes for NGF now available to the corresponding nucleic acid (26, 27).

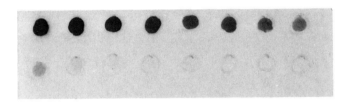

Fig. 2. Dot blots of βNGF. 2 μl samples of mouse βNGF (twofold dilutions) were applied to nitrocellulose and then incubated with affinity-purified rabbit anti-bodies to NGF. Dots were vizualised by incubation with anti-rabbit Ig antibodies conjugated with β-galactosidase followed by Bluo-Gal (a histochemical substratum for β-galactosidase, BRL). Start concentration of NGF was 10 μg/ml (top left). Highest dilution seen (although only faintly on the photograph) is 2.4 ng/ml (or 5 pg in the blot).

In the following sections some recent work to determine NGF or NGF-like substances in tissues will be described.

NGF-LIKE ACTIVITY IN THE CHICK EMBRYO

The nerve-growth activities of chick tissues as seen in bioassay have been described in a series of papers from this laboratory (11, 18-20, 28-30). A recent example illustrating the difficulties for correct estimations of NGF amounts in tissue extracts will be given here. During the course of purification of chick growth factors, gel filtration with 6 M urea in the equilibrating buffer was tested. The observation was made that the activity stimulating nerve outgrowth became indistinguishable from mouse βNGF by this procedure (20, 30). The effect has now been found to be more pronounced after some initial chromatographic steps (Belew and Ebendal, in preparation). In fact up to 100-fold increase in apparent NGF-activity compared with start fractions have now been found using new chromatographic techniques. Probably the amount of NGF in the crude extract of the chick embryo has been underestimated, possibly due to NGF-binding components preventing interaction with the cell surface receptor. Similar findings were recently made also in medium conditioned by chick heart cells (31). It is not yet known whether this is relevant to mammalian tissues and to serum, reported to carry little if any NGF (23, 32). The inactivation of NGF might be biologically

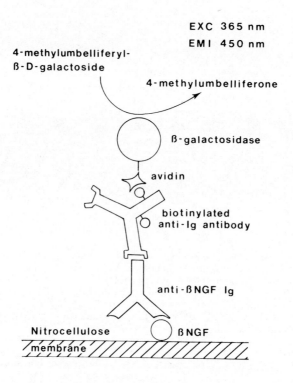

EXC 365 nm
EMI 450 nm

4-methylumbelliferyl-
ß-D-galactoside

4-methylumbelliferone

ß-galactosidase

avidin

biotinylated
anti-Ig antibody

anti-ßNGF Ig

Nitrocellulose
membrane

ßNGF

Fig. 3.   The steps used to measure ßNGF on nitrocellulose by a fluorimetric enzyme immunoassay.

important. The amounts of NGF indicated in the chick embryo would probably result in a supernormal growth if freely accessible to responsive neurons.

MEASUREMENT OF NGF BLOTTED TO NITROCELLULOSE MEMBRANES

Earlier radio or enzyme immunoassays for ßNGF were based on the binding of the protein to a solid phase via a layer of antibodies (23, 32-33). An alternative method is membranes of nitrocellulose used as blotting matrices. Initial tests with dot blots of ßNGF antibodies, visualized by enzyme-conjugated antibodies and a stained reaction product, recognize and bind to the immobilized NGF in a dose--dependent manner (Fig. 2).

This observation lead us to design an enzyme immunoassay (EIA) as illustrated in Fig. 3. The NGF sample is applied to a nitrocellulose membrane using a manifold vacuum concentrator. The membrane is then blocked with bovine serum albumin and sequentially incubated with the links shown in Fig. 3.

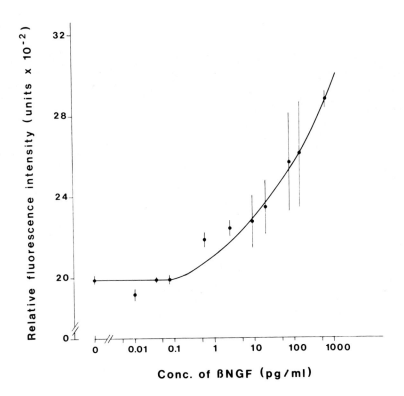

Fig. 4. Standard curve for fluorimetric determination of βNGF concentration.
Nitrocellulose membranes (pore size 0.45 μm, Bio-Rad) were soaked for 10 min in
0.02 M Tris buffer with 0.5 M NaCl (pH 7.4). The membranes were clamped in a
96-well manifold vacuum concentrator (Bio-Rad). Samples of 200 μl of βNGF,
diluted in the buffer containing bovine serum albumin (BSA) in tenfold excess
to βNGF, were applied by gravity filtration.
   After washings under vacuum, the sheet was removed and blocked for 1-2 h
with 5% (wt/vol) BSA.
   The membranes were then rinsed 3 x 10 min in washing solution containing
0.5% BSA and 0.05% (vol/vol) Tween 20 and then incubated overnight at +4°C with
affinity purified rabbit anti βNGF Ig (12 μg/ml).
   All subsequent steps were carried out at room temperature. After rinses to
remove excess Ig, the sheets were incubated for 1 h in affinity purified
biotinylated antirabbit IgG (2.5 μg/ml, Vector). Membranes were rinsed again
and exposed for 20 min to β-galactosidase conjugated to streptavidin (BRL) or
to avidin D (Vector) diluted to 2 μg/ml in 0.05 M sodium phosphate buffer
(pH 7.3) containing 1 mM $MgSO_4$. 0.2 mM $MnSO_4$, 2 mM Mg-EDTA and 1% (wt/vol) NaN_3.
After rinsing, the β-galactosidase activity was measured in cut-out blots
incubated with 0.15 mM 4-methylumbelliferyl-β-D-galactoside (Sigma) in the
phosphate buffer.
   The enzyme reaction was stopped by adding 0.15 M glycine/NaOH (pH 10.5).
Fluorescence was read in black 96-well microplates in a Microfluor reader
(Dynatech) at 450 nm (excitation 365 nm).
   Mean ± S.E.M. of quadruplicate samples are shown. Curve fitted by eye.

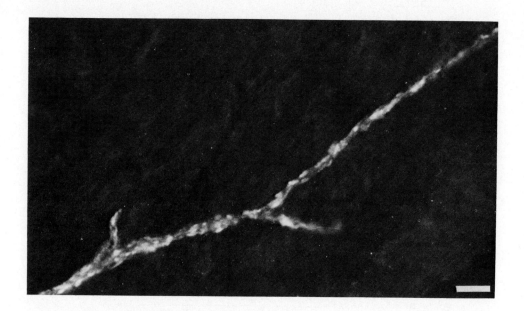

Fig. 5. Cryostat section of a 15-day-old rat fetus. NGF-like immunoreactivity is visible in a small branch of the trigeminal nerve in the maxillar region. The reactive material has a granular appearance. Antibodies as in Fig. 6 a. From ref. (35). Bar 20 µm.

Each blot is then cut out and transferred to a multiwell dish for the enzyme reaction. A fluorogenic substratum is used and the result read in a microfluorimeter scanning the multiwells.

Preliminary results using this technique indicate that very small amounts of NGF (about 0.1 pg per blot) can be detected on nitrocellulose (Fig. 4). Thus, levels of NGF reasonable to have physiological effects (above 10 pg/ml) should be

---

Fig. 6. NGF-like immunoreactivity in dorsal root ganglia (sagittal sections) of 15-day-old (13 mm) rat fetuses. Bars 50 µm.
a. Affinity purified rabbit anti-βNGF antibodies (purified on an affinity column with the standard βNGF preparation) were applied at 25 µg/ml to a 14 µm section. The second antibody, swine anti-rabbit Ig, was labelled with rhodamine. Immunoreactivity is present in spinal nerves and nerve bundles passing through the ganglia.
b. Same distribution of immunoreactivity with anti-βNGF antibodies further purified on an affinity bed prepared as described in Fig. 10.
c. Control section. The first antiserum was deprived of its antibodies against βNGF by two cycles of affinity chromatography on βNGF-Sepharose. The dilution used corresponds to 1:500 of the original serum.

367

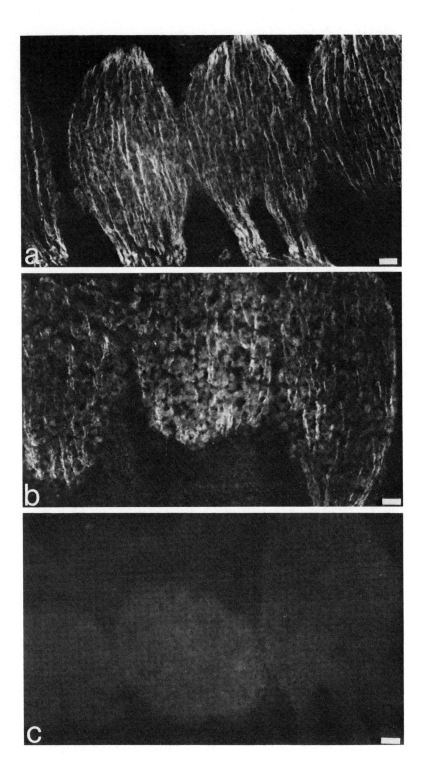

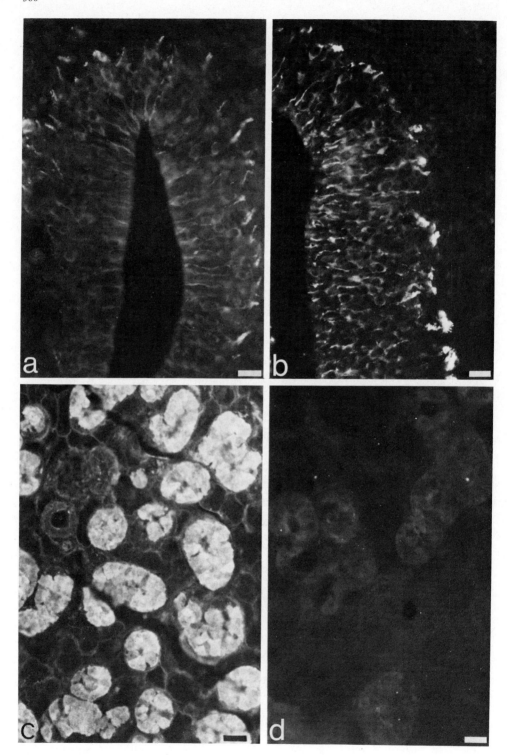

detectable using this assay. Affinity purification of NGF from tissues can be used as an initial step to lower the competition for binding sites on the nitro-cellulose membrane. Fixation of crude tissue extracts blotted to the nitro-cellulose may also be useful, as recently described (34) for another quantitative dot blot assay of neuronal proteins. One advantage of using nitrocellulose is that samples can be treated with strong detergents like sodium dodecyl sulphate (SDS), sometimes required to facilitate purification of growth factors (9, 10).

IMMUNOHISTOCHEMICAL LOCALISATION OF NGF-LIKE ACTIVITY IN THE RAT

Earlier we presented results showing positive immunoreactivity for βNGF in tissue sections of the fetal rat and in stretch preparations of the adult rat iris (35). It was found necessary to use only affinity-purified antibodies in this study since whole antiserum gave strong background fluorescence.

Immunoreactive fibres were observed already in 11-day-old rat fetuses. At Day 15 NGF-like reactivity was found along the peripheral branches of sensory and sensory-motor cranial and spinal nerves (Fig. 5). In the sensory and sympathetic ganglia densely aggregated immunoreactive material was associated with nerve bundles (Fig. 6). Cell bodies were not distinctly fluorescent but some of them showed immunofluorescence slightly above background.

Positive cell bodies were observed only in the olfactory epithelium. Both processes of these bipolar cells were immunoreactive as well as the fibres of the olfactory nerve emerging from the mucosa (Fig. 7).

At later stages (Day 21) numerous cells in the inner layer of the retina were also immunoreactive.

The iris of the adult rat was also examined. Immunoreactivity to NGF was observed along sensory and autonomic nerve bundles of normal irides and at the level of regrowing axons in irides grafted to the anterior eye chamber.

The striated ducts in adult male submaxillary glands were strongly positive (Fig. 7 c). This fluorescence was abolished by preadsorption of the antibody with NGF (Fig. 7 d).

Fig. 7. a. NGF-like immunoreactivity in the olfactory mucosa of the fetal rat. Note positive cell bodies and processes. Conditions as in Fig. 6 a.
b. Similar distribution of NGF-like immunoreactivity in the olfactory mucosa using the same first antiserum as in 6 b.
c. NGF-like immunoreactivity in the submandibular salivary gland of an adult male mouse (same antiserum as in 6 a).
d. Control. Submandibular gland of a male mouse. The anti-NGF antibodies were preadsorbed with βNGF. No reaction takes place under these conditions.
Bar 25 μm (a-d).

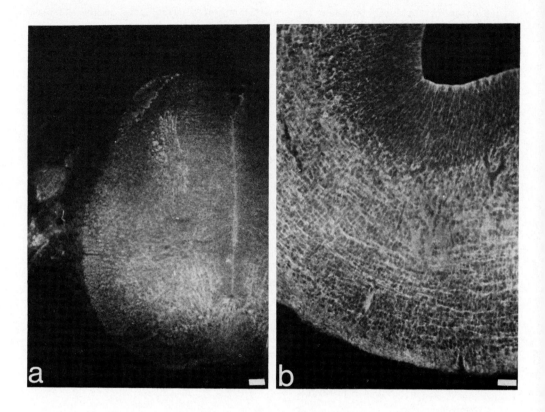

Fig. 8. Distribution of NGF-like immunoreactivity in the central nervous system of the 15-day-old rat fetus. Same antibodies as in 5 a.
    a. Spinal cord (transverse section). Bar 100 μm.
    b. Pontine flexure (sagittal section) Bar 50 μm.

Immunoreactivity was found in most areas of the central nervous system of the 15-day-old rat fetus. The immunoreactive material seemed granulated and was particularly dense along nerve bundles in the grey matter of the spinal cord (Fig. 8 a), in the ventral part of medulla oblongata and pons (Fig. 8 b), in the superficial layers of the developing tectum (Fig. 9) and cerebral cortex, in the anterior part of the forebrain, notably in the region corresponding to the olfactory bulb.

The extensive and rather unexpected NGF-like reactivity in the fetal brain is provocative. Either the fixation of the tissue induces immunoreactivity in some protein(s) not related to NGF or NGF-like substances are more widespread in the CNS than believed hitherto, indicating roles for them during brain development. It is interesting in this context that NGF has been reported to be taken up and

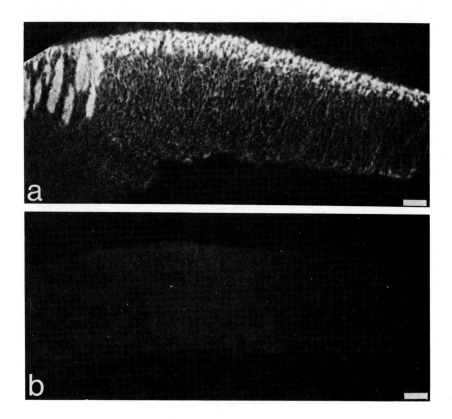

Fig. 9. NGF-like immunoreactivity in the fetal central nervous system (sagittal sections). a. Posterior commissure (left) and anterior part of the tectum.
        b. Control. Sagittal section of the tectum. The antibodies were pre-adsorbed with βNGF. No immunoreactivity observed. Bar 50 μm.

transported by cholinergic neurons in the septal and basal nuclei of the rat forebrain, affecting the level of choline acetyltransferase (36-38).

PURIFICATION OF ANTI-NGF ANTIBODIES

    βNGF is purified in our laboratory as described by Mobley *et al*. (39). NGF is eluted from the second column of CM-Sepharose with a linear gradient of NaCl (see Fig. 1 in ref. (30)), as reported to give an immunogenically pure βNGF (40). The antibodies used here were affinity purified on βNGF coupled to cyanogen bromide activated Sepharose as described (41).

    To check for possible minor contaminants in our preparations (that may lead to misinterpretations of the results from EIA and immunohistochemistry) two

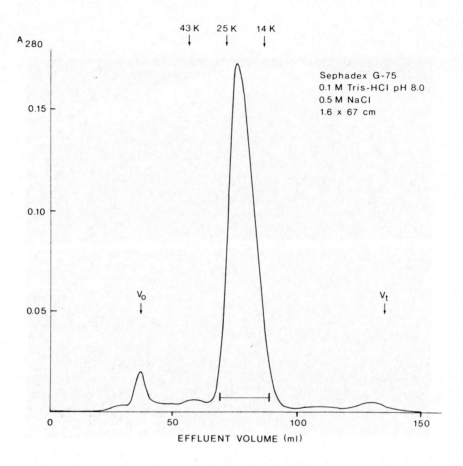

Fig. 10. Gel filtration of βNGF on Sephadex G-75 (Pharmacia). The column was equilibrated with buffer containing 0.5 M NaCl to prevent losses of βNGF due to adsorption (26). About 2 mg of protein (NGF eluted from a column of CM-Sepharose as described (30, 40)) was loaded at a flow rate of 5 ml/h. The column was calibrated with ovalbumin, chymotrypsinogen A and ribonuclease A. The apparent molecular weight of βNGF was 21,000. The peak of NGF (horizontal bar) was coupled to CNBr-activated Sepharose 4B (Pharmacia) for use in purification of antibodies to βNGF.

protocols were adopted. The first one involved a further step of gel filtration of the βNGF peak eluted by the gradient of salt. As seen in Fig. 10, βNGF elutes as an almost symmetrical peak with an apparent molecular weight of 21,000. This is in perfect agreement with the results of Furukawa *et al*. (42) and Darling *et al*. (43) from gel filtration of βNGF under conditions of high salt and low pH, respectively. An affinity bed for extracting antibodies from the antisera was prepared using only the peak fraction as indicated in Fig. 10.

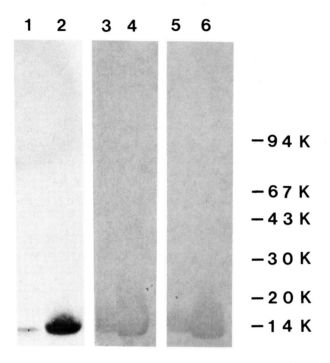

Fig. 11. SDS-gel electrophoresis and electroblots of βNGF. The NGF-peak from the column of CM-Sepharose was treated with 1% SDS and heated for 5 min in boiling water. Samples were applied to a polyacrylamide gradient gel (ranging from 2 to 16% acrylamide, top to bottom). SDS-treated marker proteins were run in other lanes as shown to the right (phosphorylase b, albumin, ovalbumin, carbonic anhydrase, trypsin inhibitor and α-lactalbumin).
Lanes 1 and 2. Staining of the gel with Coomassie Brilliant Blue R-250. The βNGF-band is seen in a 14K position. 500 ng and 4 μg, respectively, loaded on the gel.
Lanes 3-6. Electrophoretic transfer of the proteins to a nitrocellulose membrane. The membranes were incubated with anti-βNGF antibodies (from rabbits) and visualised by anti-rabbit Ig antibodies conjugated with horseradish peroxidase (substratum: 4-chloro-1-naphthol).
Lanes 3 and 4. Antibodies were against native βNGF. They were affinity-purified on βNGF coupled to Sepharose beads after a final step of gel filtration as described in the previous figure legend. 100 ng and 1 μg, respectively, loaded to the gel. Staining in the 14K position.
Lanes 5 and 6. Antibodies raised against SDS-denatured βNGF eluted from the 14K position of a polyacrylamid gel. Incubation of the nitrocellulose was made with the immunoglobulin fraction purified on a column of protein A-Sepharose. Also here, recognition of the 14K band is obvious. Amounts of NGF loaded were 100 ng and 1 μg, respectively.

374

The second approach was to SDS-denature βNGF, run the sample on a polyacrylamide gel under nonreducing conditions and use only the cut-out 14K band
containing the βNGF monomer for immunization (Fig. 11, lanes 1 and 2). In this
case the resulting antibodies were purified from serum by affinity chromatography on Protein A-Sepharose. The ability of these antibodies to block the
biological effect of NGF in a bioassay was low but they readily detected
SDS-denatured βNGF in the 14K position after electroblotting (Fig. 11, lanes
5 and 6). The NGF band was recognized also by the antibodies against native NGF
purified according to the first method outlined above (Fig. 11, lanes 3 and 4).

Only the first protocol resulted in antibodies that gave a pattern of immunoreactivity (Figs. 6 b and 7 b) similar to that based on our regular affinity-
-purified antibodies (Figs. 6 a and 7 a). Thus there is a correlation between
the ability of the antibodies to block biological activity in a bioassay and
the ability to recognize the immunoreactivity in sections described here.

ACKNOWLEDGEMENTS

Supported by the Swedish Natural Science Research Council (B-BU 4024-104,
S-FO 4024-105, B-BU 1522-102), the Swedish Medical Research Council (14X-03185,
14X-06555, 14P-5867, 25P-6326), Åke Wibergs Stiftelse, the "Expressen" Prenatal
Research Foundation, Magn. Bergvalls Stiftelse and Karolinska Institutets Fonder.
Dr. Ayer-Le Lievre received a fellowship from the French C.N.R.S. Technical
assistance was given by Mr. Bo Molin, Mrs. Annika Kylberg, Mrs. Stine Söderström,
Ms. Lena Hultgren, Ms. Anna Hultgårdh and Mrs. Barbro Standwerth. The figures
were prepared by Mrs. Vibeke Nilsson who also typed the manuscript.

REFERENCES

1. Levi-Montalcini R, Angeletti PU (1968) Physiol Rev 48:534-569.
2. Mobley WC, Server AC, Ishii DN, Riopelle RJ, Shooter EM (1977) New Engl
   J Med 297:1096-1104, 1149-1158, 1211-1218.
3. Bradshaw RA (1978) Ann Rev Biochem 47:191-216.
4. Thoenen H, Barde Y-A (1980) Physiol Rev 60:1284-1335.
5. Shooter EM, Yankner BA, Landreth GE, Sutter A (1981) Recent Progr Horm Res
   37:417-446.
6. Levi-Montalcini R (1982) Ann Rev Neurosci 5:341-362.
7. Barde Y-A, Edgar D, Thoenen H (1983) Ann Rev Physiol 45:601-612.
8. Berg DK (1984) Ann Rev Neurosci 7:149-170.
9. Barde Y-A, Edgar D, Thoenen H (1982) EMBO J 1:549-553.
10. Manthorpe M, Varon S (1984) In: Guroff G (ed) Growth and Maturation Factors.
    Vol 3. J Wiley & Sons, New York (in press).
11. Ebendal T (1979) Dev Biol 72:276-290.

12. Collins F (1984) J Neurosci 4:1281-1288.

13. Manthorpe M, Engvall E, Ruoslahti E, Longo FM, Davis GE, Varon S (1983) J Cell Biol 97:1882-1890.

14. Rutishauser U, Gall WE, Edelman GM (1978) J Cell Biol 79:382-393.

15. Edelman GM (1983) Science 219:450-457.

16. Pearce FL, Banthorpe DV, Cook JM, Vernon CA (1973) Eur J Biochem 32:569-575.

17. Harper GP, Thoenen H (1980) J Neurochem 34:5-16.

18. Ebendal T, Hedlund K-O, Norrgren G (1982) J Neurosci Res 8:153-164.

19. Ebendal T, Norrgren G, Hedlund K-O (1983) Med Biol 61:65-72.

20. Ebendal T (1984) NATO ASI Series (in press).

21. Ebendal T, Olson L, Seiger Å, Hedlund K-O (1980) Nature 286:25-28.

22. Ebendal T, Olson L, Seiger Å (1983) Exp Cell Res 148:311-317.

23. Suda K, Barde YA, Thoenen H (1978) Proc Natl Acad Sci USA 75:4042-4046.

24. Siggers DC, Rogers JG, Boyer SH, Margolet L, Dorkin H, Banerjee SP, Shooter EM (1976) New Engl J Med 295:629-634.

25. Fabricant RN, De Larco JE, Todaro GJ (1977) Proc Natl Acad Sci USA 74:565-569.

26. Scott J, Selby M, Urdea M, Quiroga M, Bell GI, Rutter WJ (1983) Nature 302:538-540.

27. Ullrich A, Gray A, Berman C, Dull TJ (1983) Nature 303:821-825.

28. Ebendal T, Jacobson C-O (1977) Exp Cell Res 105:379-387.

29. Ebendal T, Belew M, Jacobson C-O, Porath J (1979) Neurosci Lett 14:91-95.

30. Ebendal T, Olson L, Seiger Å, Belew M (1984) In: Black IB (ed) Cellular and Molecular Biology of Neuronal Development. Plenum, New York, pp 231-242.

31. Norrgren G (1984) Acta Univ Upsaliensis Abstr Uppsala Diss Fac Sci 743. 31 pp.

32. Korsching S, Thoenen H (1983) Proc Natl Acad Sci USA 80:3513-3516.

33. Furukawa S, Kamo I, Furukawa Y, Akazawa S, Satoyoshi E, Itoh K, Hayashi K (1983) J Neurochem 40:734-744.

34. Jahn R, Schiebler W, Greengard P (1984) Proc Natl Acad Sci USA 81:1684-1687.

35. Ayer-Le Lievre CS, Ebendal T, Olson L, Seiger Å (1983) Med Biol 61:296-304.

36. Gnahn H, Hefti F, Heumann R, Schwab ME, Thoenen H (1983) Dev Brain Res 9:45-52.

37. Hefti F, Dravid A, Hartikka J (1984) Brain Res 293:305-311.

38. Seiler M, Schwab ME (1984) Brain Res 300:33-39.

39. Mobley WC, Schenker A, Shooter EM (1976) Biochemistry 15:5543-5552.

40. Chapman CA, Banks BEC, Carstairs JR, Pearce FL, Vernon CA (1979) FEBS Lett 105:341-344.

41. Stoeckel K, Gagnon C, Guroff G, Thoenen H (1976) J Neurochem 26:1207-1211.

42. Furukawa Y, Furukawa S, Satoyoshi E (1984) J Biol Chem 259:1259-1264.

43. Darling TLJ, Petrides PE, Beguin P, Frey P, Shooter EM, Selby M, Rutter WJ, (1983) Cold Spring Harbor Symp Quant Biol 48:427-434.

RÉSUMÉ

Le Nerve Growth Factor (NGF) est actuellement le mieux connu des facteurs de survie neuronale. Cependant jusqu'à présent, il n'a été isolé qu'à partir de tissus exocrine dont la richesse en NGF ne semble pas avoir d'importance neurologique fonctionnelle.

Récemment, au cours de la purification progressive d'extraits de tissus embryonnaires de poulet, une activité de type NGF a été démasquée. La signification de ces résultats en tant que preuve de la présence de NGF chez l'embryon de poulet est discutée.

Des techniques appropriées à la caractérisation du NGF et d'autres facteurs ayant une activité neurotrope similaire dans des tissus adultes dénervés ou embryonnaires sont considérées. Un test microfluométrique pour le NGF, après blotting sur support de nitrocellulose, est présenté. Très sensible, ce test permet de détecter des taux de NGF inférieurs à 1 pg/ml (soit moins de 200 fg par échantillon).

Une immunoréactivité de type NGF a été observée dans le cerveau due foetus de rat de 15 jours, notamment dans la partie ventrale du bulbe et du pont, dans les couches superficielles du tectum et du cortex ainsi que dans la région olfactive de l'ébauche télencéphalique. Las techniques mises en oeuvre pour la purification par affinité des anticorps anti-NGF utilisés sont aussi décrites.

Hormones and Cell Regulation, Volume 9
INSERM European Symposium
J.E. Dumont, B. Hamprecht and J. Nunez editors
© 1985 Elsevier Science Publishers B.V. (Biomedical Division)

CHOLINERGIC DIFFERENTIATION FACTOR FOR CULTURED SYMPATHETIC NEURONS

KEIKO FUKADA

Division of Biology, California Institute of Technology, Pasadena, CA 91125, USA

Over the past ten years both in vivo and in vitro studies have shown that the phenotype adopted by a variety of neural crest derivatives can be influenced by the environment in which they develop (for a review, see Patterson, 1978; Le Donarin, 1980). Studies of the embryonic development of peripheral ganglia in chick-quail chimeras have shown that the fate of the neurons can be switched from noradrenergic to cholinergic or vice versa when their embryonic environment is experimentally altered by a suitable transplantation (Le Douarin, 1980). In fact this transmitter decision can be changed in vivo even after the cells cease migrating and form ganglia (Le Douarin et al., 1978).

These in vivo studies, however, raised the question of whether the conversion in transmitter phenotype is due to a selection of certain neural crest populations, or rather a change in transmitter metabolism of single cells. This question was answered by culture studies in which the fluid and cellular environment surrounding the neurons can be controlled experimentally. These studies showed that the alteration in transmitter phenotype is due not to a selection of certain subpopulations for survival, but to a change of phenotype at the single-cell level. First, biochemical studies on immature single sympathetic neurons grown in microcultures showed that most of the initially noradrenergic neurons could be influenced to become cholinergic (Reichardt & Patterson, 1977). More recently, a complete transition from noradrenergic to cholinergic was demonstrated in single neurons by electrophysiological recordings. The transmitter status of individual neurons on islands of heart cells was assayed over time as they developed. Both partial transitions from purely noradrenergic to dual-function and from dual-function to purely cholinergic, as well as a complete transition from purely noradrenergic to purely cholinergic, were observed (Potter et al., 1981). This switch was accompanied by changes in the cytochemistry of synaptic vesicles (Potter et al., 1981).

The culture studies also revealed the existence of a diffusible factor which controls the transmitter status of developing sympathetic neurons in culture. When these neurons are grown in the virtual absence of non-neuronal cells, they develop many of the properties expected of noradrenergic neurons in vivo, with a time course similar to that observed in vivo. They synthetize and accumulate catecholamines (CA) from tyrosine (Mains & Patterson, 1973 a,b,c), they take up, store and release CA, and form morphologically noradrenergic synapses with each other (Claude, 1973; Ree & Bunge, 1974; Landis, 1980). This noradrenergic differentiation in vitro is greatly enhanced, or stabilized, by growing the neurons under depolarizing conditions or stimulating them electrically (Walicke et al., 1977; Landis, 1980). The effect of depolarization is mediated by Ca^{++} ions (Walicke et al., 1977; Walicke & Patterson, 1981).

On the other hand, when the same sympathetic neurons are grown either in the presence of appropriate non-neuronal cells or in culture media conditioned by them (CM), or on a monolayer of p-formaldehyde fixed heart cells (Hawrat, 1980), they can be influenced to become cholinergic (Patterson & Chun, 1974, 1977). The neurons develop the ability to produce acetylcholine (ACh) and form functional cholinergic synapses on each other, and with skeletal and heart muscle cells (O'Lague et al., 1974; Nurse & O'Lague, 1975; O'Lague et al., 1978) along a time course similar to that of noradrenergic neurons in culture (Bunge et al., 1978; Patterson et al., 1978).

The CM effect is specific, since the number of neurons surviving in culture and overall neuronal growth (measured as total protein and lipid) are not affected by this factor at any concentration of CM (Patterson & Chun, 1977). In this respect, therefore, this factor is qualitatively different from another developmental signal, nerve growth factor, which also affects both survival and growth (for a review, eq. see Harper & Thoenen, 1980).

The response of the neurons to the cholinergic influence is dose-dependent (Patterson & Chun, 1977): A higher proportion of CM or a greater number of non-neuronal cells gives a higher ratio of ACh to CA synthesis, a higher incidence of cholinergic transmission and a higher proportion of synapses that lack small granular vesicles (Landis et al., 1976). If an appropriately strong cholinergic stimulus is applied, many of the noradrenergic properties are reduced: The ability of the neurons to synthesize and accumulate CA (Patterson & Chun, 1977), as well as the endogenous CA stores (Landis, 1976, 1980; Johnson et al., 1976; Johnson et al., 1980) are reduced. This is accompanied by the suppression of tyrosine hydroxylase (TH) (Swerts et al., 1983; Wolinsky & Patterson, 1983), dopamine β-hydroxylase (Swerts et al., 1983) and an isozyme of monoamine oxidase enriched in noradrenergic neurons (Pintar et al., 1981). This reciprocal change in the ability to produce transmitters (that is, increase in ACh synthesis and suppression of CA synthesis) is characteristic of cholinergic induction. It is seen even with the most purified CM material available (see below).

Recently it has been suggested that the expression of peptidergic phenotypes is coregulated with the expression of the noradrenergic/cholinergic phenotype (Kessler et al., 1984; Kessler, in press. In the presence of ganglionic non-neuronal cells, CM from the target of the neurons increases both choline acetyltransferase (CAT) and substance P (SP), and decreases TH and somatostatin

(SO). On the other hand, in the absence of non-neuronal cells, the CM increases TH and SO, but very little CAT and SP are observed. Moreover, membrane depolarization, which suppresses cholinergic development, also suppresses development of SP. It would be interesting to see whether the purified cholinergic factor also has an effect on the expression of peptidergic phenotypes.

Moreover, several lines of evidence show that CM causes other changes in neurons in addition to the alteration in transmitter status. Specific glyco-proteins and glycolipids on neuronal surfaces are also altered, as revealed by metabolic and surface labelling studies (Braun et al., 1981), and the binding of lectins (Schwab & Landis, 1981) and monoclonal antibodies (Chun et al., 1980). Another change is in the family of glycoproteins which are spontaneously released into the medium (Sweadner, 1981). These results suggest that various aspects of transmitter phenotype are regulated in concert, although it remains to be shown that the purified cholinergic factor causes all these effects.

Recently, Weber and co-workers have reported an interesting result concerning the types of neurons which this cholinergic factor may affect. They have shown that a partially purified factor from serum-containing CM which induces the cholinergic phenotype in sympathetic neurons also causes a significant increase in ACh synthesis and CAT activity in both spinal cord and nodose ganglion neurons. They suggest that the same macromolecule is involved in the regulation of neurotransmitter phenotype in these three types of neurons, despite their different embryonic origins (Giess & Weber, 1984; Swerts et al., 1984). Definitive proof for this interesting hypothesis requires either a purified factor or a monoclonal antibody against it. Furthermore, it must be determined that the CM effect is not a selection but an induction (Edgar et al., 1981), since in these cultures cholinergic neurons are not specifically labeled and cannot be distinguished from other neurons in culture.

Of related interest is the relationship of this cholinergic factor to other differentiation factors. Among these are a CAT-stimulating activity of mole-cular weight 40-45,000 daltons which affects chick ciliary ganglion neurons in culture (Nishi & Berg, 1981), and a CM factor(s) produced by skeletal muscle cells which increases the level of CAT activity of spinal cord neurons in culture (Giller et al., 1977; Brookes et al., 1980; Godfrey et al., 1980; Smith and Appel, 1983).

These studies raise several further questions: Among them are i) Does the cholinergic signal play a role in normal development in vivo? ii) What is the molecular mechanism of cholinergic induction? Recent studies on cholinergic sympathetic neurons innervating sweat glands of rat footpads have shown that the phenotypic transition from noradrenergic to cholinergic demonstrated in culture

studies is actually occurring in vivo as part of normal development: At the 7th
postnatal day the axons innervating the developing glands possess endogenous CA
histofluorescence and small granular vesicles. Thereafter these markers for
endogenous CA decrease and acetylcholinesterase staining increases (Landis,
1981, 1983; Landis & Keefe, 1983). A similar transmitter transition has been
postulated for a population of neuron-like cells in the embryonic rat gut
(Cochard et al., 1978; Teitelman et al., 1978; Jonakait et al., 1979). It still
remains to be demonstrated, however, that the transition is caused by the
cholinergic factor. Supporting this possibility is the recent study of Wolinsky
and Patterson (in press) which shows that the cholinergic inducing activity
present in rat serum does not begin to rise until 9 days postnatally. The time
course of its appearance is thus similar to the time course for the transition
occurring in cholinergic sympathetic neurons innervating rat sweat glands.

In order to determine the role of the cholinergic factor in normal in vivo
development and to understand its mechanism of action at the molecular level,
therefore, it is necessary to purify the factor to homogeneity. Purification
was first started from serum-containing CM. The partially purified activity
chromatographs on Sephadex at a molecular weight of about 45,000 daltons, is
stable to urea, guanidine and mercaptoethanol and is highly sensitive to
periodate treatment (Weber, 1981).

Since serum in the CM contains a mixture of proteins which must be purified
away from the factor, a method of obtaining an active serum-free CM was
developed. CM prepared from serum-free medium (withouth additives) was
inactive. However, CM prepared from serum-free medium supplemented with certain
hormones based on the results of Sato and co-workers (Bottenstein et al., 1979)
was more active than the serum-containing CM. This success in obtaining active
serum-free CM not only enabled an enormous purification at the outset but also
clarified two other points concerning cholinergic induction. i) The possibility that the cholinergic factor is a serum component modified by non-neuronal
cells was ruled out. ii) Another level of environmental influence on the
choice of transmitter phenotype was suggested: namely, that specific hormones
can control transmitter choice in developing sympathetic neurons indirectly via
non-neuronal cells (Fukada, 1980).

Purification of the cholinergic factor from this serum-free, hormone-
supplemented medium has been achieved about 10^5-fold with reasonable recovery by
a series of steps including ammonium sulfate precipitation, DEAE- and CM-cel-
lulose column chromatography, Sephadex gel filtration and SDS-polyacrylamide gel
electrophoresis.

The ACh/CA ratio of neurotransmitters synthesized and accumulated by 2 week-
old cultures was used as the quantitative index of the biological activity of CM

or a partially purified factor (Patterson & Chun, 1977). The estimate of 10^5-fold purification is tentative due to the difficulty in accurately measuring the amount of protein at the final step of purification, SDS-polyacrylamide gel electrophoresis. Neither the Lowry (Lowry et al., 1951) nor the Bradford (Bradford, 1976) method can be used reliably due to the low protein concentration of the material and the existence of some interfering substances in the sample eluted from the gel. Furthermore, the region where activity resides on an SDA gel is neither stained with Coomassie Brilliant Blue or silver, which can detect at least 10 ng of protein, nor iodinated with the Chloramine T, Iodo-gen or lactoperoxidase-glucose oxidase methods, which label tyrosine residues. However, it is heavily iodinated with the Bolton-Hunter reagent (Bolton & Hunter, 1973), which preferentially labels lysine residues under the conditions used. The amount of protein recovered for biological assay can then be estimated, assuming that the proteins are uniformly labeled, or using another known protein (ovalbumin) labeled in the same way as a standard (Fukada, unpublished results).

The activity resides in a slightly basic molecule based on the behaviour on ion-exchange columns. The active molecule contains protein, since it is extremely sensitive to trypsin. However, this activity is resistant to treatment with urea or SDS, although β-mercaptoethanol in the presence of SDS completely destroys the activity. Activity is recovered as a single peak of M.W. of about 45,000 daltons, both by Sephadex column chromatography (native conditions) and by SDS-polyacrylamide gel electrophoresis (denatured conditions). These results suggest that the factor may be composed of a single subunit, although it is possible that two or more subunits are connected by disulfide bonds. Moreover, these results rule out the possibility that the activity derives from a smaller component bound to the 45,000 dalton protein. Even at this purification (~10^5-fold), the two activities (that is, induction of ACh synthesis and suppression of CA synthesis) are not separable. The concentration range where this factor is active in culture is estimated to be approximately 10 ng/ml, indicating that the activity acts in a hormonal concentration range (Fukada, 1983 & unpublished results).

In this article I have described a developmental factor which may represent a mechanism by which non-neuronal cells in the local environment can influence neuronal development. I have reported on the current status of the purification of this factor, pointing out the importance of obtaining a purified factor for further studies. Progress has been slow due to the lack of a rich source of the factor, and the length of the biological assay. In order to further purify the factor, an alternative approach of raising monoclonal antibodies against it is in progress. In addition, recent advances in protein microsequencing and gene-cloning using synthetic oligonuclotide probes based on the microsequencing data

382

may be helpful in obtaining the pure factor in much larger quantities than are
presently available. This in turn should allow us to carry out the biochemical
and physiological studies necessary for understanding the role of this factor in
normal development and its mechanism of action.

REFERENCES

Bolton, AE & Hunter WM (1973) Biochem J 133: 529-539

Bottenstein, J, Hayashi, I, Hutchings, S, Masui, H, Mather, J, McClure, DB,
    Ohasa, S, Rizzino, A, Sato, G, Serrero, G, Wolfe, R, and Wu, R (1979) In:
    Jacoby, WB and Pastan, IH (eds) Methods in Enzymology. Academic Press, New
    York 45: 94-109

Bradford M (1976) Anal Biochem 72: 248-254

Braun SJ, Sweadner KJ & Patterson PH (1981) J Neurosci 1: 1397-1406

Brookes N, Burt DR, Goldberg AM & Bierkamper GG (1980) Brain Res 186: 474-479

Bunge R, Johnson M & Ross CD (1978) Science 199: 1409-1416

Chun LLY, Patterson PH & Cantor H (1980) J Exp Biol 89: 73-83

Claude P (1973) J Cell Biol 59: 57a

Cochard P, Goldstein M & Black IB (1978) Proc Natl Acad Sci USA 75: 2986-2990

Edgar D, Barde YA & Thoenen H (1981) Nature 289: 294-295

Fukada K (1980) Nature 287: 553-555

Fukada K (1983) 13th Ann Soc Neurosci Abstr. 9: 182.5, p 614

Giess MC & Weber MJ (1984) J Neurosci 4: 1442-1452

Giller EL, Neale JH, Bullock PN, Schrier BK & Nelson PG (1977) J Cell Biol 74:
    16-29

Godfrey EW, Schrier BK & Nelson PG (1980) Devl Biol 77: 403-418

Harper GP & Thoenen H (1980) J Neurochem 34: 5-16

Hawrot E (1980) Devl Biol 74: 136-151

Johnson M, Ross D, Meyers M, Rees R, Bunge R, Wakshull E & Burton H (1976)
    Nature 262: 308-310

Johnson MI, Ross CD, Meyers M, Spitznagel EL & Bunge RP (1980) J Cell Biol 84:
    680-691

Jonakait GM, Wolf J, Cochard P, Goldstein M & Black IB (1979) Proc Natl Acad Sci
    USA 76: 4683-4686

Kessler JA, Adler JE, Jonakait GM & Black IB (1984) Devl Biol 103: 71-79

Kessler JA, Brain Res (in press)

Landis SC (1976) Proc Natl Acad Sci USA 73: 4220-4224

Landis SC, MacLeish PR, Potter DD, Furshpan EJ & Patterson PH (1976)   Sixth Ann
    Soc Neurosci Abstr 2: 280, p. 197

Landis SC (1980) Devl Biol 77: 349-361

Landis SC (1981) In: Garrod D & Feldman JD, (eds.) Development of the Nervous
    System. Cambridge Univ. Press, pp 147-160

Landis SC & Keefe D (1983) Devl Biol 98: 349-372

Landis SC (1983) Federation Proc 42: 1633-1638

Le Douarin NM, Teillet M, Ziller C & Smith J (1978) Proc Natl Acad Sci USA 75:
    2030-2034

Le Douarin NM (1980) Nature, Lond 286: 663-669

Lowry O, Rosebrough N, Farr A & Randall R (1951) J Biol Chem 193: 265-275

Mains RE & Patterson PH (1973 a) J Cell Biol 59: 329-345

Mains RE & Patterson PH (1973 b) J Cell Biol 59: 346-360

Mains Re & Patterson PH (1973 c) J Cell Biol 59: 361-366

Nishi R & Berg DK (1981) J Neurosci 1: 505-513

Nurse CA &O'Lague PH (1975) Proc Natl Acad Sci USA 72: 1955-1959

O'Lague P, Obata K, Claude P, Furshpan EJ & Potter DD (1974) Proc Natl Acad Sci
    USA 71: 3602-3606

O'Lague P, Potter P & Furshpan E (1978) Devl Biol 67: 384-403

Patterson PH & Chun LLY (1974) Proc Natl Acad Sci USA 71: 3607-3610

Patterson PH & Chun LLY (1977) Devl Biol 56: 263-280

Patterson PH (1978) Ann Rev Neurosci 1: 1-17

Pintar JE, Maxwell GD, Sweadner KJ, Patterson PH & Breakfield XO (1981) 11th Ann
    Soc Neurosci Abstr 7: 273.9, p. 848

Potter DD, Landis SC & Furshpan EJ (1981) Ciba Found Symp 83: 123-138

Rees R & Bunge RP (1974) J Comp Neurol 157: 1-11

Reichardt LF & Patterson PH (1977) Nature 270: 147-151

Schwab M & Landis S (1981) Devl Biol 84: 67-78

Smith RG & Appel SH (1983) Science 219: 1079-1081

Sweadner K (1981) J Biol Chem 256: 4063-4070

Swerts JP, Le Van Thai A, Vigny A & Weber M (1983) Devl Biol 100: 1-11

384

Swerts JP, Giess MC, Mathieu C, Sauron ME, Le Van Thai A & Weber MJ (1984) In: Duprat AM, Kato AC & Weber MJ (eds) The Role of Cell Interaction in Early Neurogenesis, NATO Advanced Institutes Series. Plenum Publishing Co, London, pp. 335-344.

Teitelman G, Joh TH & Reis DJ (1978) Brain Res 158: 229-234

Walicke PA, Campenot RB & Patterson PH (1977) Proc Natl Acad Sci USA 74: 5767-5771

Walicke PA & Patterson PH (1981) J Neurosci 1: 343-350

Weber MJ (1981) J Biol Chem 256: 3447-3453

Wolinsky E & Patterson PH (1983) J Neurosci 3: 1495:1500

Wolinsky EJ & Patterson PH, J Neurosci, in press

Zurn AD (1982) Devl Biol 94: 483-498

RESUME

Des études, in vivo et in vitro, ont toutes deux montré que des neurones individuels, originaires de la crête neurale, ont la capacité de devenir soit cholinergiques soit noradrénergiques en fonction de l'environnement dans lequel ils se développent. Des études réalisées sur des cultures ont révélé l'existence de facteur(s) diffusible(s) provenant de certains types de cellules non--neuronales, qui peuvent induire une différenciation cholinergique dans des neurones sympathiques de rat se développant en culture, sans affecter leur survie ou leur croissance. Des études récentes sur des neurones sympathiques choliner-giques innervant des glandes sudoripares de pattes de rat ont montré que la transition de l'état noradrénergique à l'état cholinergique, démontré dans des études en culture, a lieu réellement in vivo en tant que part du développement normal. Afin de comprendre le rôle de ce facteur dans le développement normal et son mécanisme d'action, il est essentiel de posséder un facteur purifié. Utilisant des techniques biochimiques conventionnelles, le facteur a été purifié environ 10^5 fois, avec une récupération raisonnable, à partir de milieu de culture, sans sérum, additionné d'hormones et conditionné par des cellules cardiaques. L'activité la plus purifiée est une protéine légèrement basique d'environ 45000 daltons, qui, pense-t-on, existe sous la forme d'une seule sous-unité. La gamme de concentration à laquelle ce facteur est actif en culture est approximativement 10 ng/ml. Cette préparation possède toujours la capacité d'à la fois inhiber le développement de caractéristiques noradrénergiques et d'induire une différenciation cholinergique dans des neurones sympathiques de rats, en culture.

Hormones and Cell Regulation, Volume 9
INSERM European Symposium
J.E. Dumont, B. Hamprecht and J. Nunez editors
© 1985 Elsevier Science Publishers B.V. (Biomedical Division)

NEUROTROPHIC FACTORS

Y.-A. BARDE, and H. THOENEN

Dept. of Neurochemistry, Max-Planck Institute for Psychiatry,
D-8033 Martinsried b. Munich, FRG.

Embryonic neurons, when isolated from their normal
environment and brought into culture, will not survive
unless the culture medium (with or without serum) is supplemented with tissue extracts or medium conditioned by various
cell types (for reviews, see 1-3). Much like fibroblasts or
smooth muscle cells require specific mitogenic factors to
divide in culture, embryonic neurons require specific
factors if they are to survive in culture. Such factors can
readily be found in a variety of tissue extracts and conditioned media, and it is generally assumed that they are
present in the environment of the neurons in vivo and are
required for normal development. It has been shown in a
number of cases that these factors are proteins since they
usually are non-dialyzable, heat- and protease-sensitive.

One of the major questions remaining at this stage is
the exact role played by such factors in vivo, during
normal development. As in many other cases in the past, such
as with neurotransmitters, hypothalamic relasing factors,
protein mitogens or interleukins, it is clear that a necessary first step in the establishment of the exact role
played by these factors acting on neurons is their purification and characterization. The purification of neurotrophic factors presents two major difficulties: the first
is that in order to monitor the purification, a biological
assay is needed that determines the survival of cultured
embryonic neurons. 24 hours is usually a bare minimum of time
to decide whether or not a sample can promote the survival
of neurons. As biological activity must be measured, then
the sample must of course retain its biological activity.
Since in most cases it seems that the activity is carried by
a protein, then efficient purification techniques like HPLC

can often not be employed because of the use of solvents
that usually denature proteins to the extent that their
biological activity is lost.

The second major difficulty is linked to the large enrichment required to obtain material of reasonable purity.
Indeed, although it is quite easy to detect a biological
activity in raw extracts, the specific activity of such
extracts is usually very low. This is in contrast with the
very high specific activity of the few (three so far)
neurotrophic factors with survival activity that have been
purified (see below). At present only one factor has been
purified from a source with a low specific activity.

One of the major reasons why such projects are pursued
at all in spite of their inherent difficulties is that there
already exists one (and so far only one) example where it
could be established that one defined neurotrophic factor
does indeed play a crucial role during the normal development of the nervous system. This factor is nerve growth
factor (NGF). For reasons that are entirely unclear, the
submandibular gland of the adult male mouse contains large
amounts of NGF. The remarkable contribution of Rita Levi-
Montalcini and Stanley Cohen was not only to purify this
protein using a bioassay (more than 20 years ago), but also
to show that antibodies to NGF could specifically destroy
parts of the peripheral nervous system; those neurons that
specifically require NGF at a certain stage of development
die when NGF is made inaccessible to them after blockade of
its biological activity with antibodies (for reviews on the
physiological aspects of this molecule see 4 and 5). The
availability of pure material has also allowed the demonstration that injection of NGF would rescue neurons that are
normally eliminated during development (6,7). Thus NGF
represents an example of a well-characterized molecule (to
the extent that the mRNA translating this protein has been
sequenced, 8,9) playing a decisive role in the normal
development of specific parts of the nervous system. Well-
established targets for NGF are the neurons of the peripheral sympathetic system as well as those of the dorsal
root ganglia. Whether or not NGF plays a role in the normal
development of parts of the central nervous system is

unclear at the moment. When labeled NGF is injected into the
rat hippocampus, cholinergic cell bodies (projecting to the
hippocampus) are labeled in the medial septal nucleus and in
the nucleus of the diagonal band of Broca (10). This finding
indicates that NGF is retrogradely transported from the site
of injection, and this is a strong argument for the presence
of specific NGF receptors on these neurons. Furthermore, it
has been shown in vitro and in vivo that exposure of these
neurons and their projections to NGF leads to an increase in
the specific activity of choline acetyltransferase (11,12).
From these data, it is tempting to speculate that NGF does
play a role in the development of at least some cholinergic
neurons in the brain. However, NGF-antibodies injected into
the ventricles of newborn rats failed to prevent the increase
of choline acetyltransferase activity seen in these neurons
after birth (12). Whether or not these antibodies really get
access to the sites of production of NGF (which have not been
established yet) is not entirely clear, so that the role of
NGF in the brain during normal development is still a matter
of speculation.

The very fact that NGF has been shown to act selectively
on the peripheral sympathetic and sensory nervous system
naturally leads to the question whether or not other factors
exist that would play a role similar to that of NGF, but act
on different types of neurons.

As mentioned above, a number of in vitro experiments
do indeed suggest that there exists other proteins, different
from NGF, that are also able to keep alive embryonic neurons.
To date only two such proteins have been purified sufficiently to be reasonably certain that their action is due to
one molecule, as opposed to that of a mixture of proteins.
Both have been purified using survival and fiber outgrowth
of neurons in culture as an assay system. One is referred to
as CNTF (for ciliary neurotrophic factor) and has recently
been purified from chick eye tissue (13) using the survival
of embryonic chick ciliary neurons as bioassay. Dissection
of the eye tissues to which the ciliary neurons project has
revealed that homogenates of these structures have a high
specific activity (about 40 ng/ml protein being necessary to

388

keep alive about half the number of neurons plated for 24
h). This is the reason why a purification factor of only 400
fold (quite comparable to that needed to obtain pure NGF
from the male mouse submandibular gland) was needed to
obtain a protein that appears as one band on SDS-gel electrophoresis. Its molecular weight is 20,400 and its isoelectric
point 5.0. Its specific activity is very high: only 0.1
ng/ml is necessary to see an effect on ciliary neurons.
Perhaps somewhat surprising is the fact that this protein
not only keeps alive ciliary neurons but also sympathetic
and sensory neurons from the chick embryo. In these experiments, survival has been measured after 24 hr and it will be
interesting to see in future experiments if long term
survival is also promoted.

Aside from NGF and CNTF, only one other protein with
neuronal survival-promoting activity has been purified (14).
The source was the brain of adult pigs. This protein, presently called BDNF (for brain-derived neurotrophic factor)
has a molecular weight of approximately 12,000 and is very
basic (pI about 10). Like CNTF it appears as one band after
SDS gel electrophoresis and indeed, as for the purification
of CNTF, the last step is a preparative SDS gel electrophoresis. The activity of BDNF cannot fully be renatured
after treatment with SDS. In spite of this fact, about 15
ng/ml is enough to keep alive about 50% of the neurons
isolated from the dorsal root ganglia of the chick embryo on
a laminin substrate (Barde, unpublished results). Increasing
the concentration of the factor does not allow more neurons
to be rescued, but interestingly, the addition of NGF (at 1-
5 ng/ml), which on its own also promotes the survival of
about 50% of the neurons, allows 80-100% of the neurons
plated to be rescued, for prolonged period of time (more
than a week). Why these neurons should require two survival
factors if all of them are to survive in culture is intriguing. We hypothesize that this observation is related to
the fact that sensory neurons have a double projection: one
to the periphery, which is the one that normally sees NGF
(see 15) and one to the central nervous system, the source
of BDNF. This hypothesis is also based on the fact that

neurons not having projections of this kind such as those of the ciliary and sympathetic ganglia do not show any survival response with BDNF. On the other hand, neurons isolated from the nodose ganglia of the chick, which are in contact with CNS tissue show the same survival response as those from the dorsal root ganglia. (Lindsay and Barde, unpublished results). These neurons are not supported with NGF.

In contrast to their distinct target specificities, there are intriguing similarities between NGF and BDNF. Indeed, BDNF has essentially the same molecular weight as the monomer of NGF and both are very basic molecules. So far we failed to detect any cross-reactivity with antibodies to NGF with BDNF. Sequencing of BDNF and comparison with the known sequence of NGF will decide whether or not these two molecules belong to the same family. In this context, it should be mentioned that Ullrich et al. (9) found only a single NGF gene per haploid genome in mouse and human DNA, using high stringency hybridization conditions.

So far, nothing is known on the role played by CNTF or BDNF in vivo. Antibodies blocking the activity of these molecules are not yet available nor are the amounts of these proteins sufficient to do the necessary in vivo work.

Further work with these newly purified molecules and those that will be purified in the future should lead to a better understanding of how the nervous system is put together.

REFERENCES

1. Varon S, Adler R (1981) Adv Cell Neurobiol 2:115-163
2. Barde Y-A, Edgar D, Thoenen H (1983) Ann Rev Physiol 45:601-612
3. Berg D (1984) Ann Rev Neurosci 7:149-170
4. Levi-Montalcini R, Angeletti PU (1968) Physiol Rev 48:534-569
5. Thoenen H, Barde Y-A (1980) Physiol Rev 60:1284-1335
6. Hendry IA, Campbell J (1976) J Neurocytol 5:351-360
7. Hamburger V, Brunso-Bechtold JK, Yip JW (1981) J Neurosci 1:60-71

8. Scott J, Selby M, Urdea M, Quiroga GM, Bell GI, Rutter WJ
    (1983) Nature 302:538-541

9. Ullrich A, Gray A, Berman C, Dull TJ (1983) Nature
    303:821-825

10. Schwab ME, Otten U, Agid Y, Thoenen H (1979) Brain Res
    168:473-483

11. Honegger P, Lenoir D (1982) Dev Brain Res 3:229-239

12. Gnahn H, Hefti F, Heumann R, Schwab ME, Thoenen H
    (1983) Dev Brain Res 9:45-52

13. Barbin G, Manthorpe M, Varon S (1984) J Neurochem
    (in press)

14. Barde Y-A, Edgar D, Thoenen H (1982) EMBO J 1:549-553

15. Korsching S, Thoenen H (1983) Proc Natl Acad Sci USA 80:
    3513-3517

Résumé

    Cette brève revue résume l'état actuel de nos connais-
sances concernant les facteurs neurotrophiques. A part NGF,
seules 2 protéines ont été purifiées, l'une à partir de
certains tissus de l'oeil du poulet (CNTF) et l'autre à partir de cerveaux de porc (BDNF). Quelques problèmes propres
à la purification de ce genre de protéines sont discutés,
ainsi que la signification possible de ces protéines durant le
développement du système nerveux. Le modèle proposé s'inspire
de celui établi à partir des connaissances acquises avec NGF,
un des rares facteurs en biologie du développement, et le
seul en neurobiologie, dont le rôle physiologique est établi.

Hormones and Cell Regulation, Volume 9
INSERM European Symposium
J.E. Dumont, B. Hamprecht and J. Nunez editors
© 1985 Elsevier Science Publishers B.V. (Biomedical Division)

GLIA-INDUCED NEURITE EXTENSION AS A MODEL TO DETECT SOME BIO-
CHEMICAL INTERACTIONS BETWEEN GLIA AND NEURONS

DENIS MONARD, KAZUO MURATO, and JOACHIM GUENTHER
Friedrich Miescher-Institut, P.O. Box 2543, CH-4002 Basel,
Switzerland

Cultured mouse neuroblastoma cells have been considered as a
decent experimental model to approach the biochemical mechanisms
involved in neurite outgrowth (1). Most of the substances known to
promote neurite extension also strongly interfere with the growth
rate of those cells (2,3). A very efficient neurite outgrowth without noticeable cell growth inhibition can be promoted by incubating
the neuroblastoma cells in a medium conditioned by rat glioma cells
(4). This neurite-promoting activity is also found in the medium
conditioned by rat brain primary cultures (5). The developmental
stage at which these cultures are initiated is crucial for the production of neurite-promoting activity. Only dissociated brain cells
from rats which are at least 3 to 5 days old will release neurite-
promoting activity in the medium (5). This critical stage for an
adequate production of neurite-promoting activity coincides with
the burst of glial cell multiplication. These results represent an
experimental evidence which indicates that glial cells are able to
influence neuronal cells at the time neuronal migration ceases and
neurite outgrowth increases.

Different culture systems provide evidences that glial cells
can support neuronal survival and neurite outgrowth (6,7). Different types of glial cells could release different types of factors
which could each be specific for different types of neuronal cells
or for some of the cellular stages leading to a fully differentiated neuron. The biochemical identification and characterization
of these biologically active substances produced by glial cells is
required to approach the nature and the regulation of such glia-
neuron interactions.

The biochemical purification of the heat-sensitive, high molecular weight glia-derived factor which induces neurite extension in
neuroblastoma cells has therefore represented a key-experimental
step in our approach to understand this type of interaction between
glial and neuronal cells. Two facts have been of critical importance for the achievement of this goal. 1) The possibility to grow

and to maintain C6 rat glioma cells on microcarrier beads in serum-free medium for several days has finally allowed to produce the amount of biological activity which is required for a purification procedure. 2) The finding that serum-free medium conditioned by glioma cells contains a potent inhibitor of serine proteases as urokinase or plasminogen activator has represented another key in-formation (8). The neurite-promoting activity present in semi-pu-rified preparations is adsorbed irreversibly on urokinase-Sepharose. The use of PMSF-modified urokinase-Sepharose allows to adsorb the biological activity and to recover part of it as a single protein band upon elution with sodium chloride (Fig. 1).

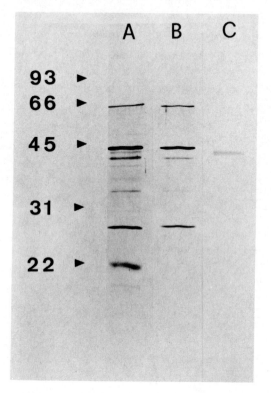

Fig. 1. Phenylmethylsulfonyl fluoride (PMSF)-urokinase affinity chromatography.

PMSF-urokinase Sepharose was prepared by coupling 10,000 Plough units human urokinase (Leo Pharmaceuticals, Denmark) to 0.7 g CH-Sepharose (Pharmacia, Sweden). The urokinase-Sepharose beads were suspended in 50 ml 50 mM Tris-HCl buffer at pH 8.0 (Tris buffer) containing 2 mM PMSF on a rocking shaker overnight at 4°C, then washed extensively with Tris buffer. Six liters of serum-free

C6 glioma-conditioned medium were concentrated 20-fold and dialysed against Tris buffer in an Amicon DC-2 hollow fiber system
equipped with a HIx50 cartridge. The concentrated conditioned
medium was the loaded on a Bio-Rad AG IX2 anion exchange column
which removed most of the proteins without adsorbing the neurite-
promoting activity. The active wash-through was loaded on an Affi-
Gel Blue (Bio-Rad, USA) column (1x12 cm). The column was rinsed
with 50 ml 0.4 M NaCl in Tris buffer, 25 ml of 1 mM adenosine-5'-
triphosphate (ATP), 1 mM reduced nicotinamide-adenine dinucleotide
(NADH) in Tris buffer, 60 ml of 0.6 M NaCl in the same buffer.
Finally, glial factor activity was eluted using 1.5 M NaCl in Tris
buffer. Following dialysis, the neurite-promoting activity was adsorbed batchwise on PMSF-urokinase-Sepharose at room temperature
for 1 hour. The beads were packed into a small column (1 cm ∅)
which was rinsed at a speed of 23 ml/h with 10 ml Tris buffer,
followed by 10 ml 20 mM NaCl in Tris buffer at 4°C. Forty mM NaCl
in Tris buffer was then required to elute simultaneously neurite-
promoting activity and urokinase inhibitory activity from such an
affinity chromatography column. The active material eluted from
Affi-Gel Blue (A), the wash through PMSF-urokinase-Sepharose (B)
and the active material eluted with 40 mM NaCl from PMSF-urokinase-
Sepharose (C) have been analysed using silver stained (9) SDS-
polyacrylamide gel electrophoresis (10). The low molecular weight
protein standards (95 Kd, 66 Kd, 45 Kd, 31 Kd, 22 Kd) from Bio-
Rad were used to calibrate the SDS-polyacrylamide gel. The results
illustrate that the 43 Kd band removed by PMSF-urokinase-Sepharose
is the only protein band found in the 40 mM NaCl eluate which contains both neurite-promoting activity and urokinase inhibitory
activity. The low recovery of this procedure does not allow to use
it for preparative purposes. This knowledge has however led to a
purification procedure which allows to purify this glia-derived
factor to homogeneity (11). The purified 43 Kd protein has both
neurite-promoting activity and urokinase (respectively plasminogen
activator) inhibitory activity. It does also form SDS-resistant
complexes with these serine proteases (11). Our present estimation
indicates that the neurite-promoting activity of this protein can
be detected at about 0.5 ng/ml, or in other terms in the range of
10^{-11} M.

The fact that the glia-derived neurite-promoting factor is a
protease inhibitor has motivated a study on the effect of differ-

ent protease inhibitors on neurite outgrowth. Only two known protease inhibitors have turned out to be as potent as the purified
glial factor. Hirudin, a thrombin inhibitor isolated from the
leech, promotes neurite outgrowth already at $5x10^{-11}M$ (8) and the
synthetic peptide D-Phe-Pro-ArgCH$_2$Cl has the same effect at the
same concentration (in preparation). The fact that many other
serine protease inhibitors tested show no or only marginal neurite-
promoting activity, even at high concentrations (8), argues for
some specific interactions between the cell-associated protease(s)
and the biologically active inhibitors. This neurite-promoting
activity of hirudin and of the synthetic peptide does further support that the biological activity of the glioma-derived factor is
due to its protease inhibitor nature. If the factor acts by interfering with a proteolytic activity associated with the neuronal
cell surface, the identification and characterisation of the proteases associated with the neuronal membrane and of their respective sensitivity to the factor is required for a further understanding of the phenomenon. Diisopropylfluorophosphate (DFP) binds
covalently to the active site of serine proteases. Tritiated diisopropylfluorophosphate (^{3}H-DFP) will therefore irreversibly label
the serine proteases present in cell membranes. The labeled proteases can then be revealed by SDS-polyacrylamide gel electrophoresis (SDS-PAGE), followed by autoradiography. This methodology
has revealed that neuroblastoma membranes contain at least six
serine proteases from distinct molecular weight (30 Kd, 32 Kd,
33 Kd, 38 Kd, 40 Kd and 44 Kd) (12). Neuroblastoma membrane proteins have also been fractionated by SDS-polyacrylamide gel electrophoresis. The fractions obtained have been tested for plasminogen activator (PA) activity using a very sensitive caseinolysis
assay. This procedure allows to detect two PA at 40 Kd and 59 Kd.
The activity measured at these molecular weights is plasminogen-
dependent. The labeling experiments indicate that the 40 Kd PA
form is revealed, as one would expect, by ^{3}H-DFP. The 59 Kd PA
form does not bind ^{3}H-DFP. This could mean that, when integrated
in the membrane, this PA form is not accessible to ^{3}H-DFP and that
its active site is only unmasked upon fractionation by SDS-PAGE.
Neuronal cells of mouse cerebellum can be cultured in serum-free
medium (13). Two PA of similar molecular weight (43 Kd and 60 Kd)
can be found associated with these neuronal cells, indicating
that PA is not only found in the neuronal cells of tumoral origin.

All these different forms of neuronal PAs are inhibited by the
glioma-derived neurite promoting factor. One could, at this stage,
tentatively conclude that the isolated neurite promoting factor
is acting by inhibiting the PAs associated with the neuroblastoma
cell membranes. However, the situation does not seem to be as
simple. The synthetic peptide D-Phe-Pro-ArgCH_2Cl mentioned above
does also induce neurite outgrowth at extremely low concentrations
($5x10^{-11}$M). This chloromethyl ketone derivative inhibits serine
proteases (preferentially thrombin or thrombin-like enzymes) by
binding covalently to the active site (14). If, in the membrane
preparation, an incubation with the cold tripeptide precedes the
incubation with ^{3}H-DFP, the active site of the protease reacting
with the tripeptide will become inaccessible to the ^{3}H-DFP and the
corresponding band will disappear from the autoradiogram. Such
experiments have indicated that this is the case for the 31 Kd
band but not for the band corresponding to the PA activity (12).

These experiments indicate that the tripeptide does not react
with the PA band labeled by ^{3}H-DFP. This is confirmed by the fact
that only high peptide concentrations which are irrelevant for
our phenomenon can marginally inhibit PA activity. We do not yet
know which protease is inhibited by hirudin. We know that the
glioma-derived neurite-promoting factor inhibits PA activity but
it is possible that it could also interfere with the protease
sensitive to the tripeptide or with one or some of the other proteases which are labeled by ^{3}H-DFP. Experiments in progress are
aiming an answer to these questions. The present results nevertheless already indicate that the inhibition of different and
distinct serine proteases associated with the neuroblastoma cell
membrane can lead to neurite outgrowth. Obviously, the inhibition
of cell surface associated proteolytic activity does not need to
be an obligatory step required for neuritic growth. In fact, many
of the different agents which induce neurite outgrowth in neuroblastoma cells (1,2,3) could act by ways which do not imply the
inhibition of proteolytic activity. For example, the neurite
outgrowth induced by an increase in cyclic adenosine monophos-
phate (cyclic AMP) could mainly be triggered by influencing the
organisation of the cytoskeleton, possibly the polymerisation of
the microtubules. Thus, the driving force would be of intracellular origin. In fact, dibutyl cyclic AMP treatment causes neurite
outgrowth and concomitantly an increase in the PA activity asso-

ciated with the cell surface (15). This could indicate that, in
this situation, cell surface modifications, such as the expression
of PA activity, are resulting from intracellular initial triggers,
one of which causes also neurite outgrowth.

Our observation that an inhibition of the proteolytic activity
associated with the cell surface can also be an initial trigger
leading to neurite outgrowth is nevertheless biologically attractive. The role of cell surface proteolytic activity for the cell
migration which takes place in many developmental systems has been
recognized (16,17). Migrating cells have more cell surface proteolytic activity than stationary differentiated cells (18,19). In
the nervous system, an increase in plasminogen activator production by cerebellar cultured cells dissociated at the time of active
granule cell migration has been reported (20). The degradation of
elements of the extracellular matrix seems to take place during
the migration of the neural crest cells (21). Neuroblastoma can
also be considered as migrating cells. In fact, such migrating
neuronal cells (granule cells, neural crest cells, neuroblastoma
cells) should be equipped with an increased cell surface proteolytic activity which would be required for the migration. Interference with this cell surface proteolytic activity would modulate,
even stop, the extent of the migration and promote cell surface
modifications which would be compatible with neurite outgrowth.

Others have presented experimental evidence that, for example
with mouse melanoma cells, certain cell surface proteins are necessary as adhesion or guidance forces for cellular migration (22).
These authors postulate that in such cases an increase in cell
surface proteolytic activity causes degradation of these surface-
associated proteins required for migration and therefore inhibit
the cellular movement. Conversely, inhibition of proteolytic
activity would cause an increase of such migration promoting proteins and therefore promote cellular migration. It is therefore
not yet excluded that, in such a situation, an inhibitor as the
glioma-derived neurite-promoting factor would rather promote cellular migration.

There is actually no evidence that such a type of regulation
plays a role during the migration of neuronal cells. In fact,
arguments against this possibility have been presented in the
studies related to the importance and function of neuronal cellular adhesion molecules (N-CAM and LI) (23,24). These surface

proteins are required for the aggregation of neuronal cells.
Recent studies have suggested that less N-CAM is available at the
cell surface at the time of migration, but that it does increase
again at the stage neuronal cells cease to migrate and organize
themselves by aggregating before sending out neurite (25). Antibodies against LI interfere with the migration of granule cells
in cultured cerebellum slices (26). In such cases it would be
of interest to consider if a proteolytic activity is responsible
for the degradation of N-CAM, respectively LI, during migration
and if the glioma-derived inhibitor would interact with such a
protease, therefore being able to decrease the turnover of N-CAM
or LI antigens.

In secreting protease inhibitory proteins as the one we have
purified, the glial cells would play an important regulatory
function for the behavior of the neuroblast cell.

REFERENCES

1. Haffke SC, Seeds NW (1975) Life Sci 16:1649-1658

2. Schubert D, Jacob F (1970) Proc Natl Acad Sci USA 67:257-254

3. Prasad KN, Hsie AW (1971) Nature New Biol 233:141-142

4. Monard D, Solomon F, Rentsch M, Gysin R (1973) Proc Natl Acad
    Sci USA 70:1894-1897

5. Schürch-Rathgeb Y, Monard D (1978) Nature 273:308-309

6. Barde YA, Lindsay RM, Monard D, Thoenen H (1978) Nature 274:818

7. Varon SA, Adler R (1981) In: Fedoroff S, Herz L (eds) Advances
    in Cellular Neurobiology 2. Academic Press, New York, pp 115-163

8. Monard D, Niday E, Limat A, Solomon F (1983) Progress in Brain
    Research 58:359-364

9. Eschenbruch M, Bürk RR (1982) Anal Biochem 125:96-99

10. Laemmli UK (1970) Nature 227:680-685

11. Günther J, Monard D, in preparation

12. Murato K, Monard D, in preparation

13. Huck S (1983) Brain Research Bulletin 10:667-674

14. Kettner C, Shaw E (1979) Throm Res 14:969-973

15. Soreq H, Miskin R, Zutra A, Littauer UZ (1983) Developmental
    Brain Research 7:257-269

16. Sherman MI, Strickland S, Reich E (1976) Cancer Res 36:4208-
    4216

17. Topp W, Hall JD, Marsden M, Teresky AK, Rifkin D, Levine AJ,
    Pollack R (1976) Cancer Res 36:4217-4223

18. Unkeless JC, Tobia A, Ossowski L, Quigley JP, Rifkin DB, Reich E (1973) J exp Med 137:85-111

19. Ossowski L, Quigley JP, Reich E (1974) J biol Chem 249:4312-4320

20. Krystosek A, Seeds NW (1981) Proc Natl Acad Sci USA 78:7810-7814

21. Thiéry JP, Duband JL, Delouvée A (1982) Develop Biol 93:324-343

22. Rieber M, Rieber MS (1981) Nature 293:74-76

23. Rutishauser U (1983) Cold Spring Harb Symp quant Biol 48:501-514

24. Schachner M, Faissner A, Kruse J, Lindner J, Meier DH, Rathjen FG, Wernecke H (1983) Cold Spring Harb Symp quant Biol 48:557-568

25. Thiéry JP, Duband JL, Rutishauser U, Edelman GM (1982) Proc Natl Acad Sci USA 79:6737-6741

26. Lindner J, Rathjen FG, Schachner M (1983) Nature 305:427-430

RESUME

Le milieu conditionné par des cellules de gliome de rat induit une élongation neuritique chez les cellules de neuroblastome de souris. Cette activité est due à une protéine d'un poids moléculaire de 43 000 daltons qui a été purifiée jusqu'à homogénéité. Cette protéine s'avère également être un inhibiteur très potent de protéases de type sérine tels que l'activateur du plasminogène ou l'urokinase. En fait, le complexe formé entre ces protéases et l'inhibiteur isolé est stable et résistant au dodecylsulfate de sodium.

Plusieurs protéases de type sérine peuvent être détectées dans les membranes de cellules de neuroblastome à l'aide de diisopropyl-fluorophosphate tritié. Parmi celles-ci, deux activités similaires à l'activateur du plasminogène sont décelables à 40 000 et 59 000 daltons. Ces deux activités sont inhibées par la protéine purifiée du milieu conditionné par les cellules gliales. Deux enzymes ayant les mêmes caractéristiques sont également présentes dans les cultures primaires de cellules neuronales dérivées de cervelet néonatal.

Une activité protéolytique accrue étant généralement mesurable à la surface de cellules en phase de migration, ces résultats suggèrent que les cellules gliales peuvent, par la sécrétion d'inhibiteurs de protéases spécifiques, moduler la migration des neuroblastes et ainsi créer des conditions favorables à l'élongation neuritique.

SELECTED ABSTRACTS OF POSTERS

SELECTION DES RESUMES DES POSTERS

Hormones and Cell Regulation, Volume 9
INSERM European Symposium
J.E. Dumont, B. Hamprecht and J. Nunez editors
© 1985 Elsevier Science Publishers B.V. (Biomedical Division)

VISUALIZATION OF EPIDERMAL GROWTH FACTOR-RECEPTOR BY CRYO-ULTRAMICROTOMY:
TWO CLONAL VARIANTS OF A431 CELLS DIFFERING IN EGF-R CHARACTERISTICS

J. Boonstra[*], L.H.K. Defize[o], P. van Maurik[o], P.T. van der Saag[o], S.W. de Laat[o], J. Leunissen[+] and A.J. Verkleij[+]

[*]Dept. of Molec. Cell Biol. and [+]Institute of Molecular Biology, University of Utrecht, Padualaan 8, 3584 CH Utrecht and [o]Hubrecht Laboratory, Uppsalalaan 8, 3584 CT Utrecht, The Netherlands.

Cryo-ultramicrotomy combined with immunogold-labeling provides a method to visualize extra- and intracelullar located antigens. Intracellular labeling being achieved with a high resolution, independently of the internalization process. We have applied this method to localize EGF-R in A431 human epidermoid carcinoma cells. Thin sections of the cells were labeled with anti EGF-R antibody, rabbit anti mouse antibody and protein A-gold complex. Dense labeling was observed on the plasma membrane and coated pits.

Intracellular, labeling was observed of coated vesicles, multivesicular bodies, lysozomes and a network of intracellular membranes. In these studies 3 different EGF-R monoclonal antibodies were used of which one (2E9) is directed against a peptide determinant at the EGF binding domain and the others are directed against external sugar moieties of the EGF-R (2D11 and 2G5). Immunofluorescence and electron microscopic studies using cryoultramicrotomy and immunogold-labeling demonstrated that 2E9 recognizes the EGF-R in all A431 cells in culture, whereas two distinct cell populations are revealed after labeling with the sugar-recognizing antibodies: 60% positive and 40% negative cells.

The two cell types were cloned (by D.J. Arndt-Jovin and T.H. Jovin) and found to have a similar morphology, growth rate and EGF-R number. A large difference was observed in basal- and EGF-induced receptor phosphorylation. These results indicate that sugars play a role in the maintenance of the spatial configuration of the receptor, thereby influencing the activity of the receptor tyrosine kinase activity.

Hormones and Cell Regulation, Volume 9
INSERM European Symposium
J.E. Dumont, B. Hamprecht and J. Nunez editors
© 1985 Elsevier Science Publishers B.V. (Biomedical Division)

HORMONE-STIMULATED Na-K PUMP IN ISOLATED RAT LIVER CELLS

T. CAPIOD, B. BERTHON and M. CLARET
Unité de Recherches INSERM U 274 - Bât. 443, Université Paris-Sud, 91405 - ORSAY, Cedex, France.

The action of noradrenaline and of the peptidic hormones vasopressin and angiotensin on the Na-K pump was studied in isolated rat liver cells. The three hormones stimulated the carrier and decreased the internal Na concentration of the cells. The actions were dose-dependent and blocked by the respective antagonists phenoxybenzamine, cyclo-tyr-(et)-val-arg-vasopressin and sar-ile-angiotensin. The stimultaneous addition of maximal doses of noradrenaline and vasopressin or angiotensin were not additive suggesting that the hormones use a common mechanism to control the Na-K pump activity.

Incubating the cells in Ca-free (EGTA, 200 µM) medium for long periods (90-120 min) to deplete internal Ca^{2+} stores increased the Na-K pump activity without increasing the internal Na content of the cells and abolished the stimulatory action of noradrenaline, vasopressin and angiotensin. The effect of a high concentration of the Ca^{2+} chelator quin2 on the resting and hormone-stimulated activity of the Na-K pump was also studied. The cells were loaded with (^{3}H)-quin2-acetoxymethyl ester for 6 min. The resulting internal content of the indicator was 3.1 nmol/mg dry wt. The cytosolic Ca^{2+} as monitored from the fluorescence emission of quin2 was 190 nM. Under these conditions, the resting level of the Na-K pump was not altered and the cells remained responsive to the three hormones. Loading to cells in the absence of external Ca^{2+} for 6 min reduced the cytosolic free Ca^{2+} concentration to 25 nM and increased the Na-K pump activity to a level close to that initiated by noradrenaline, vasopressin and angiotensin without increasing internal Na. In addition, the rat hepatocytes were no longer sensitive to the hormones. It is suggested that both the resting activity of the Na-K pump and the action of the hormones on the carrier are dependent on internal cell Ca^{2+}.

Hormones and Cell Regulation, Volume 9
INSERM European Symposium
J.E. Dumont, B. Hamprecht and J. Nunez editors
© 1985 Elsevier Science Publishers B.V. (Biomedical Division)

ADENYLATE CYCLASE IN CILIA FROM PARAMECIUM TETRAURELIA

DORIS GIERLICH and JOACHIM E. SCHULTZ

Pharmazeutisches Institut, Universität Tübingen, Morgenstelle 8, 7400 Tübingen (FRG)

The regulation of cyclic AMP levels in Paramecium tetraurelia 'in vivo' seems to involve a Donnan ratio specifically established by potassium and calcium ions (Schultz et al. FEBS Lett.167:113, 1984). Therefore, the properties and ionic regulation of a particulate adenylate cyclase (AC) localized in the excitable ciliary membrane was examined to further probe into the physiological role of cyclic AMP in this unicellular protozoan. AC can use both, MgATP and MnATP as substrates. Using Mg as metal cofactor, a highly specific activation was observed with only 3 mM K^+. While Rb-ions had a lower intrinsic activity, Na-ions were without stimulatory activity and Li-ions were inhibitory (IC_{50} = 100 mM). Maximal activation by K was dependent on the time necessary for preparation of ciliary membranes. 30% Glycerol stabilized the enzyme in a K-responsive state. Activation by K-ions was completely lost upon storage of AC at $0°C$ for 10hrs with or without the presence of protease inhibitors. K-Activation was not observed with MnATP as substrate.

Fluoride ions are known to stimulate a number of AC's. The AC from Paramecium is not affected by F-ions when MnATP is the substrate. However, using MgATP F-ions inhibited AC with an IC_{50} of 10mM. Other halogenid ions also had an inhibitory effect though higher concentrations were needed, the order of potency was J > Br = Cl. The activity of Paramecium AC was not influenced by GTP, GMPPNP. No substrates for cholera toxin and pertussis toxin mediated adenoribosylation were detected. So far, all evidence indicates that the AC in the excitable ciliary membrane from Paramecium is regulated by quite different mechanisms compared to those detected in higher systems.

Recently, solubilization of the AC from the membrane with Lubrol PX has been accomplished. Part of the K-sensitivity was retained. This enables us to further purify the enzyme to evaluate the molecular basis of its regulation and learn more about the functional role of cyclic AMP in the cilia of Paramecium.

Hormones and Cell Regulation, Volume 9
INSERM European Symposium
J.E. Dumont, B. Hamprecht and J. Nunez editors
© 1985 Elsevier Science Publishers B.V. (Biomedical Division)

THE EFFECT OF CORONARY ARTERY OCCLUSION AND REPERFUSION ON THE
ACTIVITIES OF TRIGLYCERIDE LIPASE AND GLYCEROL 3-PHOSPHATE
ACYLTRANSFERASE IN THE ISOLATED PERFUSED RAT HEART.

GUY P. HEATHERS AND ROD. V. BRUNT
Department of Biochemistry, University of Bath, Bath, Avon, U.K.

Glycerol 3-phosphate acyltransferase (GPAT) and Triglyceride lipase (TGL) were measured in homogenates from non-
ischaemic and ischaemic tissue from the isolated perfused rat
heart.   Ischaemia was produced by occlusion of the left descen-
ding coronary artery for 10 minutes.   Enzyme activites in
tissue from the non-ischaemic area were similar to the activities
measured in tissue from normally perfused hearts.   In the
ischaemic area GPAT activity was considerably reduced, while
TGL activity was markedly increased compared to activity
measured in the non-ischaemic area.   The change in activities
produced by ischaemia were prevented by pre-perfusion with
cardio-selective β_1 antagonist Atenolol.

Reperfusion of the ischaemic area, by removal of the suture,
resulted in TGL activity returning to the value measured in
tissue from normally perfused hearts.   Pre-perfusion of the
α_2 antagonist Yohimbine resulted in a prolongation of the
increased TGL activity during reperfusion.   This is consistent
with cAMP levels remaining elevated.   GPAT activity, after 10
minutes ischaemia.   The reperfusion-induced fall in GPAT
activity was prevented by pre-perfusion with the α_1 antagonist
Doxasozin.

These changes in enzyme activities show that during ischaemia
there is an increased β adrenergic drive.   On reperfusion the
β adrenergic drive is removed but an α_1 adrenergic drive
becomes apparent.

Hormones and Cell Regulation, Volume 9
INSERM European Symposium
J.E. Dumont, B. Hamprecht and J. Nunez editors
© 1985 Elsevier Science Publishers B.V. (Biomedical Division)

THE EFFECT OF ADRENERGIC AGENTS ON THE ACTIVITIES OF GLYCEROL
3-PHOSPHATE ACYLTRANSFERASE AND TRIGLYCERIDE LIPASE IN THE
ISOLATED PERFUSED RAT HEART.

G.P.HEATHERS, N.AL-MUHTASEB AND R.V.BRUNT.
Department of Biochemistry, University of Bath, Bath, Avon, U.K.

Glycerol 3-phosphate acyltransferase (GPAT) activity was
measured by the incorporation of radio-labelled glycerol 3-
phosphate into butanol-soluble products.   Triglyceride lipase
(TGL) activity was measured by the release of fatty acids from
radio-labelled triolein.   The activities of both enzymes were
measured in homogenates or 5000g supernatants from hearts
perfused with adrenergic agonists and antagonists.   Perfusion
with adrenaline or the βagonist isoprenaline produced an
increase in TGL activity and a fall in GPAT activity.   These
changes could be imitated by incubation of heart homogenates
with cAMP-dependent protein kinase.   The α_2 agonist clonidine
produced the opposite effect, thus it increased GPAT activity
and degreased TGL activity.   Methoxamine, an α_1 agonist, had
no effect on TGL activity but reduced GPAT activity.   Continuous perfusion of the α_1-antagonist atenolol reduced TGL
activity to half that found in controls but also, anomalously,
reduced GPAT activity.   No change was seen on continuous
perfusion of α_1 or α_2 antagonists.   Changes in GPAT
activity were localized mainly in the microsomal enzyme.   These
changes are consistent with both enzymes being regulated via
a cyclic-AMP dependent protein kinase system and GPAT via α_1-
adrenergic mechanisms.

Hormones and Cell Regulation, Volume 9
INSERM European Symposium
J.E. Dumont, B. Hamprecht and J. Nunez editors
© 1985 Elsevier Science Publishers B.V. (Biomedical Division)

PROTEIN PHOSPHORYLATION–DEPHOSPHORYLATION ASSOCIATED WITH THE NUCLEAR MEMBRANE
BREAKDOWN OF THE *XENOPUS LAEVIS* OOCYTE

J. HERMANN, J. BOYER, J. ASSELIN, O. MULNER, J. MAROT, R. BELLÉ and R. OZON
Laboratoire de Physiologie de la Reproduction (Groupe Stéroïdes)
Université Pierre et Marie Curie, 4, Place Jussieu, 75230 PARIS Cedex 05 FRANCE

In the course of progesterone-induced maturation of *Xenopus laevis* oocytes,
an increase in the level of phosphorylation of proteins occurs several hours
after exposure to the hormone, shortly before the germinal vesicle breakdown,
at the time of the appearance of maturation promoting factor (MPF) activity.

Partial purification of MPF activity : MPF activity was recovered from
progesterone matured oocytes using precipitation by 20% ethanol or 5% polyethylene glycol-20,000. Neither Ca^{2+} ions nor calmodulin inhibit the MPF
activity prepared in this manner. The stability of MPF extracts was improved
by adding adenosine 5'-O-(3 thiotriphosphate) (ATP-γ-S) 50 μM) to the preparation buffer containing also 2-glycerophosphate and Mg^{2+} ions. The presence of
these two compounds in addition to ATP-γ-S were always required suggesting
that phosphoprotein phosphatases may be involved in the inactivation of MPF.

Alcaline phosphatase inactivates MPF activity : microinjection of low molecular weight phospho esters such as 2-glycerophosphate, phosphotyrosine, phospho-
serine, phosphothreonine or 4-nitrophenyl phosphate were able to induce a
considerable reduction in the time course of progesterone-induced maturation
with 2-glycerophosphate being the most effective. The injection of pure alkaline
phosphatase from calf intestine (10 ng per oocyte) was able to decrease or
abolish the effect of 2-glycerophosphate. Furthermore, this enzyme totally
blocked progesterone – or MPF-induced maturation at a dose of 25 ng per oocyte.
Alkaline phosphatase might behave *in vivo* as a phosphoprotein phosphatase.

A 45 K dalton alkali stable protein is phosphorylated immediately after MPF
microinjection. This phosphoprotein which is absent in the cytoplasm of immature
oocytes may represent a molecular marker of the MPF activity.

Hormones and Cell Regulation, Volume 9
INSERM European Symposium
J.E. Dumont, B. Hamprecht and J. Nunez editors
© 1985 Elsevier Science Publishers B.V. (Biomedical Division)   407

A STUDY OF MICROTUBULE REGULATION: ZINC-STIMULATED MICROTUBULE ASSEMBLY AND
EFFECTS OF COLD ON MICROTUBULE COMPOSITION

J.E. HESKETH

Cell Biochemistry Department, Rowett Research Institute, Aberdeen, U.K.

Possible mechanisms for the control of microtubule formation and stability are
being investigated using 2 experimental systems, firstly the stimulation of
brain microtubule assembly by low zinc concentrations and secondly the
differences in the properties of microtubules which are stable or labile (i.e.
depolymerized) in the cold.

Addition of low concentrations of zinc chloride (500 µM in the presence of 1
mM EGTA; estimated free zinc 1 µM) to rat brain supernatant fluids resulted in
a stimulation of microtubule assembly. The zinc-stimulated assembly appeared
normal in terms of the microtubule structure (no sheet structures were seen),
polypeptide composition and dependence on GTP. Zinc reduced the length of the
lag phase of assembly and the critical concentration. These observations,
together with the lack of effect of microtubule 'seeds' in the presence of zinc
all suggest zinc stimulates the nucleation phase of assembly. Zinc-stimulated
assembly increased the cold-stability and the alkaline phosphatase activity of
the microtubules. The mechanism of the zinc effect is unclear but could
involve phosphorylation of MAP_2 or cold-stable specific proteins or an effect on
S100 protein.

Cold-stable (CS) and cold-labile (CL) microtubules were prepared from rat
brain by two cycles of assembly/disassembly and two periods of exposure to cold.
SDS-polyacrylamide gel electrophoresis showed differences in the polypeptide
composition of the two types of microtubule; the CS microtubules contained
certain specific minor components of approximate molecular weight, 155 KDa
(doublet), 80 KDa and 50 KDa while the CL microtubules contained more of the
high molecular weight component MAP_2. CS microtubules also exhibited a much
higher alkaline phosphatase activity. These three differences between CS and
CL microtubules, namely CS specific proteins, MAP_2 and alkaline phosphatase are
3 factors which may possibly be involved in control of microtubule stability.
The relevance of changes in alkaline phosphatase activity to cold stability is
supported by the concomitant increase in both CS and alkaline phosphatase
activity in response to zinc stimulation of microtubule assembly. Both MAP_2
and CS specific proteins have been reported to be phosphorylated and current
work is testing the hypothesis that MAP_2 phosphorylation is associated with
lability of microtubules to cold.

Hormones and Cell Regulation, Volume 9
INSERM European Symposium
J.E. Dumont, B. Hamprecht and J. Nunez editors
© 1985 Elsevier Science Publishers B.V. (Biomedical Division)

408

MICROTUBULES AND MICROTUBULE-ASSOCIATED-PROTEINS IN *XENOPUS* OOCYTE

C. JESSUS, D. HUCHON and R. OZON

Laboratoire de Physiologie de la Reproduction, Groupe Stéroïdes,
Université Pierre et Marie Curie, 4, Place Jussieu, 75230 PARIS Cedex 05,
FRANCE

In *Xenopus* oocyte, progesterone triggers the resumption of the first meiotic
cell division or meiotic maturation. Tubulin was shown to be unable to assemble
in vivo in immature oocyte ; its ability to polymerize occurs at the time of
the nuclear envelope breakdown. In order to know if tubulin and MAPs (microtubule-associated-proteins) were present at these different stages of the oocyte
development, microtubules were isolated in the presence of taxol ; different MAPs
were identified on the basis of two criteria : their affinity for microtubules
and their capacity to induce tubulin polymerization. The biological role of MAPs
in the induction of oocyte tubulin polymerization was furthermore investigated
and it was found that a ribonucleoproteic factor associated with the oocyte free
ribosomes inhibits the neurotubulin assembly. Binding experiments performed with
radioactive rat brain MAPs show that their sequestration by the ribonucleoproteic
factor in the presence of GTP accounts for the *in vitro* inhibition of tubulin
polymerization. To investigate if the binding of MAPs to oocyte ribosomes play
a biological function, we have microinjected fluorescent purified MAPs into
prophase blocked oocytes : they capped the basal nuclear envelope. In contrast,
when the fluorescent MAPs-ribosome complex is microinjected, the MAPs remain in
the cytoplasm and never reach the region underlying the nuclear envelope.

REFERENCES

1. Jessus C, Friederich E, Francon J and Ozon R (1984) Cell Diff 14:179-187

2. Jessus C, Huchon D, Friederich E, Francon J and Ozon R (1984) Cell Diff,
   in press

Hormones and Cell Regulation, Volume 9
INSERM European Symposium
J.E. Dumont, B. Hamprecht and J. Nunez editors
© 1985 Elsevier Science Publishers B.V. (Biomedical Division)

GABA DECREASES PROLACTIN mRNA IN RAT PITUITARY

J. PH. LOEFFLER, N. KLEY, C.W. PITTIUS and V. HÖLLT
Dept. Neuropharmacology, Max-Planck-Institut für Psychiatrie
Am Klopferspitz 18a, 8033 Planegg-Martinsried F.R.G.

There is evidence that γ-aminobutyric acid (GABA) inhibits the secretion of prolactin (PRL) from rat pituitary cells in vitro (e.g. Schally et al. (1977) Endocrinology 100 (3), 681-691). We report here that GABA also inhibits the biosynthesis of PRL in these cells in lowering the mRNA levels coding for this hormone.

Concentrations of PRL mRNA were measured by hybridisation of cytoplasmic mRNA (White and Bancroft (1982) JBC 257 (15), 8569-8572) to a ^{32}P-labelled cDNA probe containing 800 bases complementary to rat PRL mRNA (generous gift from J. Baxter, San Francisco).

Incubation of cultured rat anterior pituitary cells in the presence of 10^{-8} to 10^{-4}M GABA for 3 days resulted in a dose-dependent decrease in the PRL mRNA with ED_{50} of about 10^{-6}M. Maximal effects (45 to 70% decrease) were obtained with 10^{-5}M GABA. A similar inhibition was seen in the presence of the $GABA_A$ receptor agonists Muscimol and Isoguvacine, but not with the $GABA_B$ receptor agonist Baclofen. In addition, the inhibitory effect of GABA was antagonized by the $GABA_A$ receptor antagonist Bicuculline at a concentration of 10^{-5}M.

Thus these results suggest that GABA exerts an inhibitory effect on PRL biosynthesis in rat adenohypophyseal cells by an action initially mediated by $GABA_A$ receptors.

J.PH.L. is supported by INSERM postdoctoral fellowship.

Hormones and Cell Regulation, Volume 9
INSERM European Symposium
J.E. Dumont, B. Hamprecht and J. Nunez editors
© 1985 Elsevier Science Publishers B.V. (Biomedical Division)

EFFECT OF PHENYL-METHYL-SULFONYL FLUORIDE (PMSF) ON RAT LIVER CELL PROLIFERATION

M.-N. LOMBARD, J.-F. HOUSSAIS and D. ZILBERFARB
I.N.S.E.R.M. (U.22) Institut Curie, Bât. 110, Centre Univ. 91405 ORSAY (France)

Serum protease inhibitors i.e. α_2 macroglobulin can modify cell proliferation[1]. Whether or not this effect is directly related to the anti-proteolytic activity remains an open question. We report here some results obtained on cycling cells "in vitro" with a synthetic protease inhibitor, PMSF.

Rat liver cells (HAB & HH lines) grown under serum deprived conditions were partially synchronized by addition of 10 % serum. Serial pulse labelling (^{3}H-thymidine) were performed and the cells were fixed and counted after standard autoradiographic procedure. PMSF (10^{-3} M) was supplied for 2 hrs. Cells were allowed to recover in 10 % serum. Control cells were exposed to NaF (10^{-3}M) instead of PMSF or to inactivated PMSF.

	Time (hrs)		cell nr/field*	% S phase nr
HAB cells				
	0		145.9 ± 16.4	37
	23		251.1 ± 17.1	66
	37	controls	466.4 ± 30.8	61
		PMSF	265.4 ± 17.0	35
PMSF was added at 23 hrs				
HH cells				
	0		56.0 ± 2.0	
	72		79.8 ± 5.7	34
	96	controls	225.6 ± 9.3	41
		PMSF	110.7 ± 5.4	33
PMSF was added at 88 hrs				

* Average nr of 20 fields.

Cell multiplication resumes in PMSF treated cultures after 24 hrs in serum supplemented medium.

In our experimental conditions, PMSF, an alkylating reagent acting specifically on the serine residue of the active site of endoproteases, induces a reversible break down of cell proliferation.

1. Koo P.H. (1983) Ann. N.Y. Acad. Sci. 421 pp. 388-390.

Hormones and Cell Regulation, Volume 9
INSERM European Symposium
J.E. Dumont, B. Hamprecht and J. Nunez editors
© 1985 Elsevier Science Publishers B.V. (Biomedical Division)   411

CYCLIC NUCLEOTIDE-INDEPENDENT KINASE ACTIVITIES WHICH PHOS-
PHORYLATE THE RIBOSOMAL PROTEIN S6

OSKAR H. W. MARTINI and ALFONS LAWEN
Institut für Virologie und Immunbiologie der Universität Würzburg,
Versbacher Straße 7, D 8700 Würzburg (F.R.G.)

The molecular mechanisms underlying the propagation and spreading
of mitogenic signals distal of the plasma membrane level are not
known. Since mitogenic agents such as growth factors elicit rapid
changes in the pattern of protein phosphorylation in cytoplasm
and nucleus, one or several protein kinase/phosphatase systems
must be involved. The ribosomal protein S6 has been pinpointed
as a substrate for the postulated "mitogen-responsive protein
kinase(s)", because rapid incorporation of up to 5 phosphate groups
at serine sites of the S6 molecule has been demonstrated to be a
typical response phenomenon to various mitogenic stimuli. The
"mitogen-responsive S6-kinase(s)" can be expected to be cyclic
nucleotide-independent, because the cGMP- and cAMP-dependent protein kinases are known to phosphorylate S6 in only 1 or 2 sites,
respectively.

In order to avoid interference by cAMP-dependent kinases, we
used the kin$^-$ mutant of S49 cells, a murine lymphoma cell line
(kin$^-$ refers to lack of cAMP-dependent protein kinase activity),
for the first stage of our search for cXMP-independent S6-recogni-
zing kinase activities. 10 000 x g supernatants prepared from
these fast-growing cells were fractionated on DEAE-Sephacel, and
the fractions were tested for protein kinase activity using
$[\gamma-^{32}P]$ATP as phosphate donor and a variety of protein kinase substrates, including 40S ribosomal subunits, as phosphate acceptors.
A cXMP- and Ca^{++}/phospholipid-independent kinase activity with
high specificity for serine sites of the S6 protein ($<$5 % of the
phosphate transferred was bound to other proteins of the 40S subunit) could be detected. For efficient S6 phosphorylation, this
enzyme required the presence of EDTA and of higher cosubstrate
concentrations ($>$100 μM ATP) than the cXMP-dependent kinases
assayed under the same ionic conditions.

Using the same conditions of cell fractionation and assaying,
protein kinases with similar chromatographic behaviour were also
found in mouse liver and in HeLa cells.

Hormones and Cell Regulation, Volume 9
INSERM European Symposium
J.E. Dumont, B. Hamprecht and J. Nunez editors
© 1985 Elsevier Science Publishers B.V. (Biomedical Division)

MITOGEN-RESPONSIVE PROTEIN KINASE ACTIVITY OF THE 10 000 x g
SUPERNATANT OF HeLa CELLS

OSKAR H. W. MARTINI and ALFONS LAWEN
Institut für Virologie und Immunbiologie der Universität Würzburg,
Versbacher Straße 7, D 8700 Würzburg (F.R.G.)

In our search for "mitogen-responsive S6-kinases" we used HeLa
S3 cells as a further model, in addition to the S49 kin⁻ cells
studied in the accompanying poster, because HeLa cells were expected to be better suited for correlating protein kinase activity
changes with other growth parameters.

HeLa S3 cells were growth arrested by serum starvation. A
portion was then stimulated by transfer to fresh medium plus serum.
Immediately before or 30 min after this stimulation, cells were
harvested, and 10 000 x g supernatants were prepared. These were
chromatographed on DEAE-Sephacel, and the fractions of the eluate
were assayed for kinase activity using 40S ribosomal subunits
as probes (cf. accompanying poster). - Two peaks of kinase activity
recognizing the S6 protein of the 40S subunit were detected. In
eluates derived from serum stimulated cells the activity of one
of these enzymes was increased by almost one order of magnitude.
This effect of serum could be mimicked with pharmacological doses
of insulin. The response seems to be mediated by growth factor
receptors

The "mitogen-responsive S6-kinase" was partially purified and
characterized. It was insensitive to a variety of known protein
kinase regulators, including Ca^{++} and cyclic nucleotides. Its
properties (M_r = 55 kd; [K^+] optimum of 35 mM; inactive with GTP)
and its chromatographic behaviour on DEAE-Sephacel (elution in
partial overlap with a casein kinase II type activity) argue
against identity of the "mitogen-responsive S6-kinase" with either
casein kinase I or II. Rather, it may be related to an S6-kinase
activity recently observed in 3T3 cells (Novak-Hofer, I. and
Thomas, G. (1984) J. Biol. Chem. 259:5995-6000), as will be
discussed.

Mitogen-responsive S6-kinase preparations were also obtained
from chick embryo fibroblasts. The preparations derived from serum-
stimulated cells were 10 times as active as those from serum-
deprived fibroblasts.

Hormones and Cell Regulation, Volume 9
INSERM European Symposium
J.E. Dumont, B. Hamprecht and J. Nunez editors
© 1985 Elsevier Science Publishers B.V. (Biomedical Division)

HORMONE-DEPENDENT Ca^{2+} CHANNELS IN RAT HEPATOCYTES

J-P. MAUGER, J. POGGIOLI, F. GUESDON and M. CLARET

Unité de Recherhces INSERM U 274 - Bât. 443, Université Paris-
Sud, 91405 - ORSAY, Cedex, France.

In the rat liver, different hormones or agonists accelerate
glycogenolysis either by increasing the cytosolic Ca^{2+} concentration ($(Ca^{2+})_i$) (for α_1 adrenergic agonists, angiotensin and
vasopressin) or by activating the adenylate cyclase and increasing
the cAMP content of the hepatocytes (for β-adrenergic agonists and
glucagon). The increase in $(Ca^{2+})_i$ induced by the Ca^{2+} mobilizing
hormones is maximal and sustained only in the presence of extracellular Ca^{2+} ($(Ca^{2+})_o$), in keeping with the stimulatory action
of noradrenaline, vasopressin and angiotensin on the unidirectional
influx of Ca^{2+}. The measurement of Ca^{2+} influx at different $(Ca^{2+})_o$
reveals that the hormones increase both the K_m and V_{max} of Ca^{2+}
influx, suggesting that they increase the rate of transfer of
Ca^{2+} through a fixed number of Ca^{2+} channels (1).

The agonists that activate the adenylate cyclase, also increase
the $(Ca^{2+})_i$ whose maintenance also depends on external Ca^{2+}. Indeed
we have found that the β agonist isoproterenol and glucagon
increase the Ca^{2+} influx in rat hepatocytes in a dose-dependent
manner with EC_{50} of 20 nM and 0.9 nM respectively. This effect was
mimicked by the permeant cAMP analog, Bt_2 cAMP. This indicates
that isoproterenol and glucagon activate the Ca^{2+} channel by a
mechanism requiring an increase of the cAMP content of the hepatocytes. Simultaneous application of maximal concentrations of
a Ca^{2+} mobilizing hormone and isoproterenol, glucagon or Bt_2 cAMP
caused a Ca^{2+} influx which was much larger than the sum of that
seen with each agonist alone. These results suggest that cAMP
causes the activation of the Ca^{2+} channels and potentiates the
effect of Ca^{2+}-mobilizing hormones by acting on the coupling
mechanism between the hormone receptor and the Ca^{2+} channel in
rat hepatocytes.

REFERENCE

1. Mauger, J-P., Poggioli, J., Guesdon, F. and Claret, M. (1984)
   Biochem. J. 221 : 121-127.

Hormones and Cell Regulation, Volume 9
INSERM European Symposium
J.E. Dumont, B. Hamprecht and J. Nunez editors
© 1985 Elsevier Science Publishers B.V. (Biomedical Division)

414

THYROTROPIN MODIFIES THE SYNTHESIS OF ACTIN AND OTHER PROTEINS DURING THYROID CELL CULTURE.

H. Passareiro, P.P. Roger, F. Lamy, R. Lecocq, J.E. Dumont and J. Nunez.

IRIBHN, Faculty of Medicine, University of Brussels, Campus Erasme, 808 route de Lennik, 1070 Brussels, Belgium.

Primary cultures of dog thyroid cells have been used to study the effects of thyrotropin on the synthesis of proteins. The cells were cultured for 4 days in serum- and thyrotropin-free conditions and then in the presence of thyrotropin for varying periods of time (6 to 96 hours). During the culture period, in the absence of thyrotropin, the cells have an elongated flattened aspect. Exposure to thyrotropin for 6 to $\simeq$ 24 hours produces cell retraction and rounding up; after a longer period of culture the cells display an epithelial cuboidal shape. After varying periods of culture the cells were labeled with $[^{35}S]$ Methionine for 6 hours and then analyzed by mono- and bidimensional gel electrophoresis, followed by radioautography. The results showed :

1) 32 and 48 hours after exposure to thyrotropin : the synthesis of $\simeq$ 18 proteins was increased while that of $\simeq$ 14 other entities was decreased. After 6 hours of culture the labeling of 3 and 5 of these proteins was already increased or decreased respectively.

2) Some of the proteins the synthesis of which is modified in the presence of thyrotropin were identified. For instance actin synthesis was markedly decreased and such a decrease was maximal 24 to 48 hours after the addition of thyrotropin. A modification in the ratio between α and β tubulins was also observed together with very large changes at the level of a group of proteins having both the molecular weights (30-40 K) and the isoelectric points of tropomyosins.

3) Forskolin and cholera toxin caused the same qualitative and quantitative changes as thyrotropin; this suggests that the regulation by thyrotropin of the synthesis of several proteins of the thyroid cell is mediated by cAMP.

In conclusion, the data obtained in this work might help to explain the molecular mechanisms by which thyrotropin (and cAMP) trigger the marked changes in cell shape which occur during thyroid cell culture. They also indicate that one of the main effects of thyrotropin takes place at the level of several proteins which belong to the cytoskeleton and which are involved in the definition of the cytostructure of the thyroid cells.

Hormones and Cell Regulation, Volume 9
INSERM European Symposium
J.E. Dumont, B. Hamprecht and J. Nunez editors
© 1985 Elsevier Science Publishers B.V. (Biomedical Division)

INOSITOL-1,4,5TRISPHOSPHATE GENERATION BY THE CHEMOTACTIC PEPTIDE
FMet-Leu-Phe PRECEEDS CYTOSOLIC Ca^{2+} RISE IN HUMAN NEUTROPHILS

T. POZZAN[*], L. VICENTINI[§] and F. DI VIRGILIO[*]

[*]Institute of General Pathology and C.N.R. Unit for the Study of the Physiology of Mitochondria, University of Padova, Italy

[§]Department of Pharmacology, University of Milano, Italy

A number of experimental evidence (1,2,3) indicates that receptor mediated cell activation involves two distinct events: 1) Rise of cytosolic free Ca^{2+} concentration, $\left[Ca^{2+}\right]_i$; 2) Activation of protein kinase C. A key role in this process is played by phosphatidyl-inositol bysphosphate, PI (P)$_2$, breakdown operated by an endogenous phospholipase C, thought to be turned on by ligand receptor interaction. The two metabolites of PI(P)$_2$, inositol-1,4,5trisphosphate, IP$_3$, and diacylglycerol, DAG, in vitro can activate Ca^{2+} release from intracellular stores and protein kinase C respectively. IP$_3$ has been recently shown to release Ca^{2+} from stores in leaky neutrophils (5). The major problem in accepting this elegant scheme, in particular in neutrophils, is that it has been previously shown that PI breakdown depends on $\left[Ca^{2+}\right]_i$ rise (4). Furthermore in no cell type, IP$_3$ generation in the absence of $\left[Ca^{2+}\right]_i$ increase, has yet been shown. Here we demonstrate that inositol-phosphates, and in particular IP$_3$, can be generated by the chemotactic peptide FMLP in the absence of any $\left[Ca^{2+}\right]_i$ rise and even when $\left[Ca^{2+}\right]_i$ is decreased below resting level. We will show that IP$_3$ increases 3 to 5 times as soon as 15 seconds after FMLP addition, and that IP$_3$ is formed before IP$_1$ and IP$_2$.

It is therefore suggested that also in human neutrophils IP$_3$ causes the release of intracellular Ca^{2+} in response to surface receptor stimulation.

1) Rink T.J., Sanchez A. and Hallam T.J. (1983) Nature 305, 317
2) Pozzan T., Lew D.P., Wollheim C.B. and Tsien R.Y. (1983) Science 221, 1413
3) Di Virgilio F., Lew D.P. and Pozzan T. (1984) Nature 310, 691
4) Cockroft S., Bennet J.P. and Gomperts B.D. (1981) Biochem. J. 200, 501
5) Lew et al. (1984) J. Biol. Chem., in press

Hormones and Cell Regulation, Volume 9
INSERM European Symposium
J.E. Dumont, B. Hamprecht and J. Nunez editors

IN VITRO EFFECT OF L-TRIODOTHYRONINE ON THE DETERMINATION OF
THE DIFFERENTIATION OF MULTIPOTENTIAL HEMATOPOIETIC STEM CELLS.

FRANCOISE SAINTENY and EMILIA FRINDEL
UNITE 250 INSERM, INSTITUT GUSTAVE-ROUSSY, Rue Camille Desmoulins
94805 Villejuif cedex (France).

We have recently reported that circulating substances called *pluripoietins*,
produced after the injection of 20 mg Ara-C to mice, channel the differentiation
of both medullary and splenic pluripotential haemopoietic stem cells (CFU-S)
towards erythropoiesis.

Thyroid hormones are known to stimulate erythropoiesis and to enhance
the development of unipotent precursors in various species. The purpose of this
work is to determine whether thyroid hormones act at the CFU-S level in mice
and whether there is a relationship between these hormones and *pluripoietins*.

The pathway of CFU-S differentiation has been studied by histological
determination of the proportion of the different types of colonies generated by
CFU-S in the spleen of irradiated mice. Preliminary results demonstrate that
L-triodothyronine (LT3) has the capacity of inducing a preferential determination
of CFU-S differentiation towards erythropoiesis at the expense of granulopoiesis
in vitro, as do pluripoietins. Cholera toxin, which blocks the receptor gangliosides
of cell membrane abolishes the effect of LT3 on CFU-S.

These data indicate that (1) LT3 has a direct effect on the determination
of CFU-S differentiation and that (2) the integrity of cell membrane of CFU-S
or accessory cells is required for CFU-S to respond to LT3. The relationship
between LT3 and *pluripoietins* are now in the process of study.

Hormones and Cell Regulation, Volume 9
INSERM European Symposium
J.E. Dumont, B. Hamprecht and J. Nunez editors
© 1985 Elsevier Science Publishers B.V. (Biomedical Division)

FUNCTIONAL HETEROGENEITY OF β-ADRENERGIC RECEPTORS.

Yvonne SEVERNE , Dirk COPPENS and Georges VAUQUELIN.
Laboratory of Protein Chemistry, Institute of Molecular Biology,
Free University Brussels (VUB), 65 Paardenstraat, 1640 St.
Genesius-Rode, Belgium.

The β-adrenergic responsive adenylate cyclase system includes three components: the β-adrenergic receptor (R); the adenylate cyclase enzyme (C); and a coupling protein (Ns) which serves as a "shuttle" between R and C. Only part of the R can undergo functional coupling to Ns. This receptor subpopulation shows an increased affinity for agonists in the presence of Mg^{2+}. Agonists are thus able to induce a conformational transition of the R-Ns complex i.e. from a low to a high agonist affinity form of the R. The reagent N-ethylmaleimide (NEM) "freezes" this R-Ns complex, probably by alkylation of Ns. This process results in the locking-in of the receptor-bound agonists and is experimentally demonstrated by an apparent agonist/NEM mediated loss in R number.

Several experimental conditions, known to modify the total R concentration without alteration of Ns were shown not to affect the percentage of R which can undergo functional coupling: 1) homologous regulation of β1-R in rat brain by noradrenaline (via antidepressive drug or reserpine injections), 2) up- and downregulation of the β2-R in Friend erythroleukemia cells by respectively sodium butyrate and cinnarizine treatment and 3) dithiothreitol- mediated inactivation of turkey erythrocyte, Friend erythroleukemia cell and rat brain R . Our findings argue against a stoichiometric limitation in the number of regulatory components, genetically different R subpopulations, bound guanine nucleotides to Ns or reduced accessibility of part of the R to the agonists as the cause for functional receptor heterogeneity. Differences in either the R structure (differential post-translational modifications) or its membrane microenvironment (regions of facilitated coupling to Ns) are more plausible explanations (PNAS (1984) 81;4637).

Recently, investigation on the effects of the temperature on the percentage of coupling- prone R in turkey erythrocyte membranes and ghosts showed that although important in regulating the R-Ns coupling, increases in temperature above the physiological value do not allow increases in R-Ns coupling.

Hormones and Cell Regulation, Volume 9
INSERM European Symposium
J.E. Dumont, B. Hamprecht and J. Nunez editors
© 1985 Elsevier Science Publishers B.V. (Biomedical Division)

TSH CONTROLS THYROGLOBULIN (Tg) GENE TRANSCRIPTION BY A cAMP DEPENDENT
MECHANISM

B. VAN HEUVERSWYN & G. VASSART

IRIBHN, ULB, Campus Erasme, 808 route de Lennik, B-1070 Brussels, Belgium.

The precursor of the thyroid hormones, thyroglobulin (Tg) is encoded by an
exceptionnaly large gene (≈ 250 kb) and is synthetised abundantly by the thyroid
gland.

TSH has been shown to stimulate Tg gene transcription when administered to T_3
treated rats. In the present study, we have further studied the kinetics of this
effect, in vivo, and explored its mechanisms in the rat thyroid fragments
incubated in vitro.

Rats were treated during 7 days with T_3 (5 µg/100 g/day) in order to lower
their TSH plasma level. Thyroid glands were then stimulated with TSH either in
vivo or in vitro, the thyroid nuclei were prepared and transcription of the Tg
gene was measured in a "run off assay".

In vivo restimulation of thyroid gland were performed by the intravenous injection of TSH (0.2 IU/100 g). The animals were killed at different time after the
injection and their plasma level of T_3 and T_4 were measured together with Tg gene
transcription. As soon as one hour after TSH administration, T_4 was increased
(1.3 µg.dl $\pm$ 0.3 vs 0.7 µg/dl $\pm$ 0.1), there was no significant change in T_3 and
Tg gene transcription was stimulated by a factor of 3.1 (557 ppm $\pm$ 29 vs
181 ppm $\pm$ 32).

In order to investigate the mechanisms of TSH, we addressed to an in vitro
system. T_3 pretreated rats were killed and the thyroid tissue was incubated
without agent, with TSH (50 mU/ml), or with forskolin (10 µM), an agent known
to increase intracellular cAMP. cAMP levels were measured as an index of the
stimulatory effect of each agent. Tg gene transcription measured after a 150
min. incubation period was increased 1.9 and 2.8 fold by TSH and forskolin,
respectively.

In conclusion, stimulation of Tg gene transcription by TSH in T_3 treated
rats is (i) a rapid phenomenon, (ii) results from the direct effect of TSH on
the gland and (iii) is most probably mediated by cAMP dependent mechanisms.

Hormones and Cell Regulation, Volume 9
INSERM European Symposium
J.E. Dumont, B. Hamprecht and J. Nunez editors
© 1985 Elsevier Science Publishers B.V. (Biomedical Division)

PHOSPHOINOSITIDE BREAKDOWN AND REGULATION OF TRANSMITTER RELEASE FROM PC12
CELLS

L.VICENTINI, F.DI VIRGILIO, T.POZZAN and J.MELDOLESI
Dept. of Pharmacology and Pathology, Universities of Milano and Padua, Italy

Neurotransmitter and hormone-induced hydrolysis of phosphoinositides
(PI_s) results in the formation of two intracellular second messengers: inositol
1,4,5-trisphosphate (IP_3), which mobilizes intracellular Ca, ($[Ca^{2+}]_i$) and
diacylglycerol, which activates protein kinase C (PKC). We examined the role
of these two cell activation mechanisms in the control of dopamine (DA) release
in rat pheochromocytoma (PC12) cells. We find that the Ca ionophore ionomycin
induces a fast, prompt DA release which levels off at about 2 min, while
quin-2-measured $[Ca^{2+}]_i$ levels are still high. The PKC stimulators OAG and
TPA induce a lower but sustained DA release with no change in $[Ca^{2+}]_i$. Combining
these two agents causes a sinergistic and maximal DA release. The large and
persistent response obtained by simultaneous activation of Ca and PKC resembles
that elicited by black widow spider toxin, α-latrotoxin (α-LTx), which is
known to cause massive exocytotic release of neurotransmitter from nerve
terminals. We find that α-LTx, for which there are specific binding sites
on PC12 cells, in addition to inducing Ca influx (as we previously showed),
also stimulates PI_s breakdown with a dose and time course very similar to
the dose and time course for DA release. At early time points IP_3 is the
major inositol phosphate formed. We suggest that the massive effect of
α-LTx on DA release arises from the simultaneous influx of Ca and PI_s breakdown
with consequent activation of PKC. On the other hand, we show that α-LTx
is able to induce DA release in a nominally Ca free medium, even if at a
much slower rate and extent, while its effect on PI_s breakdown disappears.
In general, PI_s breakdown in PC12 cells is not a Ca-activated phenomenon,
since treatments that increase $[Ca^{2+}]_i$ such as Ca ionophore, 50 mM KCl or
gramicidin, do not induce inositol phosphate accumulation. Moreover, we find
that after depletion of $[Ca^{2+}]_i$ stores with ionomycin and in the absence of
extracellular Ca $[Ca^{2+}]_o$, muscarinic stimulation of PI_s breakdown still occurs.
So the lack of PI_s breakdown in EGTA medium is characteristic of the toxin.
The presence of $[Ca^{2+}]_o$ seems to be necessary for the coupling of the α-LTx
receptor to the PI_s hydrolyzing system. The readdition of Ca to the EGTA
medium in the presence of the toxin causes a prompt PI_s breakdown, while
ionomycin, which in these conditions causes $[Ca^{2+}]_i$ to rise is ineffectual.
In conclusion, both Ca^{2+} and activation of PKC, which is the consequence
of agonist-induced hydrolysis of PI_s, can stimulate the release of neurotransmitter by exocytosis in PC12 cells. The existence of additional control mech-
anism(s) of exocytosis is suggested by our data on α-LTx administered in
Ca^{2+}-free medium.

Hormones and Cell Regulation, Volume 9
INSERM European Symposium
J.E. Dumont, B. Hamprecht and J. Nunez editors
© 1985 Elsevier Science Publishers B.V. (Biomedical Division)

420

PARTIAL PURIFICATION OF 5,6-DIHYDROURACIL DEHYDROGENASE FROM RAT BRAIN, THE FIRST ENZYME ON THE BIOSYNTHETIC PATHWAY TO β-ALANINE?

G. WAHLER, G. LACHMANN and K.D. SCHNACKERZ

Institute of Physiological Chemistry, University of Würzburg, Koellikerstr. 2, D-8700 Würzburg, F.R.G.

β-alanine containing dipeptides, such as carnosine, anserine and N^{α}-(β-alanyl)lysine, have been isolated from excitable tissues, such as brain and muscle. Free β-alanine was found only in very small amounts in various regions of mammalian brain. Recent electrophysiological studies on neurons of the optic tectum suggest that β-alanine acts as an inhibitory neurotransmitter [1]. Until now it is still an open question whether β-alanine is carried across the blood-brain-barrier or whether it is synthesized in the brain itself. So far the synthesis of β-alanine has only been elucidated in mammalien liver via the degradative pathway of uracil [2]:

uracil $\rightleftharpoons$ dihydrouracil $\rightleftharpoons$ β-ureidopropionate $\rightarrow$ β-alanine + CO_2 + NH_3

A first indication for the synthesis of β-alanine in brain was obtained from glial cells of new-born mice in primary culture. It could clearly be shown that those cells can synthesize radioactive β-alanine from [^{3}H]uracil. Our next aim was an attempt on the purification of 5,6-dihydrouracil dehydrogenase, the first enzyme on the degradative pathway of uracil. Rat brains were homogenized and freed of mitochondria and microsomes by centrifugation. The dialyzed cytosolic extract showed dehydrogenase activity. A 95-fold purification of the dehydrogenase could be obtained by affinity chromatography on ADP-Sepharose. K_M values for uracil and thymine were found to be 0.98 and 0.4 mM, respectively. It remains to be seen whether the other two enzymes, dihydropyrimidinase and β-ureidopropionase, are also present in brain homogenates.

[1] Toggenburger, G., Felix, D., Cuénod, M. and Henke, H. (1982) J. Neurochem. 39, 176-183.
[2] Shiotani, T. and Weber, G. (1981) J. Biol. Chem. 256, 219-224.

AUTHOR INDEX

SUBJECT INDEX

tumors and, 3; 11-12

Rous sarcoma virus

cell transformation and, 18-19; 24-29

integration, 17-23; 28-29; 33-34

interferon and, 136; 140

rearrangements, 21-28

Rous sarcoma virus oncogene

duplication, 21-25; 30; 33

EGF receptor and, 44

S

Serotonin

protein kinase C and, 191

Sodium channel

activation of, 232

aldosterone and, 212-214; 219-220

biochemical properties of, 228-231

cation selectivity in, 234-238

conductance of single, 237-238

reconstitution in vesicles, 231

voltage-dependent, 236

Sodium/potassium-ATPase

aldosterone and, 214-220

cell surface expression of, 218-220

in hepatocytes, 402

monensin and, 270

Sodium/calcium countertransport

in pancreatic acinar cells, 330; 336; 340

Sodium/proton exchange system

amiloride and, 259

distribution of, 259-260

growth factors and, 270-273

internal Na^+ concentration and, 263

internal pH and, 262-263

phorbol esters and, 265; 275-277

properties of, 261-262

Somatostatin

in adrenergic neurones, 379

Stress fibers

actin capping proteins and, 137-138

definition of, 135

in non muscle cells, 135-140

vinculin and, 136; 140

Substance P

in cholinergic differentiated neurones, 318

T

Thyroglobulin (Tg)

gene transcription, 418

Thyrotropin

actin synthesis and, 414

protein synthesis and, 414

Tg gene transcription and, 418

Tight junction apparatus

aldosterone and, 220

TMB-8

calcium mobilization and, 308-309

Transcription enhancers

in immunoglobulin genes, 73; 76-79

mechanism of action of, 79-80

properties of, 70-76

Transcription promoters

gene expression and, 69; 71-72

Transfection

of methylated genes, 56; 59

Triglyceride lipase

in rat heart, 404; 405

Triiodothyronine (T_3)

erythropoiesis and, 416

Tumor

development, 3

diagnosis, 123; 125; 128